Fiber Reinforced Polymer (FRP) Composites for Construction

Fiber Reinforced Polymer (FRP) Composites for Construction

Editors

Rui Guo
Bo Wang
Muye Yang
Weidong He
Chuntao Zhang

Basel • Beijing • Wuhan • Barcelona • Belgrade • Novi Sad • Cluj • Manchester

Editors

Rui Guo
Southwest Jiaotong University
Chengdu
China

Bo Wang
Chang'an University
Xi'an
China

Muye Yang
Kyushu University
Fukuoka
Japan

Weidong He
Southwest Petroleum University
Chengdu
China

Chuntao Zhang
Southwest University of
Science and Technology
Mianyang
China

Editorial Office
MDPI AG
Grosspeteranlage 5
4052 Basel, Switzerland

This is a reprint of articles from the Special Issue published online in the open access journal *Buildings* (ISSN 2075-5309) (available at: https://www.mdpi.com/journal/buildings/special_issues/33A266AF1B).

For citation purposes, cite each article independently as indicated on the article page online and as indicated below:

Lastname, A.A.; Lastname, B.B. Article Title. *Journal Name* **Year**, *Volume Number*, Page Range.

ISBN 978-3-7258-2427-4 (Hbk)
ISBN 978-3-7258-2428-1 (PDF)
doi.org/10.3390/books978-3-7258-2428-1

© 2024 by the authors. Articles in this book are Open Access and distributed under the Creative Commons Attribution (CC BY) license. The book as a whole is distributed by MDPI under the terms and conditions of the Creative Commons Attribution-NonCommercial-NoDerivs (CC BY-NC-ND) license.

Contents

Rui Guo, Bo Wang, Muye Yang, Weidong He and Chuntao Zhang
Fiber Reinforced Polymer (FRP) Composites for Construction
Reprinted from: *Buildings* **2024**, *14*, 3216, doi:10.3390/buildings14103216 1

Rui Guo, Dan Yang, Bin Jia and Deyun Tang
Seismic Response of GFRP-RC Interior Beam-to-Column Joints under Cyclic Static Loads
Reprinted from: *Buildings* **2022**, *12*, 1987, doi:10.3390/buildings12111987 4

Marta Kałuża
Experimental Analysis of Surface Application of Fiber-Reinforced Polymer Composite on Shear Behavior of Masonry Walls Made of Autoclaved Concrete Blocks
Reprinted from: *Buildings* **2022**, *12*, 2208, doi:10.3390/buildings12122208 21

Chuntao Zhang, Yanyan Li and Junjie Wu
Mechanical Properties of Fiber-Reinforced Polymer (FRP) Composites at Elevated Temperatures
Reprinted from: *Buildings* **2023**, *13*, 67, doi:10.3390/buildings13010067 38

Jingyang Zhou, Xin Wang, Lining Ding, Shui Liu and Zhishen Wu
Numerical and Experimental Study on Large-Diameter FRP Cable Anchoring System with Dispersed Tendons
Reprinted from: *Buildings* **2023**, *13*, 92, doi:10.3390/buildings13010092 54

Naixian Li, Bin Jia, Junhong Chen, Ying Sheng and Songwen Deng
Phenomenological 2D and 3D Models of Ductile Fracture for Girth Weld of X80 Pipeline
Reprinted from: *Buildings* **2023**, *13*, 283, doi:10.3390/buildings13020283 72

Noura Khaled Shawki Ali, Sameh Youssef Mahfouz and Nabil Hassan Amer
Flexural Response of Concrete Beams Reinforced with Steel and Fiber Reinforced Polymers
Reprinted from: *Buildings* **2023**, *13*, 394, doi:10.3390/buildings13020374 91

Bo Wang, Gejia Liu and Jiayu Zhou
Properties of Concrete Columns Strengthened by CFRP-UHPC under Axial Compression
Reprinted from: *Buildings* **2023**, *13*, 596, doi:10.3390/buildings13030596 107

Rana A. Alhorani, Hesham S. Rabayah, Raed M. Abendeh and Donia G. Salman
Assessment of Flexural Performance of Reinforced Concrete Beams Strengthened with Internal and External AR-Glass Textile Systems
Reprinted from: *Buildings* **2023**, *13*, 1135, doi:10.3390/buildings13051135 124

Jesús D. Ortiz, Seyed Saman Khedmatgozar Dolati, Pranit Malla, Armin Mehrabi and Antonio Nanni
Nondestructive Testing (NDT) for Damage Detection in Concrete Elements with Externally Bonded Fiber-Reinforced Polymer
Reprinted from: *Buildings* **2024**, *14*, 246, doi:10.3390/buildings14010246 138

Junjie Wu and Chuntao Zhang
Modified Constitutive Models and Mechanical Properties of GFRP after High-Temperature Cooling
Reprinted from: *Buildings* **2024**, *14*, 439, doi:10.3390/buildings14020439 158

Qu Yu, Yu Ren, Anhang Liu and Yongqing Yang
Study on Bonding Behavior between High Toughness Resin Concrete with Steel Wire Mesh and Concrete
Reprinted from: *Buildings* **2024**, *14*, 1341, doi:10.3390/buildings14051341 180

Lore Zierul, Enrico Baumgärtel, David Sandmann and Steffen Marx
Shear Transfer in Concrete Joints with Non-Metallic Reinforcement
Reprinted from: *Buildings* **2024**, *14*, 1975, doi:10.3390/buildings14071975 . . . . **194**

Haider M. Al-Baghdadi and Mohammed M. Kadhum
Effects of Different Fiber Dosages of PVA and Glass Fibers on the Interfacial Properties of Lightweight Concrete with Engineered Cementitious Composite
Reprinted from: *Buildings* **2024**, *14*, 2379, doi:10.3390/buildings14082379 . . . . **210**

Editorial

Fiber Reinforced Polymer (FRP) Composites for Construction

Rui Guo [1,*], Bo Wang [2], Muye Yang [3], Weidong He [4] and Chuntao Zhang [5]

1. Department of Building Engineering, School of Civil Engineering, Southwest Jiaotong University, Chengdu 610031, China
2. School of Civil Engineering, Chang'an University, Xi'an 710061, China; chnwangbo@chd.edu.cn
3. Department of Civil Engineering, Faculty of Engineering, Kyushu University, Fukuoka 8190395, Japan; yang@doc.kyushu-u.ac.jp
4. School of Civil Engineering and Geomatics, Southwest Petroleum University, Chengdu 610500, China; hewd@swpu.edu.cn
5. School of Civil Engineering and Architecture, Southwest University of Science and Technology, Mianyang 621000, China; chuntaozhang@swust.edu.cn
* Correspondence: guor4867@swjtu.edu.cn

Citation: Guo, R.; Wang, B.; Yang, M.; He, W.; Zhang, C. Fiber Reinforced Polymer (FRP) Composites for Construction. *Buildings* **2024**, *14*, 3216. https://doi.org/10.3390/buildings14103216

Received: 27 September 2024
Accepted: 10 October 2024
Published: 10 October 2024

Copyright: © 2024 by the authors. Licensee MDPI, Basel, Switzerland. This article is an open access article distributed under the terms and conditions of the Creative Commons Attribution (CC BY) license (https://creativecommons.org/licenses/by/4.0/).

The development and application of new materials is one of the main driving forces of technical development in the field of civil engineering. Compared with traditional structural materials, a fiber-reinforced composite (FRP) has the advantages of being light weight, high strength, and corrosion resistant while also having designability [1,2]. Over a very long period of time, fiber composite materials have experienced a history of trials, demonstrations, development, and popularization. Now, they are widely used in existing structure reinforcements and various new structures, which can effectively improve structural performance and prolong structural life [3,4]. This Special Issue is dedicated to showcasing the latest research and development activities related to the utilization of FRP composites in construction. This Special Issue on Fiber Reinforced Polymer (FRP) Composites for Construction features 13 papers. All these contributions effectively address the main topics of this Special Issue in a targeted effort.

In line with the research topics explored in this Special Issue, the 13 published articles can be categorized into 3 major themes. The first research topic is the mechanical behavior of FRP. Paper [5] offers a comparative analysis of the flexural response of concrete beams reinforced with different types of FRPs and steel, with a specific focus on CFRP, which shows improved performance at lower reinforcement ratios. Paper [6] investigates fracture models for X80 pipeline girth welds; the 3D model outperformed the 2D model. Tensile strength was correlated with stress triaxiality. Paper [7] investigates shear loads for single non-metallic bars and concrete specimens with non-metallic reinforcements crossing two joints, showing that the behavior after exceeding adhesion was ductile in comparison to joints without reinforcements, where the behavior was brittle. Paper [8] investigates the effects of different target temperatures and cooling methods on the tensile properties of GFRP and derives the prediction equations and theoretical models for the mechanical properties after high-temperature cooling. Paper [9] conducts tests on nine specimens, comparing the superior deformation resistance of GFRP-reinforced concrete (RC) joints to RC beam-to-column joints and validating the experimental results with a proposed core zone shear capacity method. Paper [10] examines the impact of elevated temperatures on the mechanical properties of FRP composites, providing key insights into their performance post-heat exposure and laying the groundwork for fire safety considerations.

The second research topic is strengthening and rehabilitation techniques with FRP. Paper [11] assesses the use of AR-GT fabrics to increase the flexural capacity of RC beams, demonstrating significant improvements in load-carrying capacity with both internal and external layer applications. Paper [12] introduces three different strengthening methods, with the wrapped CFRP method demonstrating the most significant enhancement in carrying capacity and ductility. The paper also includes a theoretical calculation of the carrying capacity. Paper [13] introduces a dispersed-tendon cable anchor system for BFRP cables, aiming to increase anchoring efficiency, an innovation that could influence the design of future FRP anchoring systems. Paper [14] presents an experimental study on the shear behavior of masonry walls reinforced with FRP, indicating notable enhancements in strength and ductility, especially with CFRP composites.

The third research topic is the bonding properties of cementitious materials. Paper [15] investigates the interfacial properties between lightweight concrete (LWC) and engineered cementitious composites (ECC) with different polyvinyl alcohol (PVA) and glass fiber dosages under different surface roughness conditions. Paper [16] investigates the interfacial bonding behavior between high toughness resin concrete with steel wire mesh (HTRCS) and concrete and proposes an interfacial bond-slip model and a bearing capacity equation. Paper [17] examines the types, characteristics, and identification of damage and defects that were either observed or expected in EB-FRP concrete elements.

The research presented in this Special Issue encapsulates a diverse spectrum of investigative and practical advances, demonstrating the transformative impact of FRP composites across various construction scenarios. From innovative methods to robust models, these papers significantly enhance the design, assessment, and execution of construction projects. As the research unfolds, the pivotal role of FRP composites in the evolution of modern construction is both illuminated and affirmed.

Conflicts of Interest: The authors declare no conflict of interest.

References

1. Guo, R.; Hu, W.; Li, M.; Wang, B. Study on the flexural strengthening effect of RC beams reinforced by FRP grid with PCM shotcrete. *Compos. Struct.* **2020**, *239*, 112000. [CrossRef]
2. Wang, B.; Uji, K.; Wu, T.; Dai, H.; Yan, D.; Guo, R. Experimental investigation of stress transfer and failure mechanism between existing concrete and CFRP grid-sprayed PCM. *Constr. Build. Mater.* **2019**, *215*, 43–58. [CrossRef]
3. Guo, R.; Pan, Y.; Cai, L.; Shinichi, H. Bonding Behavior of CFRP grid-concrete with PCM shotcrete. *Eng. Struct.* **2018**, *168*, 333–345. [CrossRef]
4. Guo, R.; Ren, Y.; Li, M.; Hu, P.; Du, M.; Zhang, R. Experimental study on fexural shear strengthening effect on low-strength RC beams by using FRP grid and ECC. *Eng. Struct.* **2021**, *227*, 111434. [CrossRef]
5. Shawki Ali, N.K.; Mahfouz, S.Y.; Amer, N.H. Flexural Response of Concrete Beams Reinforced with Steel and Fiber Reinforced Polymers. *Buildings* **2023**, *13*, 374. [CrossRef]
6. Li, N.; Jia, B.; Chen, J.; Sheng, Y.; Deng, S. Phenomenological 2D and 3D Models of Ductile Fracture for Girth Weld of X80 Pipeline. *Buildings* **2023**, *13*, 283. [CrossRef]
7. Zierul, L.; Baumgärtel, E.; Sandmann, D.; Marx, S. Shear Transfer in Concrete Joints with Non-Metallic Reinforcement. *Buildings* **2024**, *14*, 1975. [CrossRef]
8. Wu, J.; Zhang, C. Modified Constitutive Models and Mechanical Properties of GFRP after High-Temperature Cooling. *Buildings* **2024**, *14*, 439. [CrossRef]
9. Guo, R.; Yang, D.; Jia, B.; Tang, D. Seismic Response of GFRP-RC Interior Beam-to-Column Joints under Cyclic Static Loads. *Buildings* **2022**, *12*, 1987. [CrossRef]
10. Zhang, C.; Li, Y.; Wu, J. Mechanical Properties of Fiber-Reinforced Polymer (FRP) Composites at Elevated Temperatures. *Buildings* **2023**, *13*, 67. [CrossRef]
11. Alhorani, R.A.; Rabayah, H.S.; Abendeh, R.M.; Salman, D.G. Assessment of Flexural Performance of Reinforced Concrete Beams Strengthened with Internal and External AR-Glass Textile Systems. *Buildings* **2023**, *13*, 1135. [CrossRef]
12. Wang, B.; Liu, G.; Zhou, J. Properties of Concrete Columns Strengthened by CFRP-UHPC under Axial Compression. *Buildings* **2023**, *13*, 596. [CrossRef]
13. Zhou, J.; Wang, X.; Ding, L.; Liu, S.; Wu, Z. Numerical and Experimental Study on Large-Diameter FRP Cable Anchoring System with Dispersed Tendons. *Buildings* **2023**, *13*, 92. [CrossRef]
14. Kałuża, M. Experimental Analysis of Surface Application of Fiber-Reinforced Polymer Composite on Shear Behavior of Masonry Walls Made of Autoclaved Concrete Blocks. *Buildings* **2022**, *12*, 2208. [CrossRef]
15. Al-Baghdadi, H.M.; Kadhum, M.M. Effects of Different Fiber Dosages of PVA and Glass Fibers on the Interfacial Properties of Lightweight Concrete with Engineered Cementitious Composite. *Buildings* **2024**, *14*, 2379. [CrossRef]

16. Yu, Q.; Ren, Y.; Liu, A.; Yang, Y. Study on Bonding Behavior between High Toughness Resin Concrete with Steel Wire Mesh and Concrete. *Buildings* **2024**, *14*, 1341. [CrossRef]
17. Ortiz, J.D.; Dolati, S.S.K.; Malla, P.; Mehrabi, A.; Nanni, A. Nondestructive Testing (NDT) for Damage Detection in Concrete Elements with Externally Bonded Fiber-Reinforced Polymer. *Buildings* **2024**, *14*, 246. [CrossRef]

Disclaimer/Publisher's Note: The statements, opinions and data contained in all publications are solely those of the individual author(s) and contributor(s) and not of MDPI and/or the editor(s). MDPI and/or the editor(s) disclaim responsibility for any injury to people or property resulting from any ideas, methods, instructions or products referred to in the content.

Article

Seismic Response of GFRP-RC Interior Beam-to-Column Joints under Cyclic Static Loads

Rui Guo [1], Dan Yang [1,2], Bin Jia [2,3,*] and Deyun Tang [4]

[1] School of Civil Engineering, Southwest Jiaotong University, Chengdu 610032, China
[2] School of Civil Engineering and Architecture, Southwest University of Science and Technology, Mianyang 621010, China
[3] Shock and Vibration of Engineering Materials and Structures Key Laboratory of Sichuan Province, Mianyang 621010, China
[4] China Construction First Group the Fifth Construction Co., Ltd., Beijing 100024, China
* Correspondence: jiabin216@126.com

Abstract: A total of nine specimens were constructed and tested under cyclic loads to investigate the differences in seismic behavior between glass fiber-reinforced polymer (GFRP)-reinforced concrete (RC) joints and RC beam-to-column joints. The experimental parameters included stirrup ratios, axial pressure ratios and concrete strength of the beam-to-column joints. The cyclic loading test results showed that the GFRP-RC beam-to-column joints can withstand significantly high lateral deformations without exhibiting brittle failure. Moreover, the RC beam-to-column joint exhibited significantly higher energy dissipation and residual displacement than the GFRP-RC beam-to-column joint by 50% and 60%, respectively. Finally, a shear capacity calculation method for the core zone of this kind of joint was proposed, which agreed well with the experimental results.

Keywords: glass fiber-reinforced polymer; beam-to-column; seismic performance; shear capacity

Citation: Guo, R.; Yang, D.; Jia, B.; Tang, D. Seismic Response of GFRP-RC Interior Beam-to-Column Joints under Cyclic Static Loads. *Buildings* **2022**, *12*, 1987. https://doi.org/10.3390/buildings12111987

Academic Editor: André R. Barbosa

Received: 12 October 2022
Accepted: 11 November 2022
Published: 16 November 2022

Publisher's Note: MDPI stays neutral with regard to jurisdictional claims in published maps and institutional affiliations.

Copyright: © 2022 by the authors. Licensee MDPI, Basel, Switzerland. This article is an open access article distributed under the terms and conditions of the Creative Commons Attribution (CC BY) license (https://creativecommons.org/licenses/by/4.0/).

1. Introduction

Traditional reinforced concrete (RC) frame structures absorb earthquake energy through structural deformation. However, when the residual displacement of the structures is too large, they are not conducive to earthquake relief and post-disaster reconstruction. Fiber-reinforced polymer (FRP) composites have been used as an alternative to steel reinforcement because of their favorable properties, such as their high strength-to-weight ratio, corrosion resistance, ease and speed of application, and minimal change in geometry. Aksoylu et al. [1] investigated the effect of web openings on pultruded fiber-reinforced polymer (PFRP) under compressive loads, and the results showed that pultruded profiles with carbon fiber-reinforced polymer (CFRP) wrapping are more likely to increase load carrying capacity of the structure; Mandenci and Özkılıç [2] explored the effect of porosity on the free vibration analysis of functionally graded (FG) beams with different boundary conditions using state space approach and Artificial Neural Networks (ANNs) technique, among others. Vedernikov et al. [3] demonstrated for the first time the possibility of applying large cross-sectional profiles to high-speed pultruded suitable structures.

The mechanical properties of FRP composites embedded in concrete have been widely studied, indicating that the bonding of FRP bars to concrete is sufficient and proving the feasibility of using FRP bars in place of steel reinforcement in concrete components. For example, Tavassoli et al. [4] compared the difference in seismic performance between RC columns and GFRP (glass fiber-reinforced polymer) columns. The results showed that the pier columns still had greater stiffness during a large deformation phase. Through a series of pseudo-static loading tests of RC piers with mixed configurations of steel-FRP reinforcement, Ibrahim et al. [5], Sun et al. [6], and Fahmy et al. [7] investigated whether piers with mixed reinforcement had significantly higher post-yield stiffness and significantly lower residual displacements and whether their energy dissipation capacity was consistent with that of RC piers. Kun et al. [8] conducted single shear tests on FRP-concrete bond joints with different anchorage types and found that different anchoring forms led to three different damage forms: interface delamination, FRP pull-out, and FRP fracture. The elastic-plasticity of the steel reinforcement was the main reason for the reduction in the RC members' post-yield stiffness. To overcome or reduce this shortcoming, recent studies on the seismic performance of FRP-RC frame structures are prevalent. For example, Ghomi et al. [9] and Hasaballa et al. [10] investigated the seismic performance of T-shaped GFRP-RC concrete beam-to-column joints using shear stress, reinforcement form, and concrete compressive strength as test variables, and their results showed that GFRP-RC concrete beam-to-column joints do not suffer from brittle failure and can withstand high transverse deflection. Mady et al. [11] proved that GFRP-RC concrete beam-to-column joints can reach their design capacity under cyclic loading, while their energy dissipation is significantly lower than that of RC concrete frames. Safdar et al. [12] tested three full-scale GFRP-RC T-connections under reversed cyclic loading to investigate the influence of the anchorage type at the end of the longitudinal bars of the beam on the cyclic performance of GFRP-RC T-connections. There are other ways of using FRP materials are applied in reinforcing concrete structures, such as wrapping. Gemi et al. [13] and Özkılıç et al. [14] studied the effect of FRP composite wrapping on the flexure performance of RC-filled GFRP profile hybrid beams. The ultimate load capacity, ductility, stiffness, energy dissipation capacity, and damage modes of the beams were also determined using a combination of tests and simulations.

Above all, existing studies have indicated that using FRP bars instead of steel reinforcement can further improve the strength, deformation capacity, and post-yield stiffness of frame structures. However, to date, no quantitative analysis has been performed on the seismic performance of GFRP-RC concrete frames, and there is a lack of systematic research on the calculation of the bearing capacity of the core area.

Therefore, this paper presents the pseudo-static tests on nine 1/2-scale concrete beam-to-column joints with different stirrup ratios, axial compression load ratio, and concrete strengths. In addition, the effects of different variables on their seismic performance were evaluated based on the test results. Moreover, as the utilization rate of FRP reinforcement is not considered in the existing equations for calculating the bearing capacity of the joints, a new calculation method was established for the shear bearing capacity of GFRP concrete beam-to-column joint cores based on the concept of the effective strain of GFRP bars.

2. Experimental Investigation

2.1. Materials

2.1.1. Concrete

This study used concrete design strength grades C30, C35, and C40, in adherence with the Chinese design specification of concrete structures, meaning that the design compressive strength values are 30 MPa, 35 MPa, and 40 MPa, respectively. The concrete was mixed with ordinary Portland cement (OPC), medium sand with a fineness modulus of 2.48, water, and crushed stone with a maximum size of 10 mm. The mixture proportions are tabulated in Table 1. During the process of casting concrete under continuous casting, three cubic samples (150 mm × 150 mm × 150 mm) were reserved for each strength grade concrete and then cured for 28 days under the same conditions as the corresponding joint specimens. The main mechanical properties of concrete follow standard test methods (GB/T50081-2002) [15], and the measured values of all materials are summarized in Table 2. The mean strengths of grades C30, C35, and C40 cubic specimens were 31.98 MPa, 36.05 MPa, and 40.86 MPa, respectively. Figure 1 shows the compressive test setup photo of the concrete.

Table 1. Mixture proportion of the concrete.

Grade	G$_{max}$ (mm)	Quantity (kg/m^3)			
		Water	Cement	Fine Aggregate	Coarse Aggregate
C30		218.56	383.44	720.86	1177.15
C35	10	218.75	446.43	642.86	1191.96
C40		218.94	509.17	621.18	1150.71

Note: G$_{max}$ is the maximum size of gravel.

Table 2. Material properties of concrete.

Grade	Design Strength f_{cd} (MPa)	Compressive Strength f_c (MPa)			
		Measured Values			Mean Value
C30	30	32.04	31.14	32.78	31.98
C35	35	36.56	34.38	37.21	36.05
C40	40	39.02	42.12	38.45	40.86

Figure 1. Compressive test for concrete.

2.1.2. Steel Reinforcement and GFRP Bar

Steel bars with a diameter of 6 mm, 8 mm, and 10 mm and GFRP bars with a diameter of 8 mm and 10 mm were used in this study. The GFRP bars, manufactured with glass fiber with a diameter of 36 μm, was impregnated with a suitable resin system to form a rod pattern. Its strength was determined by tensile tests (shown in Figure 2) according to the recommendations of GB/T 228.1-2010 [16]. From the tensile tests, it can be seen that the plastic contribution of steel reinforcement involved both a region of uniform deformation with all parts of the gauge length elongating to the same amount and a nonuniform region with localized deformation or necking. In the case of the GFRP bars, brittle fracture occurred in the elastic region (or after only a very small amount of plastic deformation). Furthermore, the mean values of its mechanical properties, including yield strength, ultimate strength, elongation, and elastic modulus, are summarized in Tables 3 and 4.

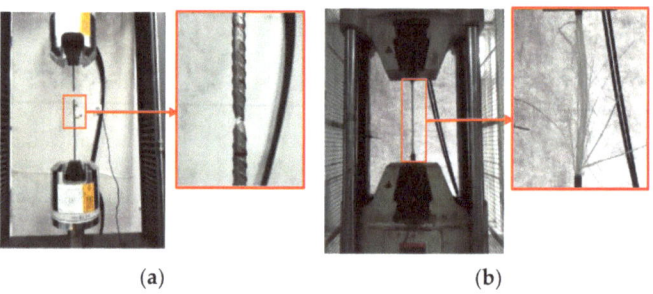

(a) (b)

Figure 2. Tensile tests for steel and GFRP bars. (**a**) Steel bars; (**b**) GFRP bars.

Table 3. Material properties of steel bars.

Diameter (mm)	Yield Strength f_y (MPa)	Ultimate Strength f_{tu}, (MPa)	Elongation (%)
6	563.87	647.19	24.03
8	486.61	581.51	22.56
10	480.46	575.16	23.84

Table 4. Material properties of GFRP bars.

Diameter (mm)	Tensile Strength f_{gt} (MPa)	Elastic Modules (MPa)
6	1481.11	5.67×10^4
8	1317.41	5.35×10^4
10	1153.71	5.03×10^4

2.2. Details of the Specimens

A total of nine joints were fabricated and tested, including eight GFRP concrete beam-to-column joints and one RC concrete beam-to-column joint as a reference specimen, and the scale ratio of all joint specimens was 1/2. The details of the test specimens are shown in Figure 3. The total length of the beam is 3000 mm, the calculated length is 2700 mm, and its cross-section is 175 mm × 250 mm. Meanwhile, to facilitate the lifting and test loading of the joint model, a certain length of loading head is reserved at the top of the column, and the distance from the lateral loading point to the fixed hinge at the base of the column is 1800 mm. The cross-section of the column is 225 mm × 225 mm.

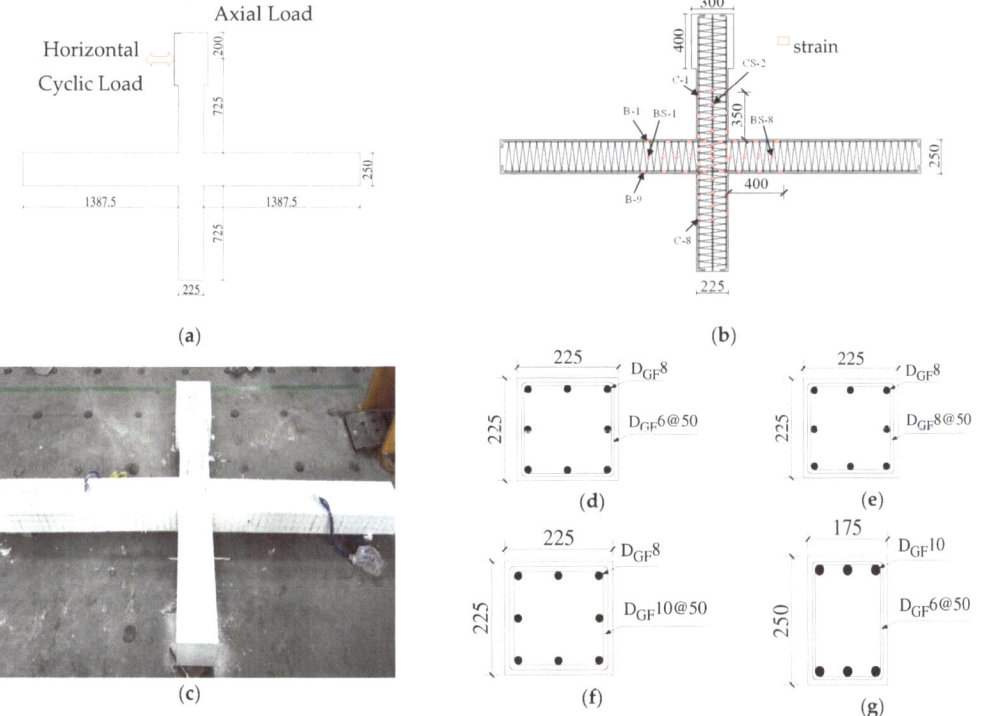

Figure 3. Details of test specimens (units: mm) (**a**) Dimensions of test specimens; (**b**) GFRP-RC specimens; (**c**) Photo of the joints; (**d**) Column BC-2; (**e**) Columns BC-5~BC-9; (**f**) Column BC-4; (**g**) GFRP-RC Beam.

The test variables in the experiments included the concrete strength, axial compression load ratio, reinforcement type (steel or GFRP bar), and stirrup ratio. Table 5 summarizes the design characteristics of the test specimens, where all joint specimens are made using concrete strengths of grade C40 and steel of HRB400 reinforcement grade. Specimen BC-1 is an RC joint model acting as a control joint with a pouring concrete strength of grade C40 and an axial pressure ratio of 0.1, and the steel stirrup diameter is 6 mm with a spacing of 50 mm. BC-2 to BC-9 are all GFRP concrete beam-to-column joint models, with beam stirrups that are GFRP spiral stirrups with a diameter of 6 mm, wherein BC-2 to BC-4 take the diameter of GFRP spiral stirrups in the column as the variable (6 mm, 8 mm, and 10 mm). BC-5 to BC-7 have different axial pressure ratios (0.2, 0.25, and 0.3) and BC-8 to BC-9 are different with respect to concrete strength (grade C30 and grade C35).

Table 5. Design characteristics of the test specimens.

Specimens	Concrete Strength	Axial Pressure Ratio	Beam		Column	
			Longitudinal Bars	Stirrups	Longitudinal Bars	Stirrups
BC-1	C40	0.1	$6 \times \Phi 10$	$\Phi 6@50$	$8 \times \Phi 8$	$\Phi 6@50$
BC-2	C40	0.1				$D_{GF}6@50$
BC-3	C40	0.1				$D_{GF}8@50$
BC-4	C40	0.1				$D_{GF}10@50$
BC-5	C40	0.2	$6 \times D_{GF}10$	$D_{GF}6@50$	$8 \times D_{GF}8$	$D_{GF}8@50$
BC-6	C40	0.25				$D_{GF}8@50$
BC-7	C40	0.3				$D_{GF}8@50$
BC-8	C30	0.1				$D_{GF}8@50$
BC-9	C35	0.1				$D_{GF}8@50$

Note: Φ means the diameter for HRB400 grade steel reinforcement; D_{GF} means the diameter for GFRP bar.

2.3. Testing Procedure

In this study, a pseudo-static loading test was carried out. As shown in Figure 4, two electrohydraulic actuators were mounted at the top of the column to the reaction frame and reaction wall. The vertical actuator provided a constant axial force and the horizontal actuator applied a cyclic load; the maximum allowable capacity of the horizontal and vertical actuators was 500 and 1000 kN, respectively. Meanwhile, the double hinge devices mounted at the end of the beams were attached to the reaction floor, and the force transducers were arranged on devices to measure the vertical reaction force at the beam end. Moreover, two rods with rolling axes were installed at the top of the column to limit the out-of-plane movement (seen in Figure 4b). To ensure that the rods did not affect the lateral translation of the specimens, it should be noted that one end of the rods was fixed to the vertical reaction frame, and the opposite end had rolling axes supporting the specimen.

Furthermore, the cyclic loads were applied by a displacement-controlled mode until the specimen fractured, which is shown in Figure 5. The cyclic loading procedure consists of several loading steps gradually increasing in lateral displacement, where the displacement increment is 5 mm per step, and three full cycles were conducted for each level. Meanwhile, strain gauges with a diameter of 3×5 mm were installed at the plastic hinge zone of the beam and column, respectively, as shown in Figure 3b. A total of 52 strain gauges were used in per specimen. In addition, a total of 14 displacement sensors were used to obtain the average curvatures of column and beam at different displacement levels. The displacement sensors were arranged in the potential plastic hinge zone and core of the beam and column.

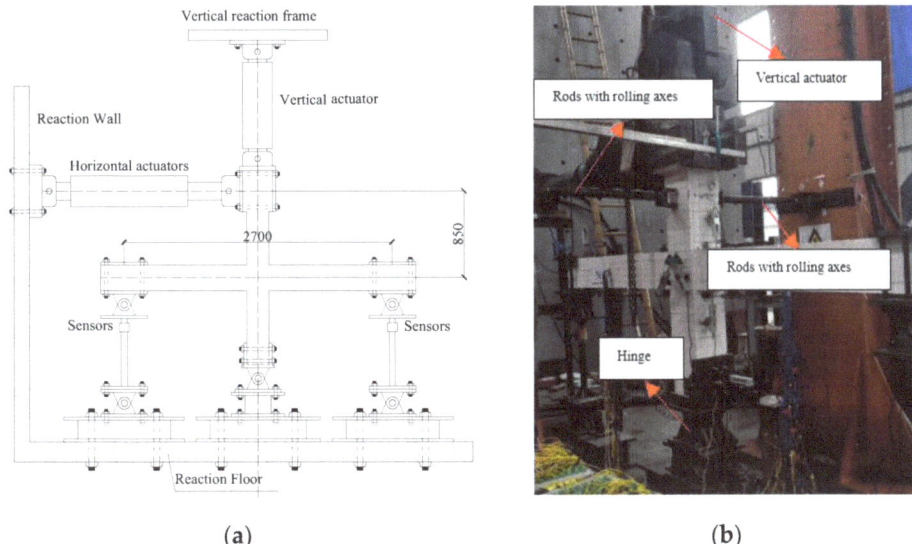

(a) (b)

Figure 4. Loading setup. (**a**) Loading setup; (**b**) Photo of setup.

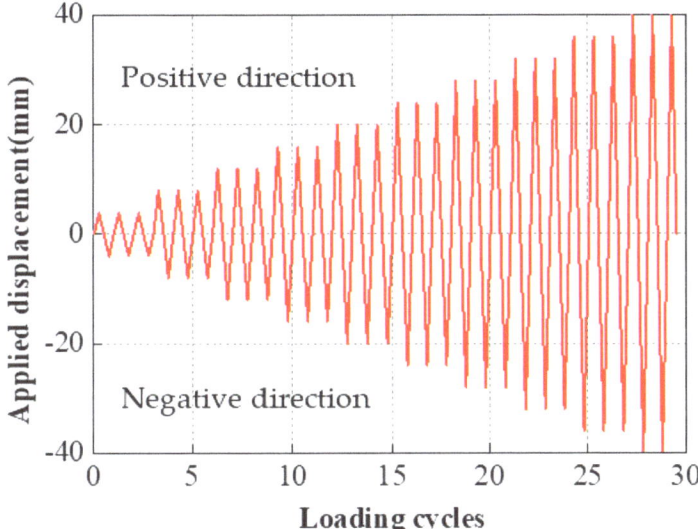

Figure 5. Loading protocol.

3. Test Results and Discussion

3.1. Crack Pattern and Failure Mode

Figure 6 shows the failure mode of each joint specimen after testing. The cracks of all specimens are mainly concentrated in the plastic hinge area of the beam; however, a few cracks occur in the non-core area of the columns, and all joint specimens undergo concrete cracking and cover concrete spalling. During the test, the first vertical bending crack appeared at the beam-to-column junction when the displacement of the column top was ±4 mm. As the cyclic load increased, a large number of cracks appeared in the plastic hinge area of the beam. Thereafter, when the cyclic load approached the peak load, cracks appeared in the beam-to-column intersection line and core area, except in the case of specimens BC-5, -6, and -7.

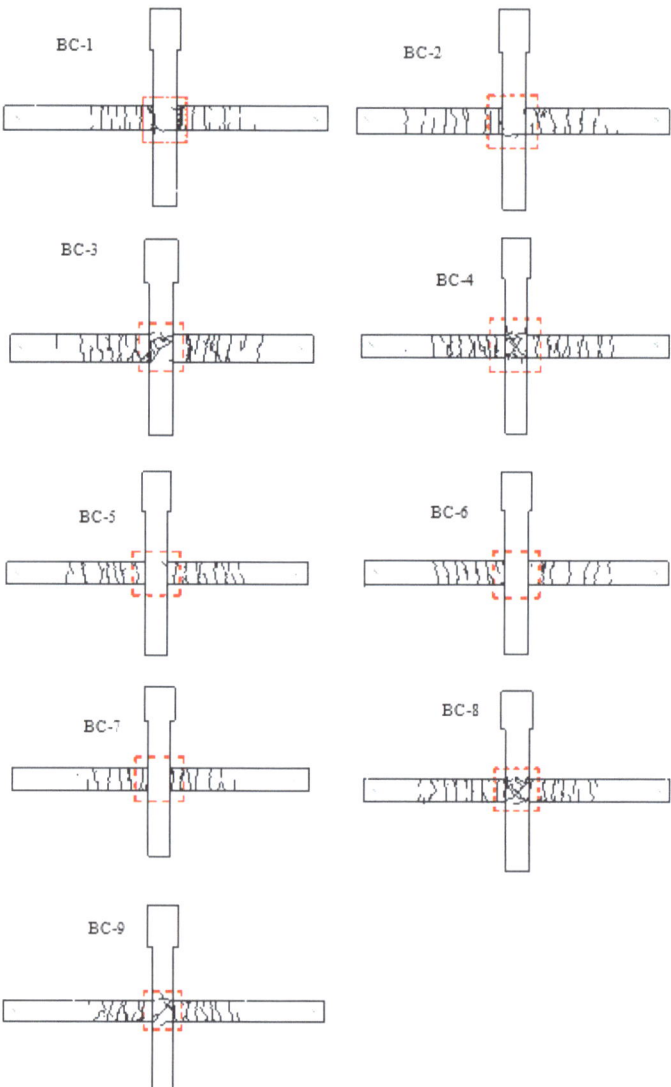

Figure 6. Crack pattern after testing.

Compared to the RC joint specimen (BC-1), the GFRP beam-to-column joints emitted a brittle sound of fiber bundle breakage when the concrete spalled. Therefore, the surface concrete of the beam end where the cracks were concentrated was chiseled away to observe the damage situation of the GFRP, as shown in Figure 7. It can be inferred that the outer rubber layer of the GFRP was pulled off first due to the cyclic load, then the fiber filament bundle gradually appeared to fracture as the load increased, and finally, the internal rubber layer fragmentation occurred in a large area. From the observed damage of the GFRP, it can be concluded that, although GFRP are brittle materials, the GFRP joints still have some energy dissipation capabilities. Meanwhile, brittle damage did not appear when the GFRP joint specimens reached a displacement angle of 5.5%, indicating that the GFRP beam-to-column joints tested can withstand relatively high lateral deformation without brittle damage.

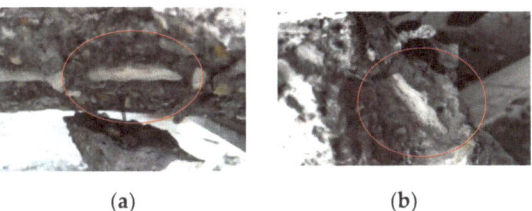

(a)　　　　　　　　　(b)

Figure 7. Observed damage of GFRP. (**a**) Outer rubber layer failure (BC-9); (**b**) Fiber bundle breakage (BC-3).

3.2. Hysteretic Load–Displacement Loops

Figure 8 shows the hysteresis load–displacement loops obtained from the cyclic loading tests. In the case of the RC joint specimen (BC-1), the total area under the hysteresis loops increases with cyclic loading after it enters the plastic phase, and a pinching phenomenon occurs at the same time. However, the hysteresis loops of specimens with different stirrup ratios and concrete strengths do not show obvious pinching phenomena. It can be speculated that the deformation of the reinforcement in the RC joint grows rapidly after yielding, and the relative slip of the concrete and reinforcement increases; however, in the case of the GFRP joints, due to the restraining effects provided by the fiber spiral hoop, the relative slip of the GFRP bars is very small after concrete failure. Therein, the maximum displacement of BC-7, which had the largest axial pressure ratio, was the smallest, and the pinching phenomenon was the most obvious among the GFRP joint specimens. In general, the energy dissipation of the GFRP joint specimens is significantly lower than that of the RC joint.

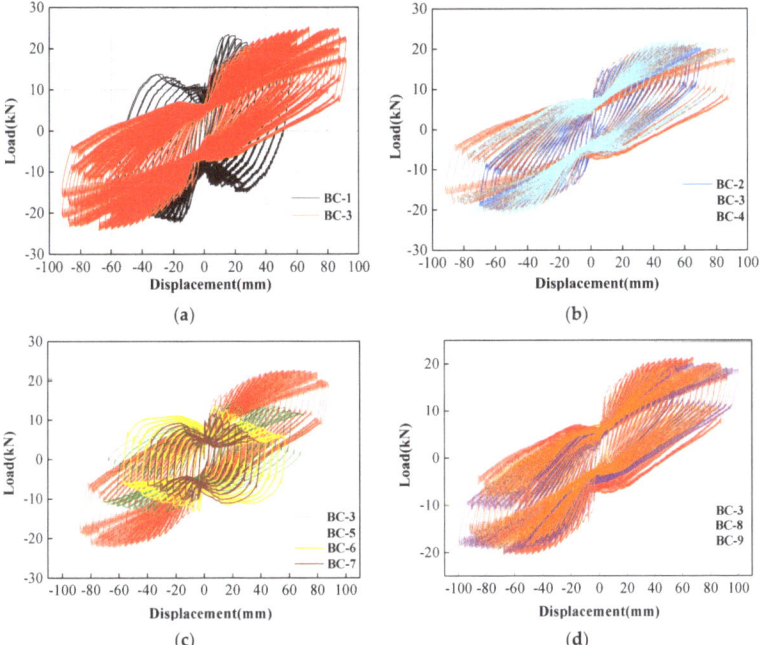

Figure 8. Hysteretic load–displacement loops. (**a**) RC vs. GFRP; (**b**) Stirrup ratio group; (**c**) Axial pressure ratio group; (**d**) Concrete strength group.

In addition, according to the displacement at the top of the column after test unloading, defined as the residual displacement, the residual displacements in the positive and negative loading directions of the RC joint specimen (BC-1) are basically equal. In contrast, the residual displacements of the GFRP joint specimens have obvious asymmetry and are much smaller, indicating that the GFRP joint has a stronger self-resetting capability than the RC joint.

Figure 9 shows the envelope load–displacement curves of all joint specimens. It seems that the bearing capacity under the same deformation of the RC joint (BC-1) is slightly larger than that of the GFRP joint specimens before the displacement of the column top reaches ±20 mm because the elastic modulus of the steel reinforcement is much larger than that of the GFRP bars (steel reinforcement: 200 GPa, GFRP bar: 26.7 GPa). However, the gap in bearing capacities among them gradually decreases with horizontal displacement. Compared to the RC joint specimen, whose bearing capacity decreases rapidly after reaching its peak point, the bearing capacities of the GFRP joint specimens decrease slowly, and the curves have obvious plateau sections. Moreover, the bearing capacities of GFRP joints increase with the stirrup ratio; meanwhile, the GFRP joint with a larger axial compression load ratio has a lower peak capacity and poor deformability. The effect of concrete strength on the bearing capacity of fiber-reinforced beam-to-column joints is not significant.

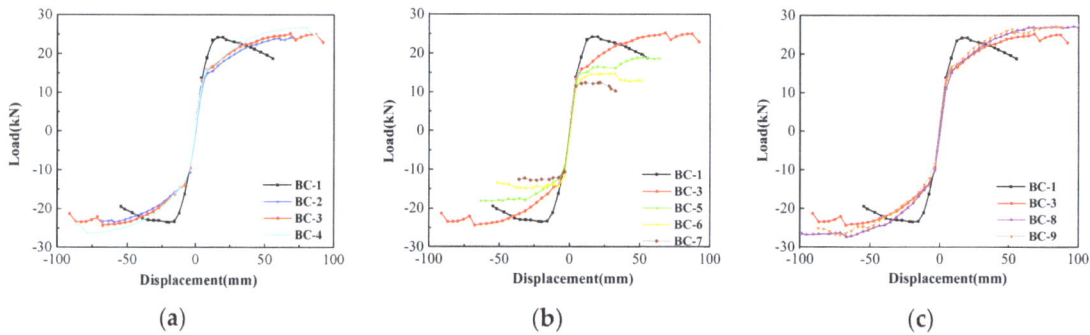

Figure 9. Envelope load–displacement curves. (**a**) Stirrup ratio group; (**b**) Axial pressure ratio group; (**c**) Concrete strength group.

Table 6 summarizes the experimental results of each specimen. The feature points of each specimen are calculated from the skeleton load–displacement curves. The yield point is obtained by graphing the farthest point method proposed by Peng et al. [17], which is the point on the curve farthest from the line connecting the origin and the peak point, and the peak point is the point on the skeleton curve of each joint model with the maximum bearing capacity. It is worth mentioning that the forward yield load of the GFRP joints is lower than that of the RC joints, but deformation exhibits the opposite trend (the deformation of the GFRP joints is significantly larger). Moreover, the yield and peak loads gradually decrease with the axial pressure ratio, and the peak loads of the joints gradually increase with the stirrup ratio. The characteristic values at each point of the beam-to-column section of GFRP bars with different concrete strengths do not significantly differ.

Table 6. Feature points of all joint specimens.

	Specimens	D_y/mm	F_y/kN	D_{max} (Drift Ratios)/mm	F_{max}/kN
Positive	BC-1	9.434	20.913	19.884 (2.34%)	24.215
	BC-2	7.664	15.240	69.944 (8.23%)	24.190
	BC-3	7.743	15.975	67.910 (7.99%)	25.288
	BC-4	7.103	15.831	79.998 (9.41%)	26.899

Table 6. Cont.

	Specimens	D_y/mm	F_y/kN	D_{max} (Drift Ratios)/mm	F_{max}/kN
	BC-5	6.119	14.775	51.990 (6.12%)	19.020
	BC-6	7.378	13.328	31.754 (3.74%)	14.840
	BC-7	3.992	11.506	21.428 (2.52%)	12.458
	BC-8	11.972	17.326	67.164 (7.90%)	27.024
	BC-9	8.046	16.747	75.584 (8.89%)	27.011
	BC-1	−12.028	−19.718	−19.882 (2.34%)	−23.475
	BC-2	−8.537	−13.238	−59.806 (7.04%)	−23.339
	BC-3	−7.998	−14.000	−67.988 (8.00%)	−24.268
	BC-4	−10.808	−14.296	−79.500 (9.35%)	−26.279
Negative	BC-5	−7.968	−12.372	−55.702 (6.55%)	−18.011
	BC-6	−3.295	−13.601	−31.940 (3.76%)	−14.668
	BC-7	−3.662	−10.578	−23.606 (2.78%)	−12.794
	BC-8	−19.508	−18.391	−67.472 (7.94%)	−27.265
	BC-9	−19.660	−17.566	−71.938 (8.46%)	−26.735

3.3. Energy Dissipation Capacity

In general, the energy dissipation capacity of the joint member can be calculated by the total area under the hysteresis loops, which is used to express the energy dissipated by the member in the process of an earthquake, as shown in Figure 10. It can be observed that all joint specimens dissipate little energy in their elastic phase; thereafter, the energy dissipation capacities increase since the specimens enter the plastic phase. Therein, the energy dissipation of the RC joint increases abruptly when the displacement of the column top reaches ±8 mm, which is significantly larger than that of the GFRP joints. However, because large plastic deformation occurs in the RC joint, the damage degree is more serious than that in the GFRP joints; therefore, the energy dissipation capacity increase rate in the RC joint becomes slower, and those of the GFRP joints still grow steadily, indicating that the GFRP joints still have a strong energy dissipation. At the same time, the hysteretic energy dissipation capacity of the GFRP joints is reduced by approximately 50% relative to that of the RC joint. Comparing the GFRP joints, it can be concluded that the energy dissipation capacities increase with the stirrup ratio due to the confinement effect on the concrete by GFRP hoops. Meanwhile, among the axial pressure ratio groups (BC-3, BC-5, BC-6, and BC-7), BC-3, having the smallest axial pressure ratio, consumes the least energy, which implies that an appropriate increase in the axial pressure ratio can lead to an increase in the energy dissipation capacities of the joint members by enhancing the aggregate interlock of concrete. On the other hand, the energy dissipation capacity of specimen BC-8 is lower than those of BC-3 and BC-9. For example, the energy dissipation capacity values of BC-8 is 5.55% and 2.22% lower than that of BC-3 and BC-9, respectively, both being under the +84 mm working condition, indicating that the energy dissipation capacity of the GFRP reinforcement beam-to-column joints increases with increasing concrete strength, but the increase is not obvious.

Furthermore, stiffness degradation is typically used to characterize the stiffness attenuation of joint members under cyclic loading, which can be calculated by the concept of equivalent stiffness, as shown by the following equation:

$$K = \frac{|F^+| + |F^-|}{|\Delta^+| + |\Delta^-|} \tag{1}$$

where F^+ and F^- are the maximum loading values in the positive and negative directions of the hysteresis loops, respectively, and Δ^+ and Δ^- are the displacement values corresponding to the maximum loading in the positive and negative directions of the hysteretic loops, respectively.

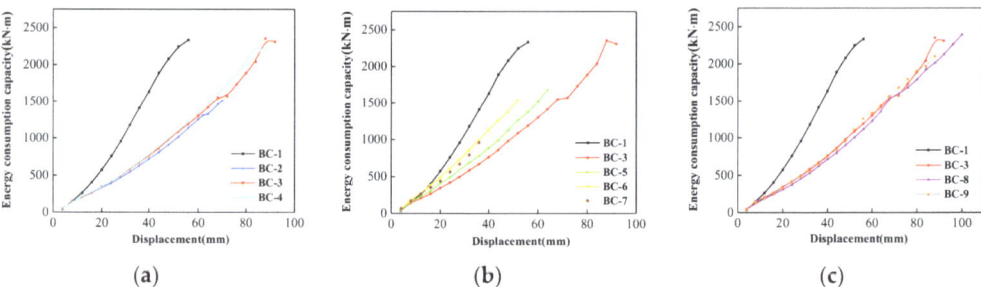

Figure 10. Energy dissipation capacities of joint specimens. (**a**) Stirrup ratio group; (**b**) Axial pressure ratio group; (**c**) Concrete strength group.

Figure 11 shows the stiffness degradation of all specimens during cyclic loading. It appears that the equivalent stiffness of BC-1 is greater than that of the GFRP joints, indicating that the longitudinal reinforcement tension affects its equivalent stiffness after entering the plastic phase due to the smaller elastic modulus of the GFRP bars. Moreover, due to the serious failure that occurred in the plastic hinge region of the BC-1 beam, the rate of stiffness deterioration of specimen BC-1 is faster than that of the GFRP joints. This comparison shows that the stiffness degradation rate of the GFRP joints increases with the axial pressure ratio, with the higher concrete strength specimen having a greater initial stiffness.

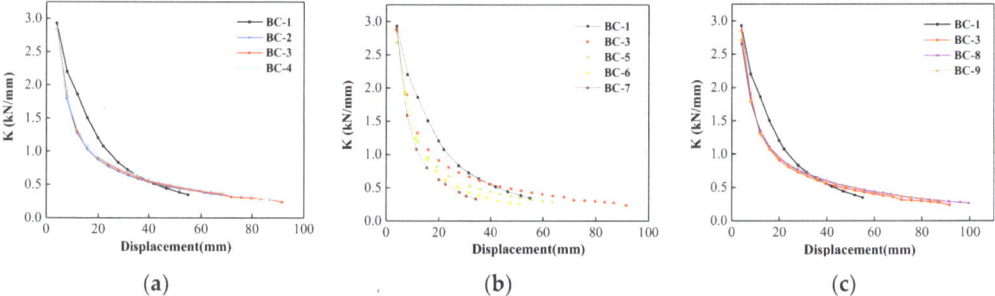

Figure 11. Stiffness degradation of joint specimens. (**a**) Stirrup ratio group; (**b**) Axial pressure ratio group; (**c**) Concrete strength group.

3.4. Stress–Strain Relationship

Strain gauges were arranged along the GFRP longitudinal bars at the end area of the beam in each specimen. Figure 12 shows that the maximum strains observed in the GFRP longitudinal bars are less than half of the rupture strain (approximately 23,000 μ), indicating that the damage of the GFRP bars takes the form of gradual softening without brittle rupture failure. Compared with the RC joint specimen, the strains of the GFRP joint specimens grow slowly without a sudden increase. The strains of the GFRP joint specimens are slightly higher before the steel reinforcement yields; thereafter, the strains of the GFRP joints are significantly lower than those of the RC joint. Furthermore, comparing the GFRP joint specimens with different variables, the GFRP longitudinal bar strain at the same displacement level gradually increases with the stirrup ratio, concrete strength, and axial compression load ratio, and the maximum utilization rate of the GFRP bars can reach 41.8%.

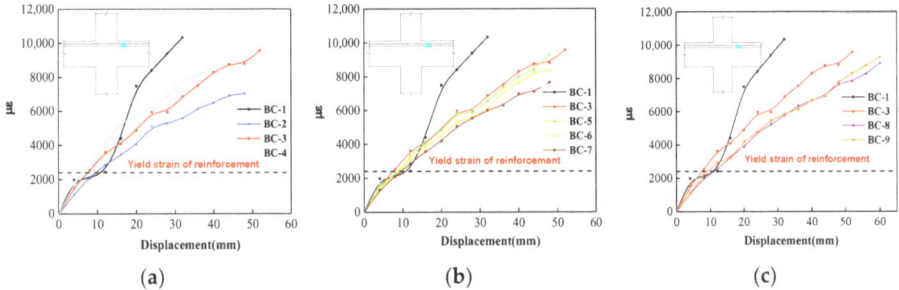

Figure 12. Maximum strains of longitudinal reinforcement. (**a**) Stirrup ratio group; (**b**) Axial pressure ratio group; (**c**) Concrete strength group.

In addition, Figure 13 shows the maximum strains of the GFRP hoops in the core area. As expected, the maximum strain in the RC joint specimen is significantly greater than those in the GFRP joint specimens, indicating that there are greater shear forces occurring in the core area of the RC joint due to the high stiffness of the longitudinal steel beam reinforcement. Comparing the GFRP joints with different variables, the maximum strains of the GFRP hoops increase with the stirrup ratio and decrease with the axial pressure ratio, and the maximum utilization rate captured in the joint stirrups exceeds 11.9% (BC-4). The GFRP hoop strains did not exceed their ultimate strains in the end. In conjunction with the final damage of the specimens, no significant damage occurs in the concrete in the core of the joint specimens, indicating that the reinforcement scheme of GFRP hoops in the core is appropriate for the joint.

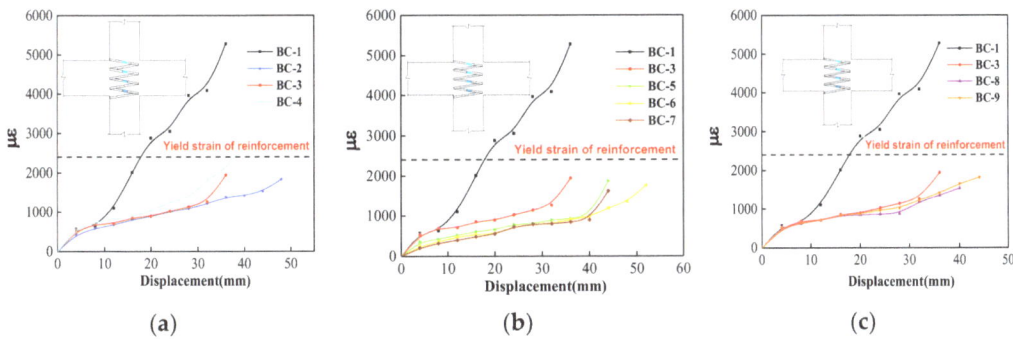

Figure 13. Maximum strains of GFRP hoops. (**a**) Stirrup ratio group; (**b**) Axial pressure ratio group; (**c**) Concrete strength group.

4. Calculation of Load Carrying Capacity

4.1. Nodal Core Shear Bearing Capacity

The reinforced concrete beam-to-column joint is very complex in terms of forces, and the core force transfer mechanism is shown in Figure 14. Its complex hysteretic behavior can be simulated by means of recent accurate and efficient models [18]. In the beam-to-column joint, there are two types of force transmission mechanisms: the inclined compression rod mechanism and the truss mechanism [19]. The diagonal compression bar mechanism (shown in Figure 14b) means that when the joint is subjected to an external load, the concrete pressure C_b at the beam end and the concrete pressure C_c at the column end will cancel a portion of the shear force of the beam-to-column section, forming a diagonal pressure field in the core area of the joint. Figure 14c shows the schematic diagram of the "truss mechanism". At the beginning of loading, the concrete in the core area bears tensile stress, which gradually increases until it reaches the tensile limit of the concrete. After the concrete cracks, the hoop reinforcement and vertical longitudinal reinforcement will bear the tensile stress, and the tensile forces T_{bs} and T_{cs}, and pressures C_{bs} and C_{cs} provided by the beam and column reinforcement will offset another part of the beam-to-column section shear force, which will be transferred to the core area to form the shear stress field and act together with the compressive stress field of the diagonal compression bar mechanism to form the truss mechanism [20].

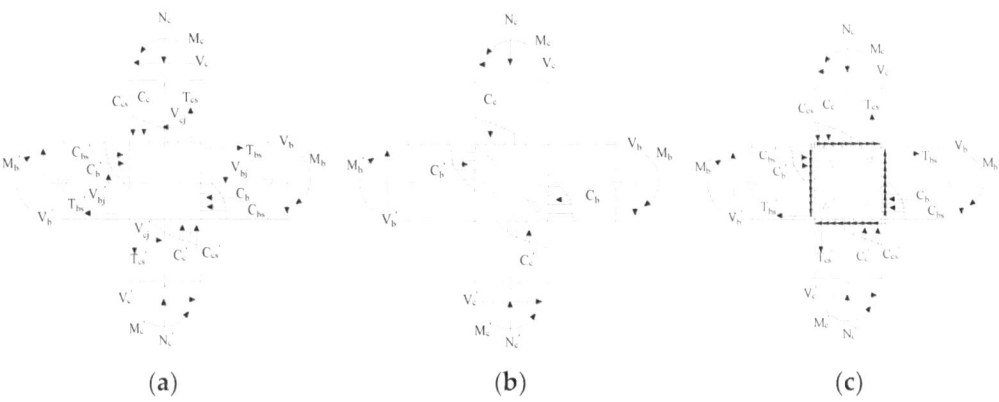

Figure 14. Shear transfer mechanism of joint core region. (**a**) Nodal force model; (**b**) Inclined lever mechanism; (**c**) Truss mechanism.

Referring to the RC concrete beam-to-column joint model and the existing diagonal compression bar model, the shear force is defined as follows [21]:

$$V_j = V_c + V_{SG} \quad (2)$$

where V_c and V_{SG} are the shear forces borne by the concrete and hoop bars, respectively:

$$V_c = \zeta f'_c b_j a_s \cos\theta \quad (3)$$

$$V_{SG} = f_{yv} \frac{A_{SV}}{s}(h_0 - a'_s) \quad (4)$$

The parameters are expressed as follows:

$$\theta = \arctan\left(\frac{h'_b}{h'_c}\right) \quad (5)$$

$$a_s = \sqrt{a_b^2 + a_c^2} \quad (6)$$

$$a_c = \left(0.25 + 0.85 \frac{N}{A_g f'_c}\right) h_c \quad (7)$$

$$\zeta \approx \frac{3.35}{\sqrt{f'_c}} \quad (8)$$

where f'_c is the concrete cylindrical compressive strength; ζ is the concrete compressive strength softening factor; b_j is the effective width of the inclined compression bar [22,23]; a_s is the height of the inclined compression bar; θ is the angle between the inclined compression bar and the horizontal axial direction; h_b' and h_c' are the outermost reinforcements between the beam-to-column cross-section distance; a_b, a_c are the beam-to-column cross-sectional pressure zone heights [24]; c_b is taken as $1/5\ h_b$, where h_b is the height of the beam cross-section; N is the column top axial pressure; A_g is the column gross cross-sectional area; h_c is the column cross-sectional height; f_{yv} is the hoop tensile strength; A_{sv} is the total area of hoop reinforcement configured in the same cross-section; s is the hoop spacing; h_0 is the effective height of the joint cross-section; and a_s' is the distance from the joint point of longitudinal compression reinforcement to the concrete edge of the same cross-section.

In summary, the hoop tensile strength is used to calculate the shear force V_{SG} borne by the hoop reinforcement. According to the proposed static test of the GFRP joint, when the GFRP joints reach their ultimate bearing capacity, the GFRP hoop does not reach its ultimate tensile strain, so the concept of the hoop utilization rate (α) is proposed in this paper. Therefore, based on the influence of the hoop ratio, axial compression load ratio, and concrete strength, the shear bearing capacity of reinforced concrete hoops can be calculated using software to carry out multiple linear regressions on the test data of eight GFRP-reinforced beam-to-column joints. The shear load capacity of GFRP joints is calculated by introducing the hoop utilization rate (α) as follows:

$$V_j = \zeta f_c' b_j a_s \cos\theta + \alpha f_{yv} \frac{A_{sv}}{s}(h_0 - a_s') \quad (9)$$

$$\alpha = 0.0025 f_c - 0.1053\mu + 2.5501\rho_{sv} - 0.052 \quad (10)$$

where μ is the axial pressure ratio and ρ_{sv} is the nodal core with the hoop ratio.

4.2. Test Verification

The values of shear bearing capacity obtained from eight GFRP-reinforced concrete beam-to-column joints in this paper are calculated according to Equation (10) and compared with the measured values. Table 7 lists the data of 22 GFRP-reinforced concrete beam-to-column joints in this paper and from other literature.

Table 7. Comparison of shear bearing capacity between theoretical and experimental values.

Specimens	Type	$b_c \times h_c$ /mm	$b_b \times h_b$ /mm	f_c /MPa	f_c' /MPa	μ	ρ_{sv}/%	V_j/kN	V_j'/kN	V_j/V_j'
BC-2	Cross-shaped	225 × 225	175 × 250	43.21	34.57	0.1	0.011	198.677	143.275	1.387
BC-3	Cross-shaped	225 × 225	175 × 250	43.21	34.57	0.1	0.020	202.909	201.938	1.005
BC-4	Cross-shaped	225 × 225	175 × 250	43.21	34.57	0.1	0.031	215.034	176.613	1.218
BC-5	Cross-shaped	225 × 225	175 × 250	43.21	34.57	0.2	0.020	228.932	201.712	1.135
BC-6	Cross-shaped	225 × 225	175 × 250	43.21	34.57	0.25	0.020	242.655	184.479	1.315
BC-7	Cross-shaped	225 × 225	175 × 250	43.21	34.57	0.3	0.020	256.732	180.252	1.424
BC-8	Cross-shaped	225 × 225	175 × 250	32.08	25.66	0.1	0.020	149.640	167.852	0.891
BC-9	Cross-shaped	225 × 225	175 × 250	35.41	28.33	0.1	0.020	165.577	176.671	0.937
G-1.3 [25]	Cross-shaped	350 × 450	350 × 450	38	30.4	0.15	0.024	809.416	968.991	0.835
G-1.8 [25]	Cross-shaped	350 × 450	350 × 450	58	46.4	0.15	0.024	1176.421	976.197	1.205
G-HT-1.0 [9]	T-shaped	400 × 350	350 × 450	47.8	38.24	0.15	0.021	803.343	738.434	1.088
G HT 1.1 [9]	T-shaped	400 × 350	350 × 450	42.2	33.76	0.15	0.021	708.375	846.611	0.837
J30-0.70 [10]	T-shaped	400 × 350	350 × 450	37.9	30.32	0.15	0.021	635.454	572.949	1.109
J30-0.85 [10]	T-shaped	400 × 350	350 × 450	32.6	26.08	0.15	0.021	545.574	490.602	1.112
J30-1.0 [10]	T-shaped	400 × 350	350 × 450	35.6	28.48	0.15	0.021	596.450	700.334	0.852
J60-0.70 [10]	T-shaped	400 × 350	350 × 450	51.3	41.04	0.15	0.021	867.281	624.406	1.389
J60-0.85 [10]	T-shaped	400 × 350	350 × 450	52.6	42.08	0.15	0.021	879.003	735.861	1.195
J60-1.0 [10]	T-shaped	400 × 350	350 × 450	52.6	42.08	0.15	0.021	879.003	881.301	0.997
H-S [26]	T-shaped	400 × 350	350 × 450	41	32.8	0.15	0.012	671.735	509.039	1.320
H-D [26]	T-shaped	400 × 350	350 × 450	31	24.8	0.15	0.012	503.956	395.719	1.274
B-S [26]	T-shaped	400 × 350	350 × 450	37	29.6	0.15	0.012	604.549	744.982	0.811
B-D [26]	T-shaped	400 × 350	350 × 450	40	32	0.15	0.012	654.871	610.418	1.073
				Mean						1.109
				Standard deviation						0.191
				Coefficient of variation						0.172

The formula for calculating the actual test value is as follows:

$$V_j' = T_l + T_r - V_c \qquad (11)$$

where V_j' is the measured shear force; T_l and T_r are the joint left and right beam end tensile steel tensions; and V_c is the column end shear force. The mean value of the ratio between the calculated and measured values of 22 GFRP-reinforced concrete beam-to-column joints is 1.109, the standard deviation is 0.191, and the coefficient of variation is 0.172.

5. Conclusions

To improve the post-earthquake restorability of the RC frame structures, an alternative solution was proposed that uses GFRP bars instead of steel bars in beam-to-column joints. The cyclic response of GFRP was experimentally investigated and the shear capacity of this joints was proposed. The following conclusions can be drawn based on the experimental and numerical results:

1. Under the cyclic displacement load, the damage of the GFRP-RC interior beam-to-column joints is mainly concentrated in the plastic hinge zone at the end of the beam, which is in line with the design concept of a "strong column and weak beam". At a 5.5% displacement drift ratio, the GFRP-RC interior beam-to-column joints did not show brittle damage, indicating that the joints can withstand significantly large lateral drift ratios.
2. Compared to RC beam-to-column joints, GFRP-RC interior beam-to-column joints have a slow increase in load capacity with increasing drift, while it can reach its design capacity. The use of GFRP bars instead of steel bars in concrete beam-to-column joints can significantly reduce the residual displacement of beam-to-column joints, but their energy dissipation capacity is also reduced.
3. The energy dissipation capacity of the GFRP-RC joints increases with increasing the axial load ratio. However, a large axial load ration can lead to large residual displacement. Thus, a lower axial load ratio is recommended to improve the self-centering capacity of the GFRP-reinforced concrete frames.
4. It is advisable to reduce the axial pressure ratio (less than 0.3) of GFRP-RC interior beam-to-column joints to improve the post-earthquake functionality of GFRP-reinforced concrete frames.
5. A shear capacity calculation method for the core zone of GFRP-RC beam-to-column joints was proposed, which agreed well with the experimental results.

The GFRP beam-to-column joints have high self-centering capacity and weak energy dissipation capacity. Therefore, this manuscript suggests that a certain amount of reinforcement can be allocated in the core area to increase the energy dissipation capacity. In addition, the GFRP and steel bars can be used simultaneously to improve the seismic performance of beam-to-column joints. More studies are needed for the response of the joints under dynamic loads, and the response of the joints need to be simulated by means of hysteretic models [27,28].

Author Contributions: Conceptualization, R.G., D.Y., B.J., and D.T.; Data Curation, D.Y.; Writing—Original Draft Preparation, D.Y.; Writing—Review and Editing, R.G. All authors have read and agreed to the published version of the manuscript.

Funding: This research was funded by the Open Fund Project of Shock and Vibration of Engineering Materials and Structures Key Laboratory of Sichuan Province grant number [No.20kfgk05]. Funder: Shock and Vibration of Engineering Materials and Structures Key Laboratory of Sichuan Province grant number.

Institutional Review Board Statement: The study did not require ethical approval.

Informed Consent Statement: The study did not involve humans.

Data Availability Statement: Some or all data and models generated or used during the study are available from the corresponding author by request.

Conflicts of Interest: The authors declare no conflict of interest.

References

1. Aksoylu, C.; Özkılıç, Y.O.; Madenci, E. Compressive Behavior of Pultruded GFRP Boxes with Concentric Openings Strengthened by Different Composite Wrappings. *Polymers* **2022**, *14*, 4095. [CrossRef] [PubMed]
2. Madenci, E.; Özkılıç, Y.O. Free vibration analysis of open-cell FG porous beams: Analytical, numerical and ANN approaches. *Steel Compos. Struct.* **2021**, *40*, 157–173.
3. Vedernikov, A.; Gemi, L.; Madenci, E. Effects of high pulling speeds on mechanical properties and morphology of pultruded GFRP composite flat laminates. *Compos. Struct.* **2022**, *301*, 116216. [CrossRef]
4. Tavassoli, A.; Liu, J.; Sheikh, S. Glass fiber-reinforced polymer-reiforced circular columns under simulated seismic loads. *ACI Struct. J.* **2015**, *112*, 103–114. [CrossRef]
5. Ibrahim, A.I.; Wu, G.; Sun, Z. Experimental study of cyclic behavior of concrete bridge columns reinforced by steel basalt-fiber composite bars and hybrid stirrups. *J. Compos. Constr.* **2017**, *21*, 4016091. [CrossRef]
6. Sun, Z.; Wu, G.; Wu, Z.; Zhang, M. Seismic behavior of concrete columns reinforced by steel-FRP composite bars. *J. Compos. Constr.* **2011**, *15*, 696–706. [CrossRef]
7. Fahmy, M.F.; Wu, Z.; Wu, G.; Sun, Z. Post-yield stiffnesses and residual deformations of RC bridge columns reinforced with ordinary rebars and steel fiber composite bars. *Eng. Struct.* **2010**, *32*, 2969–2983. [CrossRef]
8. Kun, D.; Zhipeng, J.; Song, Y.; Derun, D.; Yang, L. Experimental study on bearing capacity of FRP-concrete bonding interface with mechanical end anchorage. *J. Build. Struct.* **2020**, *41* (Suppl. 1), 399–405. (In Chinese)
9. Ghomi, S.; El-Salakawy, E. Seismic performance of GFRP-RC exterior beam-column joints with lateral beams. *J. Compos. Constr.* **2016**, *20*, 04015019. [CrossRef]
10. Hasaballa, M.H.; El-Salakawy, E. Shear capacity of Type-2 exterior beam-column joints reinforced with GFRP bars and stirrups. *J. Compos. Constr.* **2016**, *20*, 04015047. [CrossRef]
11. Mady, M.; El-Ragaby, A.; El-Salakawy, E. Seismic behavior of beam-column joints reinforced with GFRP bars and stirrups. *J. Compos. Constr.* **2011**, *15*, 875–886. [CrossRef]
12. Safdar, M.; Sheikh, M.N.; Hadi, M.N. Cyclic Performance of GFRP-RC T-Connections with Different Anchorage and Connection Details. *J. Compos. Constr.* **2022**, *26*, 04022022. [CrossRef]
13. Gemi, L.; Madenci, E.; Özkılıç, Y.O. Effect of Fiber Wrapping on Bending Behavior of Reinforced Concrete Filled Pultruded GFRP Composite Hybrid Beams. *Polymers* **2022**, *14*, 3740. [CrossRef] [PubMed]
14. Özkılıç, Y.O.; Aksoylu, C.; Yazman, Ş. Behavior of CFRP-strengthened RC beams with circular web openings in shear zones: Numerical study. *Structures* **2022**, *41*, 1369–1389. [CrossRef]
15. GB/T50081-2019; Standard for Test Methods of Concrete Physical and Mechanical Properties. China Architecture & Building Press: Beijing, China, 2019. (In Chinese)
16. GB/T228.1-2021; Metallic Materials-Tensile Testing-Part 1: Method of Test at Room Temperature. China Architecture & Building Press: Beijing, China, 2021. (In Chinese)
17. Peng, F.; Hanlin, Q.; Lieping, Y. Discussion and definition on yield points of materials, members and structures. *Eng. Mech.* **2017**, *34*, 36–46. (In Chinese)
18. Vaiana, N.; Sessa, S.; Marmo, F. A class of uniaxial phenomenological models for simulating hysteretic phenomena in rate-independent mechanical systems and materials. *Nonlinear Dyn.* **2018**, *93*, 1647–1669. [CrossRef]
19. Tasligedik, A.S. Shear Capacity N-M Interaction Envelope for RC Beam-Column Joints with Transverse Reinforcement: A Concept Derived from Strength Hierarchy. *J. Earthq. Eng.* **2020**, *26*, 1–31. [CrossRef]
20. Paulay, T.; Park, R.; Priestley, M.J.N. Reinforced concrete beam-column joints under seismic actions. *ACI Struct. J.* **1978**, *75*, 585–593.
21. Danying, G.; Ke, S.; Shunbo, Z. Calculation method for bearing capacity of steel fiber reinforced high-strength concrete beam-column joints. *J. Build. Struct.* **2014**, *35*, 71–79. (In Chinese)
22. Hwang, S.J.; Lee, H.J. Analytical model for predicting shear strengths of interior reinforced concrete beam-column joints for seismic resistance. *ACI Struct. J.* **2000**, *97*, 35–44.
23. Hwang, S.J.; Lee, H.J. Analytical model for predicting shear strengths of exterior reinforced concrete beam-column joints for seismic resistance. *ACI Struct. J.* **1999**, *96*, 846–857.
24. Paulay, T.; Priestley, M.N.J. *Seismic Design of Reinforced Concrete and Masonry Buildings*; John Wiley & Sons: New York, NY, USA, 1992; pp. 210–239.
25. Ghomi, S.K.; El-Salakawy, E. Effect of joint shear stress on seismic behaviour of interior GFRP-RC beam-column joints. *Eng. Struct.* **2019**, *191*, 583–597. [CrossRef]
26. Hasaballa, M.H.; El-Salakawy, E. Anchorage Performance of GFRP Headed and Bent Bars in Beam-Column Joints Subjected to Seismic Loading. *J. Compos. Constr.* **2018**, *22*, 04018060. [CrossRef]

27. Vaiana, N.; Sessa, S.; Rosati, L. A generalized class of uniaxial rate-independent models for simulating asymmetric mechanical hysteresis phenomena. *Mech. Syst. Signal Process.* **2021**, *146*, 106984. [CrossRef]
28. Yi-Kwei, W. Method for Random Vibration of Hysteretic Systems. *J. Eng. Mech. Div.* **1976**, *102*, 249–263.

Article

Experimental Analysis of Surface Application of Fiber-Reinforced Polymer Composite on Shear Behavior of Masonry Walls Made of Autoclaved Concrete Blocks

Marta Kałuża

Department of Civil Engineering, Silesian University of Technology, Akademicka 5 St., 44-100 Gliwice, Poland; marta.kaluza@polsl.pl

Abstract: This paper presents the results of an experimental study of the shear behavior of masonry walls made of aero autoclaved concrete (AAC) blocks strengthened by externally bonded fiber-reinforced polymer (FRP) composites. Fifteen small wall specimens were constructed and tested in a diagonal compression scheme. Two types of composite materials—carbon- and glass-reinforced polymers—were arranged in two configurations of vertical strips, adopted to the location of the unfilled head joints. The effect of the strengthening location and strengthening materials on changes in the strength and deformability parameters are discussed and the failure process of unstrengthened walls is also presented. The placement of the composite on unfilled head joints proved to be a better solution. Carbon-fiber-reinforced polymer (CFRP) strips provided a threefold increase in stiffness, a 48% increase in load-bearing capacity and a high level of ductility in the post-cracking phase. Glass-fiber-reinforced polymer (GFRP) strips offered a 56% increase in load-bearing capacity but did not change the stiffness of the masonry and provided relatively little ductility. Placing the composite between unfilled joints was only reasonable for CFRP composites, providing a 35% increase in load-bearing capacity but with negligible ductility of the masonry.

Keywords: AAC blocks; FRP strengthening; shear behavior; masonry walls; diagonal compression

1. Introduction

Aero autoclaved concrete blocks (AAC) are commonly used to erect load-bearing walls in low-rise buildings, as well as the infill walls in a frame system [1,2]. The popularity of this material is mainly due to its very good physical parameters, particularly its excellent thermal insulation properties, relatively high fire resistance and low density (resulting in the weight of the elements made of AAC) [3,4]. The second positive aspect is the widespread workability and very large variety of available products. This material can be easily processed (cutting on site), transported, and it provides a fast and simple technique for erecting walls with thin horizontal joints and unfilled vertical joints (a huge advantage in terms of the time investment required). Unfortunately, this technology makes walls made of AAC blocks sensitive to shear forces.

Enhancement of the shear parameters of existing masonry walls can be carried out through the application of externally bonded nonmetallic materials, such as fiber-reinforced polymer (FRP) composites [5–8]. This material has long been used to effectively strengthen reinforced concrete elements [9–12], primarily due to its very good strength parameters and the corrosion resistance of the composites [13–15]. The FRP system uses laminates or fabrics reinforced with high-strength carbon, glass or other nonmetallic fibers. These materials are glued to the surface of the elements using systemic epoxy adhesives.

There are numerous studies available in the literature describing the high effectiveness of this method in terms of increasing the load-bearing capacity of the masonry walls made of ceramic or stone elements subjected to static in-plane shearing [16,17]. However, the epoxy resins used here (adhesive layer) significantly deteriorate the diffusivity of such strengthened walls. Therefore, in order to ensure at least a partial diffusion of water vapor from the strengthened masonry elements, the composite is arranged in one-way strips, a grid setup or diagonal configuration (in the direction of the tensile stresses in the wall being sheared). An important finding of the study conducted by Valluzi et al. [18] was that the diagonal configuration of the FRP strips is more efficient, in terms of shear capacity, than the grid setup; however, the grid arrangement offers a better stress redistribution. Similar conclusions were reached by Kalali and Kabir [19], and Bui et al. [20]. Santa-Maria et al. [21] indicated that horizontally placed CFRP strips were more effective in crack propagation than a diagonal arrangement, which, in contrast, increased energy dissipation. It has been proven that the less stiff FRP material (with lower E-modulus) appeared to be more effective in terms of the ultimate strength and stiffness increase in the masonry panels [18,22]. Luccioni and Rougier [23] compared the impact of different CFRP configurations in retrofitted and repaired solid clay walls, indicating that the strip arrangement of FRP in repairing techniques presents the same benefits as FRP retrofitting. Noteworthy are the studies of two research teams, Kwiecień et al. [24,25] and Umair et al. [26], which successfully attempted to eliminate the major disadvantage of the FRP solution, namely the delamination of the strengthening due to the low stiffness of the adhesive and the composite itself, compared to the relatively highly deformable masonry substrate. In the study [24], flexible polymer joints were used to achieve more ductile behavior of the strengthened structures with a simultaneous increase in their load-bearing capacity. Umair et al. [26] proposed the use of a combination of different FRP materials (CFRP, AFRP and GFRP strips) and PP bands (polypropylene), which are characterized by a high tensile failure strain. Such a combination of materials has not only increased the initial strength and deformation capacity, but also the residual strength of masonry wall panels. The positive effects of the PP band are also presented by Sathiparan and Meguro [27].

AAC blocks are unusual masonry materials that are not very well recognized in research. There are very few studies reporting the influence of FRP strengthening system on the load-bearing capacity of such walls. The first experiments were initiated by a team led by Kubica [28,29]. Their studies indicated the positive influence of strengthening in the form of vertical FRP strips, providing a significant (30–75%) increase in load-bearing capacity and deformability of the walls, depending on the composite used. Saad et al. [30] presented the results of testing two walls made of AAC blocks and subjected to lateral static loads. The walls were strengthened in a single grid configuration (vertical and horizontal strips) using CFRP fabrics, which wrapped the entire wall. An almost two-and-a-half fold increase in the load capacity and stiffness of the strengthened walls was indicated. Interesting conclusions from explosion tests, which were conducted on 10 cm thick panels made of AAC, were presented in the work of the team lead by Wang [31]. The panels were strengthened using CFRP sheets. Such operations made it possible to achieve an increase in mechanical properties and excellent anti-blast resistance of strengthened panels. In addition to these singular studies, it is also possible to find some analysis of the impact of full-surface strengthening applications using various types of glass [32] and basalt [33] meshes, systemic PBO materials [34] or highly ductile concrete cover [35,36]. However, this research involves a different type of wall strengthening based on TRM systems.

Taking into account the results of the studies shown in [28,29] and the conclusions derived from the work by [18,22], this paper presents a detailed analysis of the influence of the vertical strips made of CFRP and GFRP materials on the behavior of small masonry walls. The most feasible way of laying the FRP material was chosen, i.e., vertical strips covering (type a) or not covering (type b) the unfilled head joints. The specimens were tested in a diagonal compression scheme according to recommendation in [37]. The first part of the paper presents the results of the laboratory tests and analyzes the behavior of the walls in the pseudo-elastic phase (until full load-bearing capacity is reached) and post-cracking phase, depending on the location and type of the strengthening material used. The second part includes discussion of the failure process of the unstrengthened and selected strengthened walls and comparative analysis of the impact of strengthening mode, with respect to the unstrengthened walls. Such a detailed qualitative analysis of the issue and describing the failure process of unstrengthened and strengthened walls made of AAC blocks has not been performed before. The information provided in the cited studies is fragmentary and does not allow for a proper recognition of the changes that occur in walls strengthened using this method.

The investigation presented in this paper is the first part of a large research program to identify an effective method for enhancing the shear parameters of AAC block walls and to develop a simple calculation method that takes into account the increase in load-bearing capacity and deformability of such walls.

2. Materials and Testing Procedure

2.1. Characteristic of the Masonry Walls

Laboratory tests were carried out on small masonry walls according to the recommendations in Rilem Lumb 6 [37]. The dimensions of the elements were 805 × 900 mm, with a thickness of 240 mm. Each wall consisted of four rows of one and a half AAC blocks. The dimensions of a single block were 200 × 600 × 240 mm. The walls were made using the typical erecting technique for AAC blocks, i.e., with thin bed joints (up to 3 mm) and unfilled head joints. Figure 1 presents the AAC block and the preparation of the specimens.

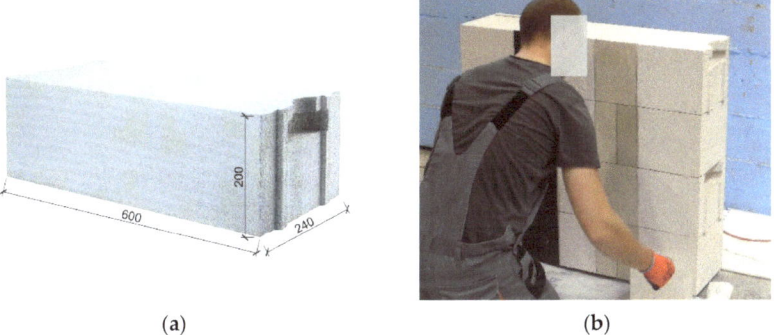

Figure 1. The view of element: (**a**) AAC block; (**b**) masonry wall before the application of strengthening.

The compressive strength of AAC blocks was determined according to EN 772-1:2000 [38]. The normalized mean compressive strength of masonry units are specified in Table 1. The standard deviation is given in brackets.

Table 1. Main strength parameters of masonry components.

Materials	Compressive Strength (N/mm^2)	Flexural Strength (N/mm^2)	Density (kg/m^3)
AAC blocks	4.65 (0.49)	-	600
Mortar	16.91 (1.74)	4.57 (0.51)	-

The mortar used for the thin bed joints was tested in accordance with the recommendations in EN 1015-11 [39]. According to very limited data obtained from the mortar manufacturer, the base of the mortar is Portland cement, dust from the Portland cement production and calcium hydroxide, with their mass content below 40%, 1% and 3%, respectively. The flexural strength of the mortar was determined on typical beams of 40 × 40 × 160 mm, and then the compressive strength of the beam halves was determined. The strength parameters are specified in Table 1.

2.2. Characteristics of the FRP Strengthening

The walls were strengthened with two types of FRP composite, produced by S&P Company. The carbon-fibers-reinforced polymer (CFRP) was C-Sheet 240, with a weight of fibers 200 g/m², and the glass-fiber-reinforced polymer (GFRP) used was G-Sheet AR 90/10. The sheets were glued to the masonry surface using two-component epoxy resin dedicated to a given strengthening system—S&P Resin 55. Two types of FRP configuration were adopted: (a) the FRP strips held together the unfilled vertical joints and (b) the strips were placed in the areas between the unfilled vertical joints. The strengthening arrangement is shown in Figure 2. In the case of the carbon sheets, the strengthening strips were 150 mm wide, while the glass sheets were 200 mm wide. The FRP strengthening was made on both sides of the wall. The parameters of the sheets and the epoxy—according to the manufacturer's data—are summarized in Table 2.

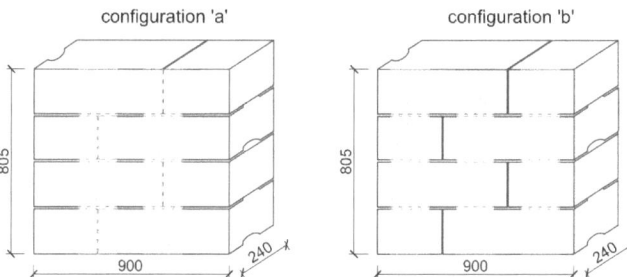

Figure 2. Two arrangements of the FRP strengthening.

Table 2. Main strength parameters of FRP strengthening.

Material	Ultimate Stress (N/mm²)	E-Modulus (kN/mm²)	Ultimate Strain (%)	Longitudinal Fiber Fraction (%)
C-Sheet 240	3800	240	1.55	100
G-Sheet AR	≥2400	73	4.50	90
Resin 55	≥100 (in compression)	≥3.20	1.73	-

2.3. Experimental Program

In total, 15 single ACC walls were made and tested. The experimental program was divided into three main groups: unstrengthened elements, wallets strengthened using CFRP sheets and GFRP sheets. In each series, three elements were tested and a summary of all the tested wallets is given in Table 3.

Table 3. Summary and designations of all test series.

Specimens	Number of Specimens	Type of Strengthening	Description
Y-US	3	unstrengthened	wallets without strengthening
Y-CFRP-a	3	FRP strengthening	walls strengthened with carbon strips in arrangement 'a'
Y-CFRP-b	3	FRP strengthening	walls strengthened with carbon strips in arrangement 'a'
Y-GFRP-a	3	FRP strengthening	walls strengthened with glass strips in arrangement 'b'
Y-GFRP-b	3	FRP strengthening	walls strengthened with glass strips in arrangement 'b'

2.4. Testing Protocol

The walls were tested in a diagonal tension scheme according to the RILEM Lumb 6 [37] standard. The loading was applied using a manually activated hydraulic jack, placed on the upper edge of the panel. The speed load was approximately 0.15 kN/s.

During the test, the applied load and the diagonal displacements were measured and recorded by two linear variable displacement transducers (LVDT) on each side. The base measurement of the LVDT was 900 mm. In two of the models, one unstrengthened and one from the Y-CFRP-a series, an optical measurement of the displacement of the wall surface was performed using the ARAMIS measurement system. Figure 3 shows the test setup, model with the traditional measurement system (inductive gauges) and the Y-CFRP-a series model with optical measurement, which were prepared for testing.

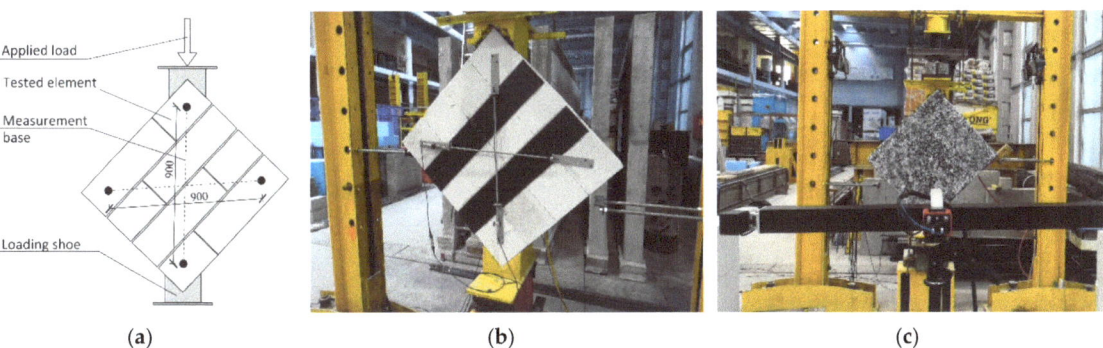

Figure 3. Masonry wallets prepared for tests: (**a**) test setup; (**b**) model with traditional measurement system; (**c**) model with surface prepared for optical measurements.

2.5. Analyzed Strength and Deformation Parameters

2.5.1. Shear Strength

Traditionally, diagonal compression testing of masonry walls is used to determine the tensile strength of the elements, which results in the failure mode—diagonal cracking. Such an assumption is made by assuming that the panel starts to collapse at its center when the principal tensile stress attains its maximum value. Therefore, the strength parameter determines the principal tensile stress in the center of the panel, assuming its isotropic elastic properties.

The strength parameter is calculated as:

$$S_{dt} = 0.707 \frac{P_i}{A_n} \tag{1}$$

where P_i is the load value and An is the net area of the panel section determined using all dimensions of the masonry wall (*h*—height, *l*—length, *t*—thickness) as:

$$A_n = \frac{l+h}{2} \cdot t \tag{?}$$

2.5.2. Shear Deformation Parameters

In addition to determining the strength, the deformation analysis is important for defining the deformation capacity of the panels. Based on the measurements of the elongation along the diagonals, the angular strain (shear strain) is calculated as:

$$\gamma = \frac{\Delta V + \Delta H}{l_g} \tag{3}$$

where ΔV is the vertical shortening, ΔH is the horizontal lengthening and l_g is the gauge length equal to 900 mm. To improve readability, the stress–strain relationship diagrams were made independently for each type of strengthened wall; an identical scale was used for easier comparison.

The displacements of the wall in a vertical (compression) and horizontal (tensile) direction are also presented.

An important parameter that describes the stiffness of the masonry in the elastic phase is the shear modulus (G). The G modulus is defined as the secant modulus between 10 and 40% of the maximum shear stress:

$$G = \frac{\sigma_{0.4} - \sigma_i}{\gamma_{0.4} - \gamma_i} \quad (4)$$

where $\sigma_{0.4}$ and $\gamma_{0.4}$ are the stress and strain at 40% of the maximum load, respectively. The initial stress and strain (σ_i and γ_i) were taken at a load level of 10% of the maximum diagonal load [40].

In many cases of the analyzed strengthened walls [20,41], the first cracks were followed by a phase of so-called 'pseudo-ductile' behavior. The pseudo-ductility coefficient (μ_d) best characterizes this phase of the structure's operation. It describes the behavior of the wall in the post-elastic or post-peak phase. This coefficient is a ratio between the ultimate and elastic strain. The ultimate strain corresponds to the largest strain experienced during the test or, in the case of ambiguous identification of the moment of damage, the strain at a level of shear stress 20% below the peak, if the stress–strain diagram continues with a descending branch [41]. The definition of elastic strain (sometimes called cracking or yielding strain) was adopted, depending on the behavior of the masonry in the elastic phase. The elastic strain can be defined differently [41]: at the bend-over point where the stress–strain curve tends to be flat (yielding strain) or when the shear strain amounted to 70% or 75% of the peak load. In this paper, the ultimate strain (γ_u) corresponds to a strain value at a level of shear stress 20% below the maximum shear stress (or, if the failure occurs faster, at the ultimate load); the elastic strain (γ_{cr}) is taken at the moment when the first crack appears.

$$\mu_d = \frac{\gamma_u}{\gamma_{cr}} \quad (5)$$

3. Results

It should be emphasized that the behavior and mode of failure of the tested walls are related to the adopted method of testing (diagonal compression) and do not fully reflect the work of the actual wall. The test stand for determining the tensile strength of a masonry wall in a diagonal compression scheme does not limit the possibility of displacement of the wall in the plane, which, in the actual structure, is ensured by further wall fragments or perpendicular walls. Additionally, the specimens are rotated with no contact to the ground. This is most evident in unstrengthened walls, as their failure mode differs from the actual mode.

3.1. Unstrengthened Wallets

In unstrengthened elements (Y-US) tested under diagonal compression, only one phase was distinguished—up until the cracking. The appearance of the crack was identified in a state of complete damage to the tested models. Therefore, the failure was characterized by one wide open crack with an almost diagonal orientation. No previous cracking was observed and the damage itself appeared suddenly and proceeded rapidly. The element split into two independent pieces (Figure 4a,b). The crack runs through unfilled joints and through masonry elements. It can also be seen that the bond capacity of the thin joint is insufficient as, each time, there was a detachment of the masonry elements in the plane of their horizontal connection.

(a) (b)

Figure 4. Damage of the unstrengthened wallets: (**a**) element no 1; (**b**) element no 2.

The characteristics of the tested models were determined by the stress–strain relationship, which was linear in each case, with a proportional increase in stress as a function of the strain. So, the elements exhibited pseudo-elastic behavior, ending with the brittle failure of the wall. The stress–strain relationships of unstrengthened walls are shown on the graphs, along with the characteristics of the strengthened masonry walls, which allow a better assessment of the impact of strengthening.

Table 4 summarizes the values of the ultimate load (which is both the cracking load and the load-bearing capacity of the wall), the recalculated stresses, the strains and the shear modulus. The table does not include a pseudo-ductile coefficient, as the unstrengthened walls did not exhibit a ductile stage.

Table 4. Specific values characterized the unstrengthened walls.

Specimens	Cracking ≅ Load-Bearing Capacity			G Modulus (GPa)
	Force (kN)	Stress (MPa)	Strain (‰)	
Y-US-s.1-1	78.28	0.270	1.331	260
Y-US-s.1-2	76.40	0.264	1.297	192
Y-US-s.1-3	75.88	0.262	1.228	268
Mean value	**76.85**	**0.265**	**1.285**	**240**

3.2. Walletes Strengthened Using FRP Materials

3.2.1. Characterization of Walls Strengthened in Configuration 'a'

Walls strengthened using FRP strips covering unfilled head joints (configuration 'a') behaved similarly, regardless of the strengthening material used. Three phases of the element's operation can be distinguished: the period up until the cracking (phase I), reaching full load-bearing capacity (phase II) and the final damage (phase III). These phases are clearly visible in the stress–strain relationship shown in Figure 5a,b. Figure 6 shows the build-up of the displacement in the direction of the main stresses (compression—axial direction and tension—horizontal direction) as a function of the applied load. In addition, Table 5 summarizes the shear stress values that correspond to the characteristic points, namely: cracking, load-bearing capacity and ultimate damage. Table 6 lists the deformation parameters of the strengthened walls.

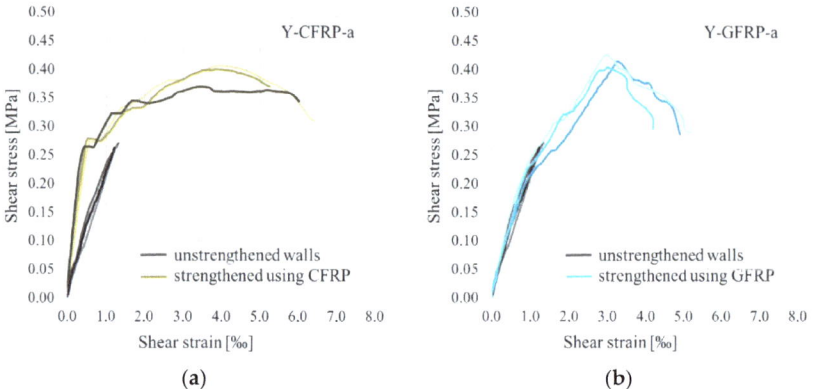

Figure 5. The stress–strain relationships for the wallets strengthened in configuration 'a' and unstrengthened walls: (**a**) the elements with carbon strips; (**b**) the elements with glass strips.

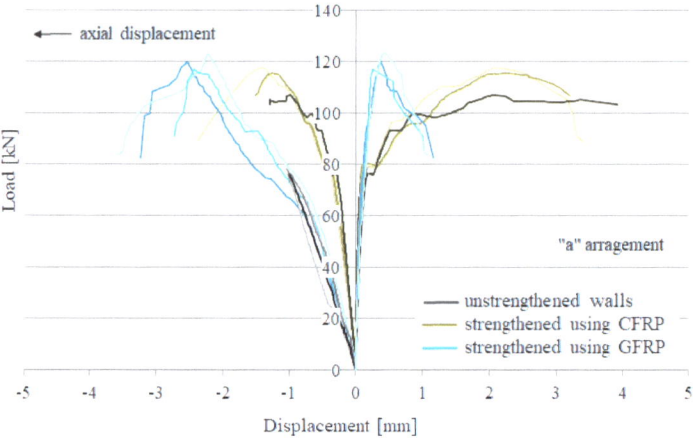

Figure 6. Horizontal and vertical displacement of the tested walls—configuration 'a'.

Table 5. Specific forces and stresses characterizing the walls strengthened in configuration 'a'.

Specimens	Cracking		Load-Bearing Capacity		Damage	
	Force (kN)	Stress (MPa)	Force (kN)	Stress (MPa)	Force (kN)	Stress (MPa)
Y-CFRP-a-1	78.97	0.273	117.80	0.406	89.29	0.308
Y-CFRP-a-2	80.59	0.278	115.41	0.398	107.03	0.369
Y-CFRP-a-3	76.54	0.264	106.99	0.369	103.53	0.357
Mean value	**78.70**	**0.272**	**113.40**	**0.391**	**99.95**	**0.345**
Y-GFRP-a-1	61.34	0.212	119.90	0.414	82.77	0.286
Y-GFRP-a-2	68.66	0.237	123.31	0.426	83.92	0.290
Y-GFRP-a-3	67.40	0.233	116.99	0.404	85.55	0.295
Mean value	**65.80**	**0.227**	**120.07**	**0.414**	**84.08**	**0.290**

Table 6. Deformation parameters characterized the walls strengthened in configuration 'a'.

Specimens	Shear Strain (‰)			G Modulus (GPa)	Pseudo-Ductility Coefficient
	Cracking	Load Capacity	Damage		
Y-CFRP-a-1	0.574	3.936	6.421	533	10.8
Y-CFRP-a-2	0.526	3.654	5.243	581	10.0
Y-CFRP-a-3	0.451	3.452	5.799	727	12.9
Mean value	0.517	3.681	5.821	613	11.2
Y-GFRP-a-1	0.964	3.240	4.892	226	4.9
Y-GFRP-a-2	0.923	2.976	5.184	284	5.0
Y-GFRP-a-3	1.006	2.984	4.169	230	4.1
Mean value	0.964	3.067	4.748	247	4.7

In the first phase, the elements exhibited a pseudo-elastic behavior that ended in cracking. Within this range, the modulus of elasticity was determined. It can be seen that the CFRP sheets provided much higher wall stiffness (2.5 times higher than GFRP sheets) which is related to a significant reduction in structural deformation at cracking. The strain of the walls of series Y-CFRP-a was only 0.517‰, while the cracking strain of the walls strengthened with GFRP strips was over 80% higher. A positive aspect of the use of CFRP sheets—in comparison with the application of GFRP sheets—was also the increase in cracking forces, which led to extension of the uncracked condition.

The second phase, ending in reaching full load-bearing capacity, varied depending on the material used. Carbon sheets provided a certain level of ductility to the structure; the deformation increased much faster than the loads. Therefore, it can be considered that the structure exhibited elasto-plastic behavior, which is confirmed by the high pseudo-ductility coefficient (Table 6). The GFRP sheets prevented the uncontrollable growth of deformation in the structure (the AAC blocks were gradually moving apart in the area of the unfilled joints) and, therefore, we do not observe the ductility effect. The value of the pseudo-ductility coefficient is smaller than 5, which confirmed the above. Eventually, a similar load-bearing capacity was obtained in both test series, with a slight advantage in favor of glass sheets. The GFRP sheets also provided less deformation at maximum force, but it was only 15% less.

In both configurations, the FRP materials provided the post-peak phase, i.e., the post-failure capacity. Both graphs (Figure 5a,b) show a descending branch of the curves.

3.2.2. Characterization of Walls Strengthened in Configuration 'b'

The application of the FRP sheets in the areas between unfilled head joints (configuration 'b') significantly changed the behavior of the elements, depending on the type of strengthening material. Figure 6 shows the successive work phases of the walls strengthened with CFRP (Figure 7a) and GFRP (Figure 7b) composites. Tables 7 and 8 contain the relevant quantities (force, stress, strain, shear modulus and coefficient) determined at characteristic points—cracking, load-bearing capacity and failure.

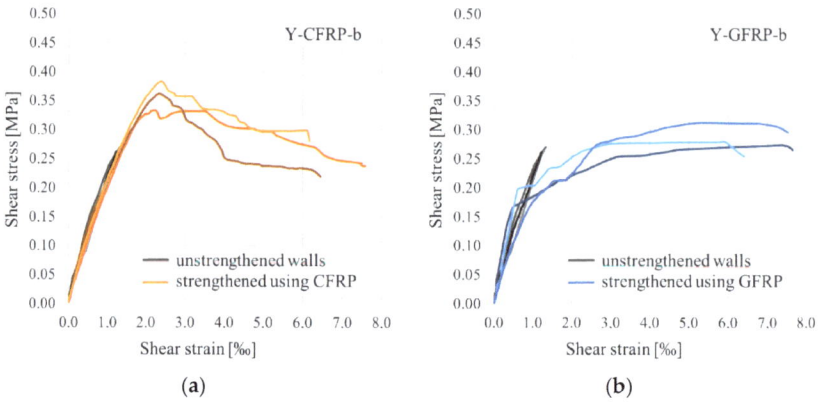

Figure 7. The stress–strain relationships for the wallets strengthened in configuration 'a' and unstrengthened walls: (**a**) the elements with carbon strips; (**b**) the elements with glass strips.

Table 7. Specific forces and stresses characterizing the walls strengthened in configuration 'b'.

Specimens	Cracking		Load-Bearing Capacity		Damage	
	Force (kN)	Stress (MPa)	Force (kN)	Stress (MPa)	Force (kN)	Stress (MPa)
Y-CFRP-a-1	96.12	0.332	96.12	0.332	68.96	0.238
Y-CFRP-a-2	104.82	0.362	104.82	0.362	63.63	0.220
Y-CFRP-a-3	111.19	0.384	111.19	0.384	81.31	0.281
Mean value	**104.04**	**0.359**	**104.04**	**0.359**	**71.30**	**0.246**
Y-GFRP-b-1	48.76	0.168	79.36	0.274	65.91	0.227
Y-GFRP-b-2	57.60	0.199	80.77	0.279	50.40	0.174
Y-GFRP-b-3	49.12	0.169	90.47	0.312	85.66	0.296
Mean value	**51.82**	**0.179**	**83.53**	**0.288**	**67.32**	**0.232**

Table 8. Deformation parameters characterizing the walls strengthened in configuration 'b'.

Specimens	Shear Strain (‰)			G Modulus (GPa)	Pseudo-Ductility Coefficient
	Cracking	Load Capacity	Damage		
Y-CFRP-b-1	3.197	3.197	7.586	198	2.0
Y-CFRP-b-2	2.367	2.367	6.456	240	1.5
Y-CFRP-b-3	2.406	2.406	6.174	220	2.0
Mean value	**2.657**	**2.657**	**6.739**	**219**	**1.8**
Y-GFRP-b-1	0.526	7.369	7.647	450	15.1
Y-GFRP-b-2	0.637	5.942	6.436	317	9.9
Y-GFRP-b-3	0.938	5.317	7.527	188	8.0
Mean value	**0.700**	**6.209**	**7.6203**	**318**	**-**

The use of CFRP sheets resulted in a significant increase in cracking forces—which should be considered to be positive—but the appearance of the cracks was equivalent to reaching the full load-bearing capacity of the walls. Up to this point, the walls exhibited pseudo-elastic behavior; the elasto-plastic phase was not recorded here. The stiffness of the walls did not change, compared to the unstrengthened walls. In contrast, the lack of stabilization of unfilled joints and the use of GFRP sheets caused a significant acceleration of the cracking; the cracking force was twice as small as in the walls of the Y-CFRP-b series. In the pseudo-elastic phase observed in these walls, a highly variable stiffness modulus was noted. The cracking was followed by a process of rapid strain increase, accompanied by a minimal increase in bearing capacity (Figure 7b)—the clear ductile phase appeared.

In this strengthening configuration, the load-bearing capacity of elements from the Y-GFRP-b series was almost identical to the unstrengthened elements, while the use of CFRP sheets provided 25% higher load-bearing capacity than GFRP sheets and allowed for a significant reduction in strain at this point. The average deformation (shear strain) of the walls with CFRP was 2.657‰, while it was as high as 6.205‰ when GFRP sheets were used.

Generally, the post-peak phase was only observed in the walls strengthened with CFRP sheets. However, due to a relatively fast decrease in force in this phase and high cracking force, the value of the pseudo-ductile coefficient—determined for a 20% decrease in maximum force—was very small (less than 2, see Figure 8). In walls from the Y-GFRP-b series, no post-peak phase was observed but the walls were characterized by very high pseudo-ductility coefficients (Figure 8). However, it should be kept in mind that the load-bearing capacity of these walls was achieved just before the damage was slightly higher than that of unreinforced walls. Thus, the high level of safety of the masonry cannot be analyzed here, since its plasticity occurs at very low forces.

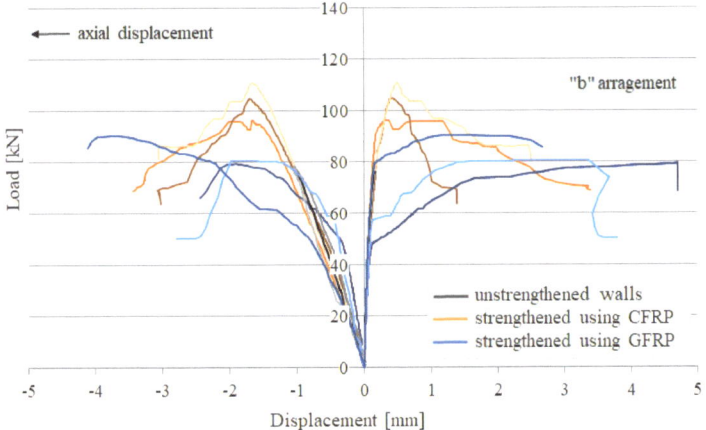

Figure 8. Horizontal and vertical displacement of the tested walls—configuration 'b'.

3.2.3. Failure Mode of Strengthened Walls

The final failure pattern of the walls with FRP strips glued on unfilled joints (configuration 'a') was the delamination of the strengthening materials from the masonry surface (Figure 9a,b). However, the precursory phenomenon of the delamination was different. In the case of the CFRP sheets, there was diagonal cracking of the masonry panels; then, delamination of the composite occurred due to the gradual widening of the cracks. Figure 9a clearly shows diagonal cracks running through the masonry and a detached part of the CFRP sheet (along with pieces of the masonry element). The intensity of the damage was very large and the walls were destroyed in their entirety. In the case of GFRP sheets, the deformation of the unfilled head joints took place, which led to 'tensioning' of the composite. Finally, through the use of a rigid epoxy adhesive, the outer layer of the masonry block was cut almost on the plane of the wall (delamination). Figure 9b shows a widened vertical joint, as well as the delamination of the GFRP sheets nearby.

Figure 9. Failure pattern observed in walls strengthened in configuration 'a': (**a**) the elements with carbon strips; (**b**) the elements with glass strips.

The application of the strengthening between unfilled joints resulted in a different failure pattern, depending on the type of FRP material. The masonry with CFRP strips did not show diagonal failure; almost no cracks were observed on the masonry surface (Figure 10a). The damage occurred due to the separation of larger pieces of masonry—cracks in the masonry plane—along with the composite adhered to them. In the walls of Y-GFRP-b series, the excessive deformation of the unfilled head joints is clearly visible (Figure 10b). This led to the cracks in the planes of the vertical joints and led to the element breaking into vertical fragments, separated by strengthened pieces.

Figure 10. Failure pattern observed in walls strengthened in configuration 'b': (**a**) the elements with carbon strips; (**b**) the elements with glass strips.

4. Discussion

4.1. Failure Initiation and Analysis

The use of an optical strain measurement system made it possible to determine the points initiating the failure of the unstrengthened masonry walls. Due to the unexpected and sudden damage to this type of specimen, it was not possible to capture the failure process with the naked eye. Figure 11 shows the successive steps in the process of strain growth and strain concentration. It is clearly visible that, in the initial phase of loading (about 50% of the maximum force), the strains increase in the areas of unfilled head joints (blue zone in Figure 11a). Subsequently, the widening of the head joints initiated the loss of adhesion within the thin bed joints, leading to a sudden growth of deformation in this area (Figure 11b). Thus, a strong strain concentration is observed at the crossing of these two joints (Figure 11c) and this is the point of masonry damage (Figure 11d).

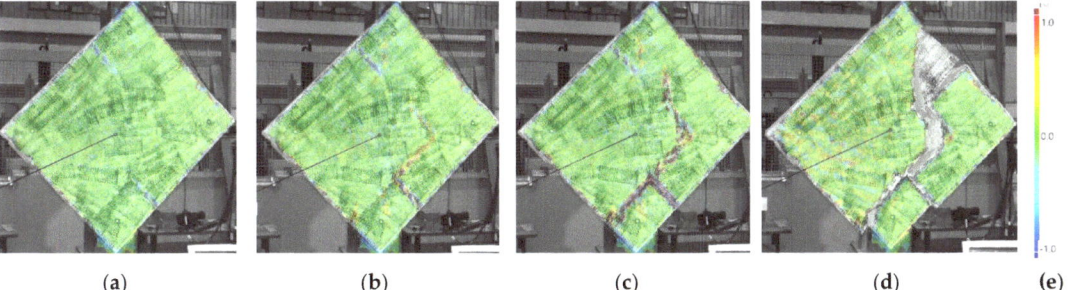

Figure 11. The process of strain growth in unstrengthened wall of Y-US series: (**a**) beginning of deformations in vertical joints; (**b**) loss of adhesion in bed joints; (**c**) deformation just before failure; (**d**) damage of wall; (**e**) deformation scale.

As described in Section 3.2.3, the use of strengthening influenced the masonry failure; the type of strengthening material was crucial. Strengthening with CFRP sheets—regardless of their arrangement—eliminated the strain concentration in the joints (initially, unfilled joints and, then, thin joints), making the masonry become a more homogeneous and uniform material. This is evident when observing the surface deformation of the Y-CFRP-a series element, where an optical strain measurement system was used. Figure 12 shows the process of strain growth in the element with CFRP strips in configuration 'a'. Figure 12a shows the location of the first strain concentration, which occurred at the force, causing cracking/failure of the unstrengthened walls (about 78 kN). There is no noticeable strain increase within the unfilled joins and thin bed joints. With successive loading, the diagonal character of the areas with intense color (the growth of the deformation) become more and more pronounced (Figure 12b). The places where delamination of the CFRP material begins also become clear (Figure 12c). At this level (about 95% of the maximum force), only a slight increase in deformation in the unfilled joint can be seen noticed. However, the deformation values are much smaller than in the other areas. Figure 12d shows the masonry just before failure, where the areas of the highest strain concentration, and, therefore, damage, are visible.

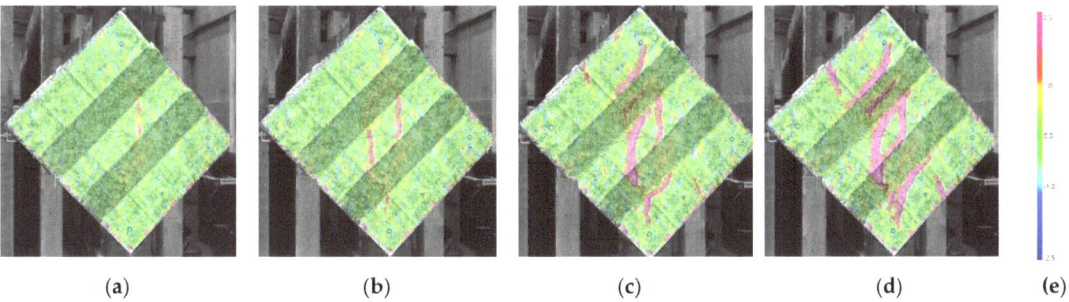

Figure 12. The process of strain growth in strengthened wall of Y-CFRP-a series: (**a**) deformations at force causing destruction of the Y-US series wall; (**b**) diagonal character of deformation; (**c**) beginning of CFRP delamination; (**d**) deformation just before failure; (**e**) deformation scaleIn walls strengthened with CFRP sheets (both configurations), the location of the composite detachment was arbitrary and did not depend on the location of unfilled joints. The relatively stiff CFRP composite effectively integrated the wall, preventing the joints from deformation. This ensured uniform masonry operations and relatively high safety of use.

In the case of the application of a much more deformable strengthening material (GFRP sheets), an excessive deformation of the unfilled joints was not avoided. However, the strengthening location prevented—in different ranges—the masonry from very rapid damage or from falling apart, which was observed in the case of unstrengthened elements.

4.2. Comparison Analysis

A quantitative comparison of the behavior of the strengthened walls was made by summarizing the significant values for all the series tested. Figure 13a–c show the relative increases in load capacity, cracking load and strain at cracking, respectively, in accordance with the unstrengthened walls (value = 1.00). The dashed pattern denotes elements in which the appearance of the first cracks was equivalent to full load-bearing capacity. Figure 14a,b show the values of shear modulus and pseudo-ductility coefficient, respectively. In Figure 14b the dashed pattern refers to the average value, which is debatable due to the very large discrepancies in the partial results.

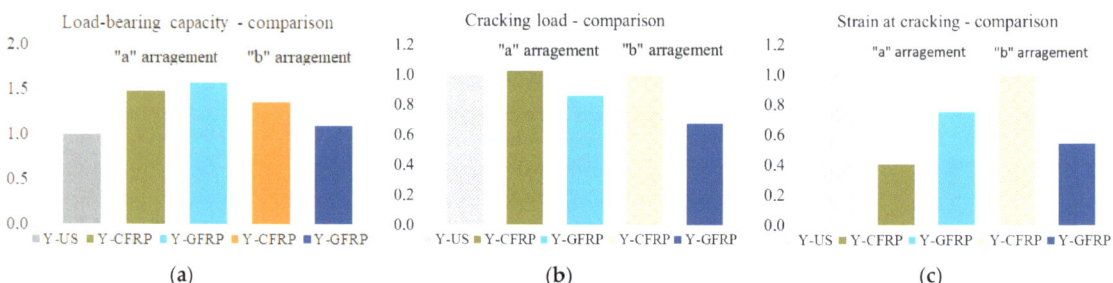

Figure 13. Relative changes in significant quantities for all tested series in a range of: (**a**) load-bearing capacity; (**b**) cracking loads; (**c**) strains at cracking.

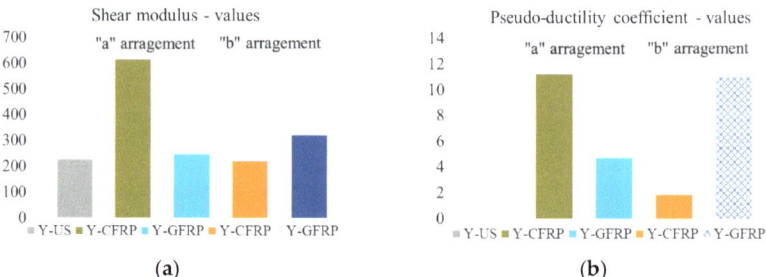

Figure 14. Comparison of significant values for all tested series in a range of: (**a**) shear modulus; (**b**) pseudo-ductile coefficient.

The strengthening made using CFRP strips in configuration 'a' was found to be the most effective solution. The strengthening combination provided a high level of cracking force, a significant increase in stiffness, an almost 50% increase in load-bearing and a high ductile coefficient (very desirable in terms of the occurrence of dynamic actions). This type of strengthening eliminated an important drawback of the technology of erecting these walls (unfilled joints initiate the failure process) by ensuring the uniformity of the structure. The application of CFRP sheets in the areas between the unfilled joints also has a positive effect, however, significantly smaller. This strengthening is characterized by the smallest load-bearing capacity, pseudo-ductility coefficient and the fact that the appearance of cracking is equivalent to reaching the load-bearing capacity of the wall results in a negligible safety reserve in the post-cracking phase.

The GFRP sheets in configuration 'a' offered high load-bearing capacities; however, they did not provide a clear elasto-plastic phase, nor a sufficient level of ductility. The use of GFRP outside of unfilled joints turned out to be completely ineffective. This solution results in the fast cracking and a negligible increase in the load-bearing capacity. The panels exhibit ductile behavior, which indicates a failure process; however, all this takes place at relatively low loads (equal to the failure loads of unreinforced masonry) and very high deformation of the structure.

Obtaining better characteristics of masonry walls by using CFRP materials (instead of GFRP) is the opposite of the trend presented in the literature [18,22]. This phenomenon is due to the specification of the masonry walls, i.e., relatively large masonry units and atypical erection technology. Excessive deformation of the unfilled joints, which leads to the insufficient adhesion in thin joints, is responsible for the failure of the unstrengthened walls. The stiff composite limits the excessive separation of unfilled joints (in configurations 'a'), significantly reducing the deformation when the first cracks appear (Figure 13c) and changing the distribution of cracks in the entire structure (Figure 12b).

5. Conclusions

The enhancement of the shear capacity of masonry walls is a necessary action wherever there are horizontal forces acting in the plane of the wall. A good example of this is areas exposed to the seismic actions or influence of mining operations. The problem is particularly important for walls made with unfilled head joints, including those made of AAC blocks, which have poor resistance to any shear forces.

In the literature, a number of examples of shear strengthening of walls using FRP materials can be found but mostly concerning walls (mainly ceramic) with all solid joints. On this basis, it was deemed worthwhile to study the surface strengthening of very popular AAC block walls in an attempt to select an effective strengthening system due to the materials used for this purpose. Therefore, a series of laboratory tests on small walls made of AAC blocks were performed in accordance with the recommendations of the standard [37]. This assumed them to be representative in terms of recognition of the issue. Based on the results of the laboratory tests, the following conclusions were made:

(1) Analysis of the failure process in unstrengthened AAC masonry walls identified the critical points in the structure that initiate its final damage. These were unfilled head joints in which displacement of adjacent blocks occurred, resulting in overloading and subsequent destruction of the bed joints.

(2) The application of CFRP sheets—regardless of their arrangement—changed the behavior of the masonry, which now worked as an almost homogeneous material. There was no deformation of the unfilled head joints. This provided positive effects, in terms of the crack delay, an increase in stiffness (more than two times higher than in the unstrengthened walls) and load-bearing capacity by 48% (with strips on unfilled joints) and 35% (with strips between the vertical joints). In the first case, the failure was in the form of diagonal cracking with a final sheet detachment; in the second case, there was a splitting in the wall plane of the entire specimens.

(3) The use of much deformable GFRP sheets did not avoid the excessive deformation of the unfilled head joints. At the same time, with strips applied to unfilled joints, the load capacity of the specimens increased by 56% and, in the case of GFRP strips located between head joints, by only 9%. In the first case, there was delamination of the sheets after large mutual displacements of the blocks. In the second, there were pronounced cracks parallel to the sheets (in the line of the head joints).

(4) The advantage of application of CFRP sheets was revealed primarily in the greater ductility and stiffness of such strengthened walls, which seems to be valuable in the case of dynamic loads (e.g., seismic/parasismic effects). In typical situations of quasi-static loads (e.g., uneven settlement or the effect of continuous mining deformations), the aspect of ductility is less important and, here, a clear advantage of using GFRP

strengthening is their price; the GFRP sheets are about four times cheaper than CFRP sheets in presented configurations.

(5) The tests performed were preliminary and recognizable, and, so, quantitative analyses of the results should be regarded as indicative. Nevertheless, the qualitative analysis is fully reliable, because the tests were carried out on wall fragments with the actual layout of the joints and the real strengthening intensity. The superiority of a strengthening system directly applied to unfilled head joints over strengthening applied in a random arrangement (here, the most unfavorable one was between the head joints) can clearly be seen.

Funding: This publication was performed with financial support by the project BK-225/RB6/2022 in Silesian University of Technology, Department of Structural Engineering.

Data Availability Statement: Not applicable.

Conflicts of Interest: The authors declare no conflict of interest. The funders had no role in the design of the study; in the collection, analyses, or interpretation of data; in the writing of the manuscript.

References

1. Fudge, C.; Fouad, F.; Klingner, R. Autoclaved Aerated Concrete. In *Developments in the Formulation and Reinforcement of Concrete*, 2nd ed.; Mindess, S., Ed.; Woodhead Publishing: Sawston, UK, 2019; pp. 345–363. ISBN 978-0-08-102616-8.
2. Wittmann, F.H. Advances in Autoclaved Aerated Concrete. In Proceedings of the 3rd RILEM International Symposium, Zürich, Switerland, 14 October 1992; pp. 1–374.
3. Jerman, M.; Keppert, M.; Výborný, J.; Cerný, R. Hygric, Thermal and Durability Properties of Autoclaved Aerated Concrete. *Constr. Build. Mater.* **2013**, *41*, 352–359. [CrossRef]
4. Aroni, S. *Autoclaved Aerated Concrete—Properties, Testing and Design*; Rilem Technical Committees; CRC Press: Boca Raton, FL, USA, 1993.
5. Coccia, S.; Di Carlo, F.; Imperatore, S. Masonry Walls Retrofitted with Vertical FRP Rebars. *Buildings* **2020**, *10*, 72. [CrossRef]
6. Babatunde, S.A. Review of Strengthening Techniques for Masonry Using Fiber Reinforced Polymers. *Compos. Struct.* **2017**, *161*, 246–255. [CrossRef]
7. de Lorenzis, L. Strengthening of Masonry Structures with Fibre-Reinforced Polymer (FRP) Composites. In *Strengthening and Rehabilitation of Civil Infrastructures Using Fibre-Reinforced Polymer (FRP) Composites*; Hollaway, L.C., Teng, J.G., Eds.; Woodhead Publishing: Sawston, UK, 2008; pp. 235–266. ISBN 978-1-84569-448-7.
8. Tinazzi, D.; Nanni, A. *Assessment Of Technologies Of Masonry Retrofitting With FRP*; Center of Infrastructure Engineering Studies, University of Missouri-Rolla: Rolla, MO, USA, 2000.
9. Abdel-Jaber, M.S.; Walker, P.R.; Hutchinson, A.R. Shear Strengthening of Reinforced Concrete Beams Using Different Configurations of Externally Bonded Carbon Fibre Reinforced Plates. *Mater. Struct.* **2003**, *36*, 291–301. [CrossRef]
10. Kałuża, M.; Ajdukiewicz, A. Comparison of Behaviour of Concrete Beams with Passive and Active Strengthening by Means of CFRP Strips. *ACEE* **2008**, *1*, 51–64.
11. Hollaway, L.C.; Teng, J.G. *Strengthening and Rehabilitation of Civil Infrastructures Using Fibre-Reinforced Polymer (FRP) Composites*; Woodhead Publishing Series in Civil and Structural Engineering: Sawston, UK, 2008.
12. Singh, S. *Analysis and Design of FRP Reinforced Concrete Structures*; McGraw-Hill Education—Europe: New York, NY, USA, 2015; ISBN 0-07-184789-8.
13. Mosallam, A. Composites: Construction Materials For The New Era. In *Advanced Polymer Composites for Structural Applications in Construction*; Hollaway, L.C., Chryssanthopoulos, M.K., Moy, S.S.J., Eds.; Woodhead Publishing: Sawston, UK, 2004; pp. 45–58. ISBN 978-1-85573-736-5.
14. Grellmann, W.; Seidler, S. Mechanical Properties of Polymers. In *Polymer Testing*, 2nd ed.; Grellmann, W., Seidler, S., Eds.; Carl Hanser Verlag: Munich, Germany, 2013; pp. 73–231. ISBN 978-1-56990-548-7.
15. Bakis, C.E.; Bank Lawrence, C.; Brown, V.L.; Cosenza, E.; Davalos, J.F.; Lesko, J.J.; Machida, A.; Rizkalla, S.H.; Triantafillou, T.C. Fiber-Reinforced Polymer Composites for Construction—State-of-the-Art Review. *J. Compos. Constr.* **2002**, *6*, 73–87. [CrossRef]
16. Saghafi, M.H.; Safakhah, S.; Kheyroddin, A.; Mohammadi, M. In-Plane Shear Behavior of FRP Strengthened Masonry Walls. *APCBEE Procedia* **2014**, *9*, 264–268. [CrossRef]
17. Capozucca, R.; Magagnini, E. Experimental Response of Masonry Walls In-Plane Loading Strengthened with GFRP Strips. *Compos. Struct.* **2020**, *235*, 111735. [CrossRef]
18. Valluzzi, M.R.; Tinazzi, D.; Modena, C. Shear Behavior of Masonry Panels Strengthened by FRP Laminates. *Constr. Build. Mater.* **2002**, *16*, 409–416. [CrossRef]
19. Kalali, A.; Kabir, M.Z. Experimental Response of Double-Wythe Masonry Panels Strengthened with Glass Fiber Reinforced Polymers Subjected to Diagonal Compression Tests. *Eng. Struct.* **2012**, *39*, 24–37. [CrossRef]

20. Bui, T.-L.; Si Larbi, A.; Reboul, N.; Ferrier, E. Shear Behaviour of Masonry Walls Strengthened by External Bonded FRP and TRC. *Compos. Struct.* **2015**, *132*, 923–932. [CrossRef]
21. Santa-Maria, H.; Alcaino, P.; Luders, C. Experimental Response of Masonry Walls Externally Reinforced with Carbon Fiber Fabrics. In Proceedings of the 8th U.S. National Conference on Earthquake Engineering, San Francisco, CA, USA, 18 April 2006; Volume 1402, p. 11.
22. Marcari, G.; Manfredi, G.; Prota, A.; Pecce, M. In-Plane Shear Performance of Masonry Panels Strengthened with FRP. *Compos. Part B Eng.* **2007**, *38*, 887–901. [CrossRef]
23. Luccioni, B.; Rougier, V.C. In-Plane Retrofitting of Masonry Panels with Fibre Reinforced Composite Materials. *Constr. Build. Mater.* **2011**, *25*, 1772–1788. [CrossRef]
24. Kwiecień, A.; Kubica, J.; Stecz, P.; Zając, B. Flexible Joint Method (FJM)—A New Approach to Protection and Repair of Cracked Masonry. In Proceedings of the First European Conference on Earthquake Engineering and Seismology, Geneva, Switerland, 3–8 September 2006; Volume 282.
25. Kwiecień, A. Polimerowe Złącze Podatne—Innowacyjna Metoda Naprawy i Konserwacji Obiektów Zabytkowych. *Wiadomości Konserw.* **2009**, *26*, 234–244.
26. Umair, S.; Numada, M.; Amin, M.; Meguro, K. Fiber Reinforced Polymer and Polypropylene Composite Retrofitting Technique for Masonry Structures. *Polymers* **2015**, *7*, 963–984. [CrossRef]
27. Sathiparan, N.; Meguro, K. Shear and Flexural Bending Strength of Masonry Wall Retrofitted Using PP-Band Mesh. *Constr. J.* **2013**, *14*, 3.
28. Kubica, J.; Kałuża, M. Diagonally Compressed AAC Block's Masonry—Effectiveness of Strengthening Using CRFP and GFRP Laminates. In Proceedings of the 8th International Masonry Conference, Dresden, Germany, 4 July 2010; International Masonry Society: Dresden, Germany; pp. 1–10.
29. Kubica, J.; Galman, I. Comparison of Two Ways of AAC Block Masonry Strengthening Using CFRP Strips—Diagonal Compression Test. *Procedia Eng.* **2017**, *193*, 42–49. [CrossRef]
30. Saad, A.S.; Ahmed, T.A.; Radwan, A.I. In-Plane Lateral Performance of AAC Block Walls Reinforced with CFPR Sheets. *Buildings* **2022**, *12*, 1680. [CrossRef]
31. Wang, B.; Wang, P.; Chen, Y.; Zhou, J.; Kong, X.; Wu, H.; Fan, H.; Jin, F. Blast Responses of CFRP Strengthened Autoclaved Aerated Cellular Concrete Panels. *Constr. Build. Mater.* **2017**, *157*, 226–236. [CrossRef]
32. Kałuża, M. Effectiveness Of Shear Strengthening Of Walls Made Using AAC Blocks—Laboratory Test Results. *Arch. Civ. Eng.* **2020**, *66*, 33–44, LXVI. [CrossRef]
33. Kałuża, M. The Influence of FRCM System with a Basalt Mesh on the Shear Ptoperties of AAC Masonry Walls. In *Brick and Block Masonry—From Historical to Sustainable Masonry*; Taylor & Francis Group: London, UK, 2020; pp. 1–10.
34. Drobiec, Ł.; Jasiński, R.; Mazur, W.; Jonkiel, R. The effect of the strengthening of AAC masonry walls using FRCM system. *Cem. Wapno Beton.* **2020**, *25*, 376–389. [CrossRef]
35. Lyu, H.; Deng, M.; Ma, Y.; Yang, S.; Cheng, Y. In-Plane Cyclic Tests on Strengthening of Full-Scale Autoclaved Aerated Concrete Blocks Infilled RC Frames Using Highly Ductile Concrete (HDC). *J. Build. Eng.* **2022**, *49*, 104083. [CrossRef]
36. Deng, M.; Zhang, W.; Yang, S. In-Plane Seismic Behavior of Autoclaved Aerated Concrete Block Masonry Walls Retrofitted with High Ductile Fiber-Reinforced Concrete. *Eng. Struct.* **2020**, *219*, 110854. [CrossRef]
37. RILEM. LUMB6Diagonal Tensile Strength Tests of Small Wall Specimens. In *RILEM*; Technical Recommendations for the Testing and Use of Construction Materials: London, UK, 1994.
38. *EN 772-1*; Methods of Test for Masonry Units—Part 1: Determination of Compressive Strength. CEN/TC 125: Brussels, Belgium, 2011.
39. *EN 1052-1*; Methods of Test for Masonry—Part 1: Determination of Compressive Strength. CEN/TC 347: Brussels, Belgium, 1998.
40. Borri, A.; Corradi, M.; Sisti, R.; Buratti, C.; Belloni, E.; Moretti, E. Masonry Wall Panels Retrofitted with Thermal-Insulating GFRP-Reinforced Jacketing. *Mater. Struct.* **2016**, *49*, 3957–3968. [CrossRef]
41. Babacidarabad, S.; De Caso, F.; Nanni, A. URM Walls Strengthened with Fabric-Reinforced Cementitious Matrix Composite Subjected to Diagonal Compression. *J. Compos. Constr.* **2014**, *18*, 04013045. [CrossRef]

Article

Mechanical Properties of Fiber-Reinforced Polymer (FRP) Composites at Elevated Temperatures

Chuntao Zhang [1,2,3,*], Yanyan Li [1,2] and Junjie Wu [1,2]

[1] Shock and Vibration of Engineering Materials and Structures Key Laboratory of Sichuan Province, Mianyang 621010, China
[2] School of Civil Engineering and Architecture, Southwest University of Science and Technology, Mianyang 621010, China
[3] Department of Mechanical Engineering, University of Houston, Houston, TX 77204, USA
* Correspondence: chuntaozhang@swust.edu.cn

Abstract: Many materials are gradually softened with increasing temperatures in the fire, which will cause severe damage. As a new fiber-reinforced polymer (FRP) composite, the change in mechanical properties of nanometer montmorillonite composite fiber-reinforced bars or plates at elevated temperatures has not been investigated. To obtain a more comprehensive study of the mechanical properties of FRP composites at high temperatures, experimental research on the nanometer montmorillonite composite fiber material under the tensile rate of 1 mm/min was conducted at target temperatures between 20 °C and 350 °C. Finally, the failure mode of the FRP composites after the tensile test was analyzed. The results demonstrate that the elevated temperatures had a major impact on the residual mechanical properties of fiber-reinforced polymer (FRP) composites when the exposed temperatures exceeded 200 °C. Below 200 °C, the maximum decrease and increase in the fracture load of fiber reinforced polymer (FRP) composites were between −34% and 153% of their initial fracture load. After exposing to temperatures above 200 °C, the surface color of fiber-reinforced polymer (FRP) composites changed from brown to black. When exposed to temperatures between 200 and 300 °C, the ultimate load of fiber-reinforced polymer (FRP) composites significantly increased from 731.01 N to 1650.97 N. Additionally, the stress−strain behavior can be accurately predicted by using the proposed Johnson−Cook constitutive model. The experimental results studied in this research can be applied to both further research and engineering applications when conducting a theoretical simulation of fiber-reinforced polymer (FRP) composites.

Keywords: fiber-reinforced polymer (FRP) composites; elevated temperatures; mechanical properties; reduction factor; constitutive model

Citation: Zhang, C.; Li, Y.; Wu, J. Mechanical Properties of Fiber-Reinforced Polymer (FRP) Composites at Elevated Temperatures. *Buildings* **2023**, *13*, 67. https://doi.org/10.3390/buildings13010067

Academic Editor: Xin Wang

Received: 9 November 2022
Revised: 2 December 2022
Accepted: 22 December 2022
Published: 27 December 2022

Copyright: © 2022 by the authors. Licensee MDPI, Basel, Switzerland. This article is an open access article distributed under the terms and conditions of the Creative Commons Attribution (CC BY) license (https://creativecommons.org/licenses/by/4.0/).

1. Introduction

Generally speaking, the matrix resin of FRP typically consists of thermosetting resin and thermoplastic resin. In the fire environment, with the increase of temperature, the mechanical properties of composite materials mainly experience a three-time decrease. When the temperatures increase to the glass transition temperature T_g of the resin matrix, it softens and enters the rubber state from the glass state. The ability of the resin matrix to transfer shear stress between reinforced fibers decreases, resulting in the first significant decrease in the mechanical properties of FRP. When the temperature is further increased to the resin decomposition temperature T_d (about 300–400 °C), the resin matrix of FRP is gradually decomposed and carbonized and the toxic smoke is released, resulting in the second significant decrease in the mechanical properties of FRP. When the temperature continues to be increased, the resin matrix begins to burn and the combustion process releases more heat, resulting in the second significant decrease in the mechanical properties of FRP.

Studies on fiber-reinforced polymer (FRP) composites at high temperatures have been conducted. The bonding strength of the concrete matrix between carbon and glass fiber sheet changes after being exposed to temperatures of 20, 50, 65, and 80 °C, according to research by Leone et al. [1]. In addition, the test results demonstrated that the concrete matrix's transition temperature from shearing failure to cohesion failure was 65 °C. Salloum [2] conducted an axial compression test on FRP-strengthened cylinders (Φ: 100 cm; height: 200 mm) after exposing them to temperatures of 100 and 200 °C and were left at each target temperature for 1, 2, and 3 h separately. The test results indicated that external-bonded FRP materials' reinforcing efficacy was sensitive to high temperatures. At temperatures 2.5 times Tg, the ultimate capacity of concrete specimens enhanced with FRP was 25% less than at ambient temperatures. The above scholars mainly focused on the mechanical behavior of concrete elements reinforced with FRP. There are studies about the mechanical behavior of FRP as a standalone material at elevated temperatures. Pultruded carbon fiber-reinforced polymer (P-CFRP) specimens and CFRP tensile specimens manufactured with a hand lay-up method were subjected to a series of tests by Nguyen et al. [3,4] at temperatures that reached 700 °C. According to their findings, hand lay-up specimens' ultimate tensile strength and Young's modulus were reduced by 50% at 350 °C and 30% at 600 °C. Additionally, they demonstrated that the thermomechanical strength was lower than the residual strength for P-CFRP samples at the same degree of applied temperature. One of the pioneering studies regarding the behavior and characteristics of FRP materials at high temperatures that are utilized in industrial domains, such as the automotive, marine, and aerospace industries, was performed by Mouritz and Mathys [5]. At high temperatures, Shenghu and Zhishen [6] performed a series of tension tests on single-layer FRP sheets composed of GFRP, CFRP, and basalt-fiber reinforced polymer (BFRP). Among all the tested fiber-reinforced sheets, they concluded that the CFRP sheets had the highest strength and the GFRP sheets had the lowest strength [7]. At around 55 °C, all of the sheets' tensile strength significantly decreased, but no further substantial decline occurred as the temperature increased. The CFRP sheets had the highest residual strength, with almost 69% of their initial tensile strength. However, there still lacks the work of establishing the constitutive model to better predict the mechanical behavior of FRPs at elevated temperatures. In this research, we proposed a constitutive model based on the experimental results of FRPs at elevated temperatures to fill the research gap.

Currently, steel is a hot topic of high-temperature research. To more accurately evaluate the fire resistance of steel structures, a variety of experimental studies on the mechanical properties of different steels at high temperatures have been conducted [8–12]. After subjecting high-strength steels of S460, S690, and S960 to fire, Qiang et al. [13,14] performed tensile tests to investigate the residual mechanical properties after the fire. Test results demonstrated that the mechanical properties of the tested steels were affected by heating temperatures below 600 °C. In contrast to this research, Gunalan and Mahendran [15] demonstrated that the residual mechanical properties of high-strength steels decreased noticeably when the target temperature exceeded 300 °C. Chicw et al. [16] investigated high-strength S690 steel that suffered from the RQT process and found that the yield strength of S690 steel that suffered from the RQT process declined more slowly than that without the RQT procedure. Additionally, earlier research [17–26] has demonstrated that when exposure temperatures rose above a particular value, substantial changes in the residual mechanical properties of low-carbon steels and HSS were observed. According to previous studies [27–29], there is a critical temperature of various steels in the post-fire mechanical properties. When the exposure target temperatures do not exceed the critical temperature, the post-fire mechanical properties of various steels remain basically unaffected. However, when exposed to temperatures above the critical temperature, the post-fire mechanical properties change significantly, irrespective of the cooling methods.

In recent years, nanometer montmorillonite composite fiber materials have gradually been used in building structures. As a new fiber-reinforced polymer (FRP) composite, the change in mechanical properties of nanometer montmorillonite composite fiber-reinforced bars or plates has not been investigated. Therefore, the mechanical behavior of nanometer montmorillonite composite fiber-reinforced plates subjected to different temperatures was studied. The experimental results provided in this paper can be applied to both further research and engineering applications when conducting theoretical analysis and numerical simulation of nanometer montmorillonite composite fiber-reinforced polymer (FRP) composites. In addition, this research is a part of the larger experimental program that aims to examine the mechanical characteristics and behavior of nanometer montmorillonite composite fiber-reinforced polymer (FRP) composites in various situations.

2. Test Program

2.1. Test Specimens

In this test, physical and mechanical properties of the nanometer montmorillonite composite fiber material, including the density ρ, Barcol hardness, fiber volume fraction, insoluble content of resin, water absorption, glass transition temperature T_g, tensile strength (main fiber direction) f_{tm}, tensile strength (secondary fiber direction) f_{ts}, compressive strength (main fiber direction) f_{cm}, compressive strength (secondary fiber direction) f_{cs}, and shock resistance are provided in Table 1. Furthermore, the decomposition temperature of the FRP-reinforced bonding colloid was less than 310 °C.

Table 1. Physical and mechanical properties of the nanometer montmorillonite composite fiber material.

Performance	Performance Index
ρ (kg·m^{-3})	$\leq$2000
Barcol hardness (HBa)	$\geq$50
Fiber volume fraction (%)	$\geq$70
Insoluble content of resin (%)	$\geq$90
Water absorption (%)	$\leq$1.0
T_g (°C)	$\geq$290
f_{tm} (MPa)	$\geq$400
f_{ts} (MPa)	$\geq$10
f_{cm} (MPa)	$\geq$100
f_{cs} (MPa)	$\geq$15
Shock resistance (kJ·m^{-2})	$\geq$240

To better comprehend how high temperatures affect composites made of FRP, experiments were conducted. The size and the form of the FRP specimens are shown in Figure 1, and the specimens were fabricated in a thickness of 5 mm, which followed the specifications outlined in GB/T 228.1-2010 [30]. The experiments included 24 specimens and 8 target temperatures. Three specimens were loaded at each temperature to reduce the test error.

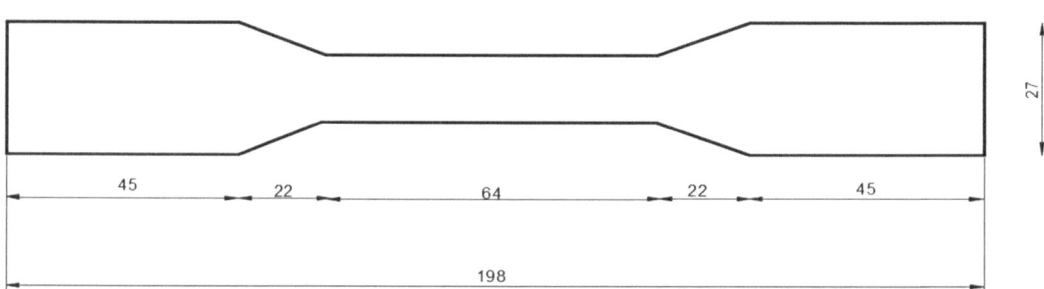

Figure 1. The FRP tensile coupon specimen (mm).

2.2. Test Details

To simulate different fire accidents, the ambient temperature and seven target temperatures—50 °C, 100 °C, 150 °C, 200 °C, 250 °C, 300 °C, and 350 °C—were considered herein. During high-temperature tensile tests, the specimen was first heated to a predetermined temperature at a heating rate of 15 °C/min. To ensure uniform temperature over the entire gauge length, the specimen was kept for about 30 min at each target temperature, and the elevated temperatures remained unchanged based on the thermocouples in the test equipment. Then, the specimen was loaded until it failed, during which the target temperatures were unchanged since the specimen was still in the test equipment. Both the displacement and engineering strain of the FRPs were output by the computer. As shown in Figure 2, this experiment was conducted by an ETM series electronic universal testing machine, and the displacement control method was used to test the specimens at a constant rate of 1 mm/min until fracture, which conformed to the requirements of GB/T 228.1-2010 [30]. Both the displacement and engineering strain of the FRPs were output by the computer. Based on the experimental results, the mechanical properties of fiber-reinforced polymer (FRP) composites were discussed.

 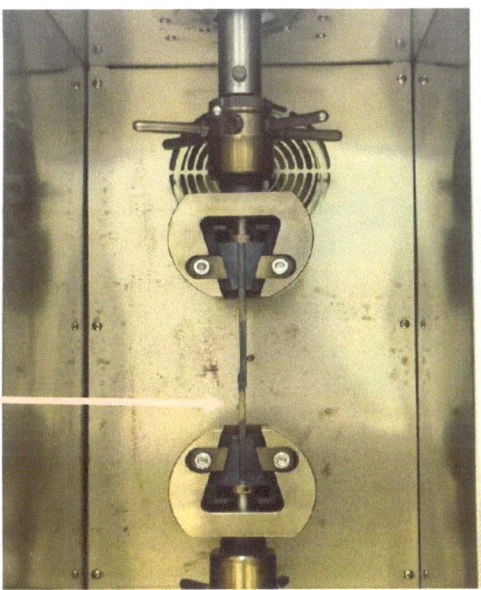

Figure 2. Test machine.

3. Experimental Results

3.1. Load−Displacement Curves

The tensile load−displacement curves of FRP composites at elevated temperatures of 50–350 °C were obtained and compared with the as-received state, as illustrated in Figure 3. When exposed to temperatures below 200 °C, the tensile load−displacement curves of fiber-reinforced polymer (FRP) composites exhibited minor differences, and the size and shape of tensile load−displacement curves were similar compared with the initial state, as shown in Figure 3a–d. In contrast, when exposed to temperatures of 300 °C, the tensile load−displacement curves of FRP composites exhibited significant differences when compared to the FRP composites in their as-received state, as illustrated in Figure 3f. Moreover, when exposed to temperatures of 250 °C, the tensile load of fiber-reinforced polymer (FRP) composites slightly increased when compared to that of the FRP composites at ambient temperature with the increased displacement, as illustrated in Figure 3e. However, the tensile load of FRP composites decreased sharply with the increased displacement when exposed to temperatures of 350 °C, which indicated that the tensional strength and ductility increased significantly at those target temperatures, as shown in Figure 3e,f. The reason for this phenomenon is that the matrix bonded by the fiber resin changed with the increase of temperature and the glue or epoxy resin was softened at high temperatures. Therefore, the critical temperature of FRP composites was 200 °C, and the ultimate bearing temperature of FRP composites was 300 °C.

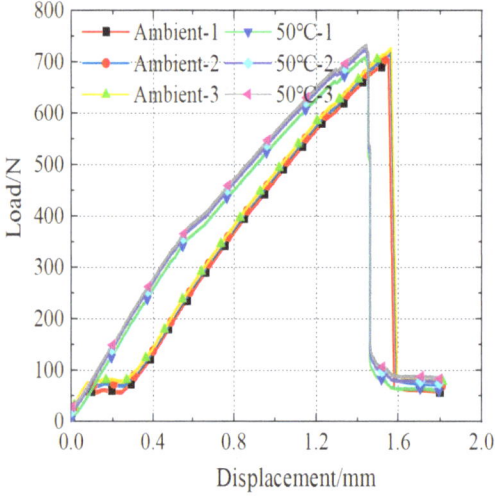

(a) Comparison between ambient temperature and 50 °C

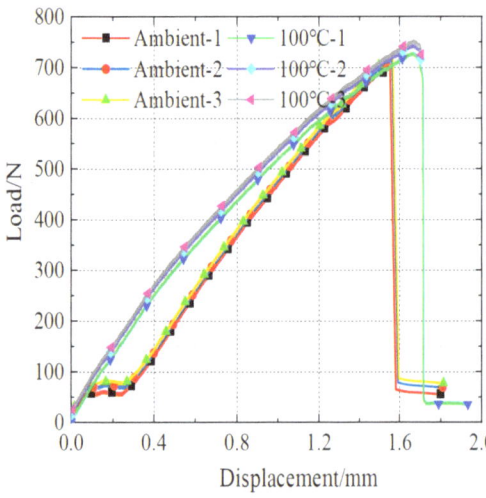

(b) Comparison between ambient temperature and 100 °C

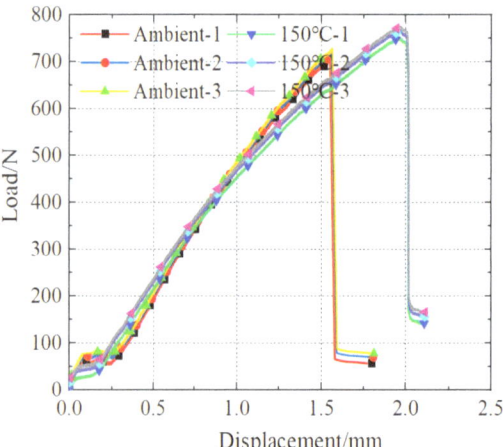

(c) Comparison between ambient temperature and 150 °C

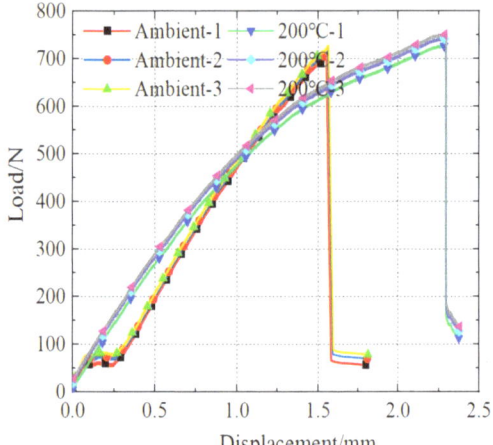

(d) Comparison between ambient temperature and 200 °C

Figure 3. *Cont.*

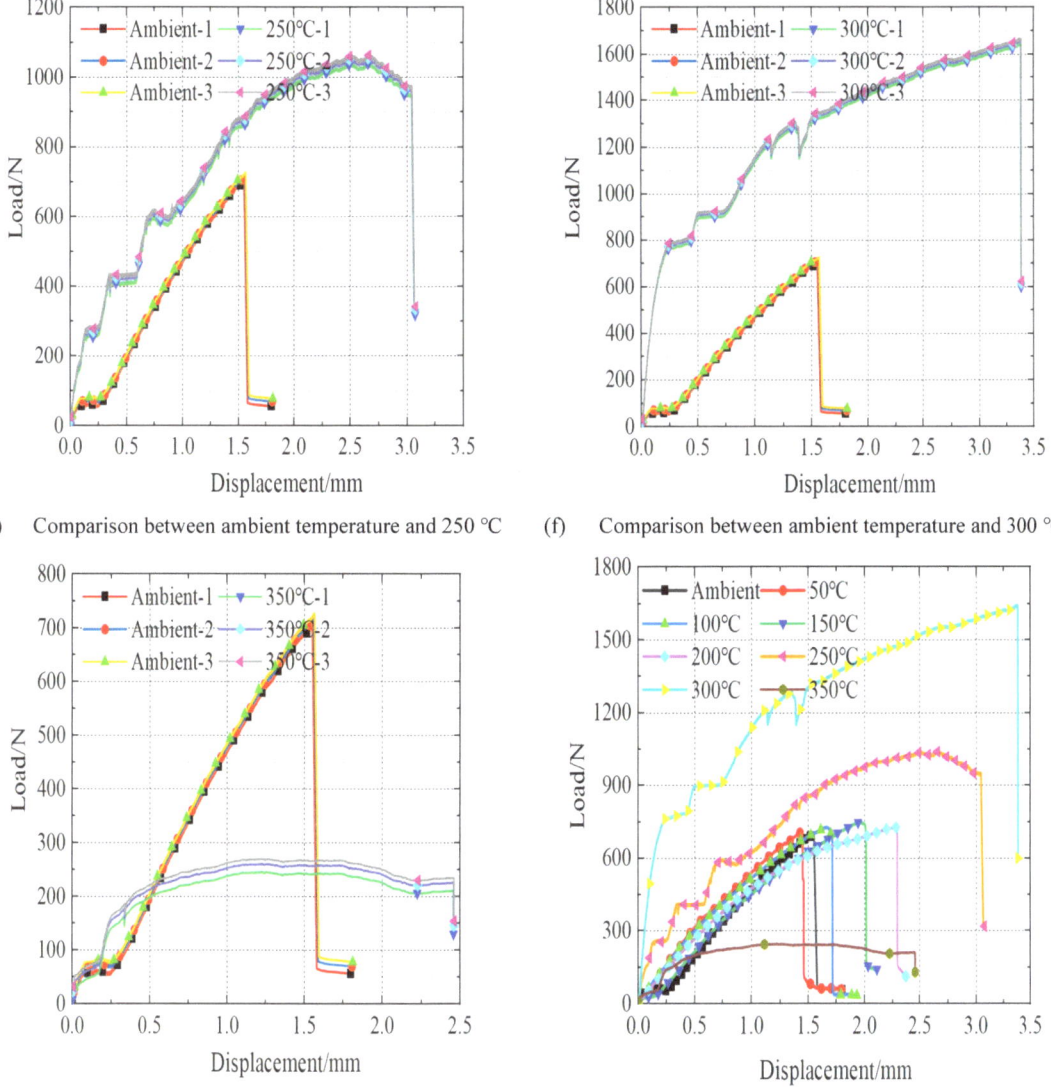

Figure 3. Load−displacement curves.

3.2. Visual Observations

Figure 4 exhibits the visual observations of fractured specimens at different elevated temperatures. The surface color of the FRP composites was significantly affected by the elevated temperatures. The surface color of FRP composites at the ambient temperature was fully brown, and it gradually changed to black when exposed to elevated temperatures between 50 °C and 200 °C. After exposure to temperatures above 200 °C, the surface color of fiber-reinforced polymer (FRP) composites changed to fully black. It is worth noting that the fibers on the surface of the fiber composite material were shed after exposure to temperatures above 200 °C. Moreover, with the increase in temperature, the phenomenon of spalling at the center fracture position of fiber-reinforced polymer (FRP) composites was more obvious.

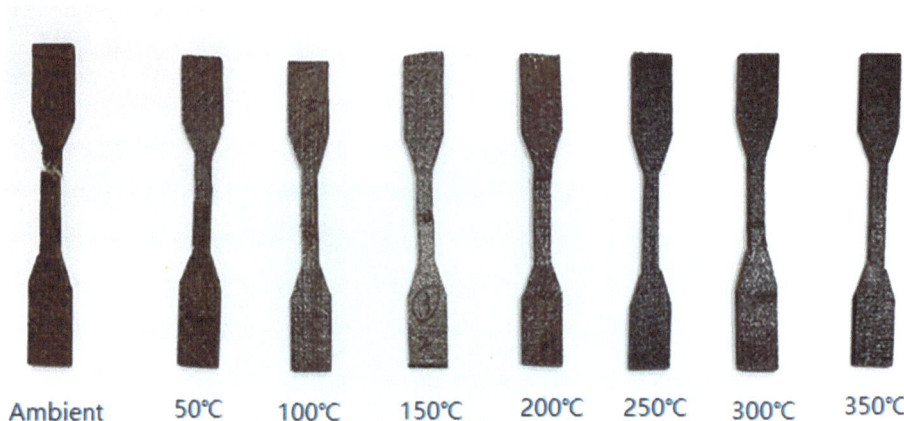

Figure 4. Specimens after experiencing elevated temperatures and tensile loading.

4. Discussion of Results

4.1. Ultimate Load

The maximum load that a structure or component can sustain is referred to as the ultimate load, and the component will enter an unstable state if the maximum load is reached. The ultimate load and residual factors of FRP composites at high temperatures are shown in Table 2. The variations in the reduction factor of the ultimate load are illustrated in Figure 5a.

The ultimate load of the specimens remained basically unchanged when exposed to temperatures below 200 °C, and the variation in the reduction factors of ultimate load did not exceed 6% when compared to the initial ultimate load of the FRP composites. However, when exposed to temperatures between 200 and 300 °C, the ultimate load of the FRP composites significantly increased from 731.01 N to 1650.97 N and increased by 133.51% of the initial ultimate load, which indicates that the FRP composites experienced a strengthening process. The reason for this phenomenon is that the matrix bonded by the fiber resin changed with the increase in temperature, when exposed to temperatures below 200 °C. The mechanical properties slightly increased, especially when the exposure temperatures are between 200 and 300 °C. This could be attributed to that the bonding effects of the nanometer montmorillonite and the fiber material were most obvious, which led to the increment of ultimate load. Notably, the ultimate load of FRP composites significantly decreased from 1650.97 N to 252.24 N and reduced 64.32% of the initial ultimate load when exposed to the temperature of 350 °C. This could be attributed to the fact that the bonding of the nanometer montmorillonite and the fiber material was softened at this temperature and the resin matrix entered the rubber state from the glass state, in which the transition temperature Tg was nearly 300 °C based on the test results.

Table 2. Ultimate loads and residual factor sof FRP composites at elevated temperatures.

Temperature (°C)	Ultimate Load (N)				Residual Factor ($F_{u,T}/F_{u,20}$)			
	Group-1	Group-2	Group-3	Average	Group-1	Group-2	Group-3	Average
20	706.87	710.24	703.92	707.01	1.00	1.00	1.00	1.00
50	709.99	708.34	705.67	708.00	1.00	1.00	1.00	1.00
100	727.13	720.61	725.85	724.53	1.03	1.02	1.03	1.02
150	752.58	750.37	755.60	752.85	1.06	1.06	1.07	1.06
200	730.25	728.26	734.52	731.01	1.03	1.03	1.03	1.03
250	1036.51	1031.27	1049.58	1039.12	1.47	1.46	1.48	1.47
300	1643.84	1658.31	1650.77	1650.97	2.33	2.35	2.33	2.34
350	244.75	253.67	258.31	252.24	0.35	0.36	0.37	0.36

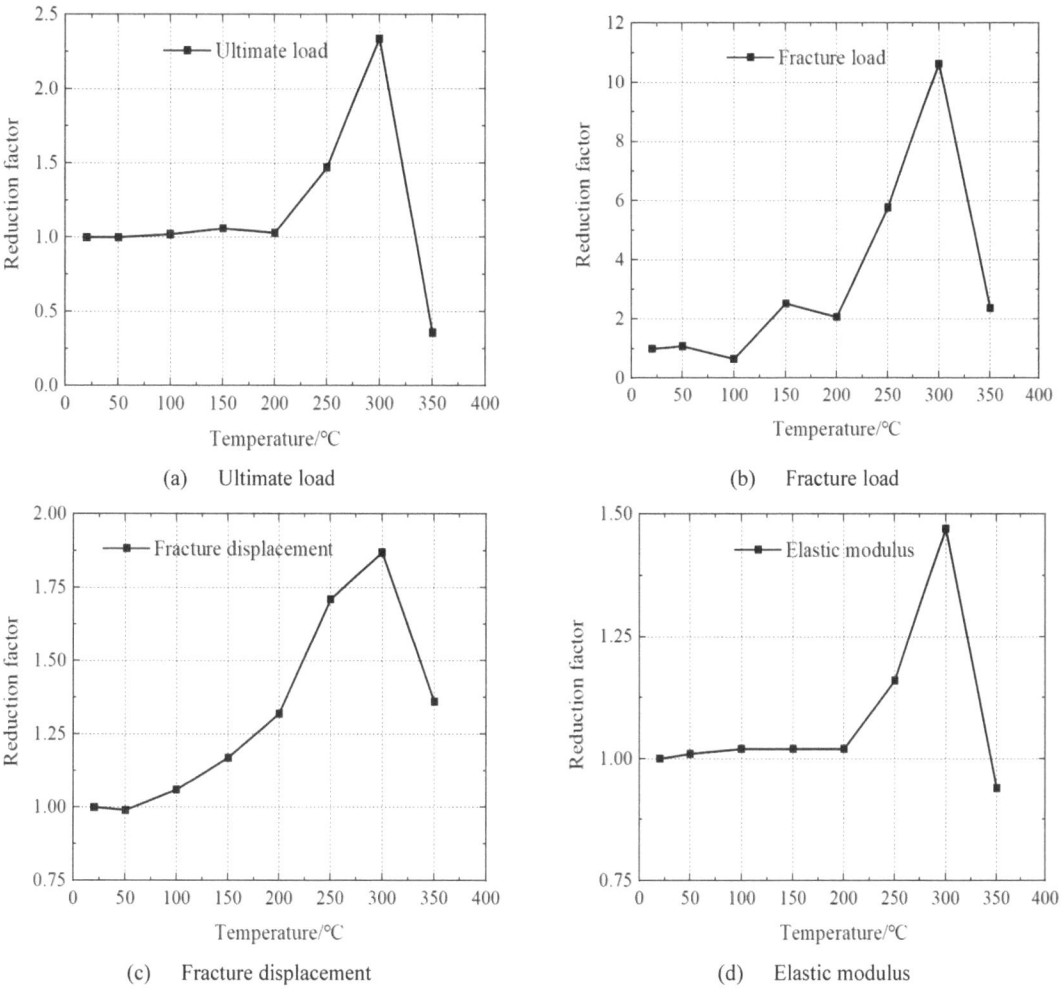

Figure 5. Residual mechanical properties of FRP composites at elevated temperatures.

4.2. Fracture Load

The critical load at which a material fails when subjected to continuous loading is referred to as the fracture load. The fracture load and the residual factor of FRP composites at high temperatures are listed in Table 3. The residual fracture load factors are plotted in Figure 5b.

When exposed to temperatures below 200 °C, the fracture load of fiber-reinforced polymer (FRP) composites remained basically the same as the initial state. The maximum decrease and increase in the fracture load of the FRP composites were between −34% and 153% of their initial fracture load, as depicted in Table 3. However, the maximum fracture load of the FRP composites was 541.35 N and was 963.26% of their initial fracture load when exposed to temperatures between 200 and 300 °C, which is consistent with the phenomenon of ultimate load. This could be attributed to the fact that the bonding effects of the nanometer montmorillonite and the fiber material were most obvious at those temperatures, which led to the increment of fracture load. Furthermore, the fracture load of the FRP composites significantly decreased from 597.55 N to 133.62 N when exposed to temperatures of 350 °C. This could be attributed to the fact that the bonding of the nanometer montmorillonite and the fiber material was softened at this temperature and the resin matrix entered the rubber state from the glass state, in which the transition temperature Tg was nearly 300 °C based on the test results.

Table 3. Fracture loads and residual factors of the FRP composites at elevated temperatures.

Temperature (°C)	Fracture Load (N)				Residual Factor ($Ff_{,T}/Ff_{,20}$)			
	Group-1	Group-2	Group-3	Average	Group-1	Group-2	Group-3	Average
20	56.88	58.51	53.22	56.20	1.01	1.04	0.95	1.00
50	60.25	61.38	62.51	61.38	1.07	1.09	1.11	1.09
100	36.36	38.52	36.33	37.07	0.65	0.69	0.65	0.66
150	142.31	139.82	144.33	142.15	2.53	2.49	2.57	2.53
200	113.23	118.33	120.87	117.48	2.01	2.11	2.09	2.07
250	319.33	325.41	330.57	325.10	5.68	5.79	5.88	5.78
300	600.87	581.34	610.45	597.55	10.69	10.34	10.86	10.63
350	130.47	133.91	136.47	133.62	2.32	2.38	2.43	2.38

4.3. Fracture Displacement

The fracture displacement is the displacement that corresponds to the fracture load. Table 4 lists the fracture displacement and the residual factor of the FRP composites at high temperatures. The residual fracture displacement factors are plotted in Figure 5c.

Table 4. Fracture displacements and residual factors of the FRP composites at elevated temperatures.

Temperature (°C)	Fracture Displacement (mm)				Residual Factor ($X_{f,T}/X_{f,20}$)			
	Group-1	Group-2	Group-3	Average	Group-1	Group-2	Group-3	Average
20	1.80	1.81	1.81	1.81	1.00	1.00	1.00	1.00
50	1.79	1.80	1.79	1.79	0.99	1.00	0.99	0.99
100	1.93	1.92	1.91	1.92	1.07	1.06	1.06	1.06
150	2.11	2.11	2.12	2.11	1.17	1.17	1.17	1.17
200	2.37	2.38	2.40	2.38	1.31	1.32	1.32	1.32
250	3.07	3.10	3.12	3.10	1.70	1.72	1.73	1.71
300	3.38	3.37	3.36	3.37	1.87	1.87	1.86	1.87
350	2.46	2.44	2.47	2.46	1.36	1.35	1.37	1.36

Contrary to what was discussed above, the fracture displacement and the residual factor of the FRP composites at high temperatures gradually increased with the increasing temperature. Particularly when the FRP composites were exposed to 300 °C, the maximum increase in fracture displacement was 1.56 mm and 87% of their initial fracture displacement. This could be attributed to the fact that the bonding effects of the nanometer montmorillonite and the fiber material were most obvious at those temperatures, which led to the mass increment of ductility, causing the fracture displacement to increase. It is worth noting that when exposed to temperatures above 300 °C, the fracture displacement of the FRP composites decreased from 3.37 mm to 2.46 mm. This could be attributed to the fact that the bonding of the nanometer montmorillonite and the fiber material was softened at this temperature and the resin matrix entered the rubber state from the glass state, in which the transition temperature Tg was nearly 300 °C based on the test results.

4.4. Elastic Modulus

The elastic modulus is termed as the ratio of engineering stress to engineering strain in the elastic deformation stage during the tensile process. The elastic modulus and residual factors of the FRP composites at high temperatures are listed in Table 5. The residual elastic modulus factors are depicted in Figure 5d.

Table 5. Elastic moduli and residual factors of FRP composites at elevated temperatures.

Temperature (°C)	Elastic Modulus (MPa)				Residual Factor (E_T/E_{20})			
	Group-1	Group-2	Group-3	Average	Group-1	Group-2	Group-3	Average
20	799.21	802.34	803.11	801.55	1.00	1.00	1.00	1.00
50	810.35	815.24	808.65	811.41	1.01	1.02	1.01	1.01
100	812.77	817.65	815.37	815.26	1.01	1.02	1.02	1.02
150	813.38	815.48	816.74	815.20	1.01	1.02	1.02	1.02
200	820.39	819.35	821.22	820.32	1.02	1.02	1.02	1.02
250	925.33	925.49	929.64	926.82	1.15	1.15	1.16	1.16
300	1173.65	1182.37	1188.29	1181.44	1.46	1.48	1.48	1.47
350	750.24	758.41	749.59	752.75	0.94	0.95	0.94	0.94

The elastic modulus of the FRP composite specimens remained basically unchanged when exposed to temperatures below 200 °C, and the variation in the residual factors of elastic modulus did not exceed 2% when compared to the initial ultimate load of FRP composites. However, when exposed to temperatures between 200 and 300 °C, the elastic modulus of the FRP composites significantly increased from 820.32 MPa to 1181.44 MPa and increased by 47.39% of the initial elastic modulus, which indicated that the FRP composites experienced a strengthening process. This could be attributed to the fact that the bonding effects of the nanometer montmorillonite and the fiber material were most obvious at those temperatures, which led to the increment of elastic modulus. Notably, the elastic modulus of FRP composites significantly decreased from 1181.44 MPa to 752.75 MPa and reduced by 6.1% of the initial elastic modulus when exposed to the temperature of 350 °C. This could be attributed to the fact that the bonding of the nanometer montmorillonite and the fiber material was softened at this temperature and the resin matrix entered the rubber state from the glass state, in which the transition temperature Tg was nearly 300 °C based on the test results.

5. Constitutive Modeling

5.1. Johnson–Cook Model

Johnson and Cook initially proposed the Johnson–Cook model in 1983 [31,32]. The various stress–strain relationships of metallic materials in situations of large deformation, high strain rates, and elevated temperature could be properly described by this model. It has been frequently utilized, since it was first introduced due to its simple form. This constitutive model was expressed as follows:

$$\sigma(\varepsilon^p, \dot{\varepsilon}, T) = \left[A + B(\varepsilon^p)^n\right]\left[1 + C\ln\left(\frac{\dot{\varepsilon}}{\dot{\varepsilon}_R}\right)\right]\left[1 - \left(\frac{T - T_R}{T_m - T_R}\right)^m\right] \quad (1)$$

where n is the constant coefficient of strain hardening, C is strain rate strengthening coefficient, m is thermal softening coefficient, A is the nominal yield stress (MPa) in the tensile process, B is the strain hardening constant (MPa), and σ and ε are the engineering stress and plastic strain, respectively, $\dot{\varepsilon}_R$ and T_R are the reference strain rate and reference deformation temperature, respectively, T_m is the melting temperature of the various metallic materials, and T is the experimental temperature in the test. The three terms in the constitutive model, read from left to right, represent the effects of heating of elevated temperatures, strengthening of strain rate, and strain hardening of flow stress [33,34]. In this study, the reference temperature and strain rate in this experiment were T_R = 293 K and $\dot{\varepsilon}_R$ = 0.005 s^{-1}, respectively. Under this experimental circumstance, A = 9.43 MPa and T_m = 1300 K.

5.1.1. Identification of Parameters B and n

Under the deformation rate and temperature $\dot{\varepsilon} = \dot{\varepsilon}_R = 0.005$ s^{-1}, $T = T_R = 293$ K. Equation (1) was transformed to the following:

$$B\varepsilon^n + A = \sigma \qquad (2)$$

The effects of thermal softening and strain rate strengthening are neglected. By transforming Equation (2) and dividing Equation (2) by the natural logarithm into both sides, Equation (2) was changed to the following:

$$n \ln \varepsilon + \ln B = \ln(\sigma - A) \qquad (3)$$

Figure 6 depicts the relationship of $\ln\varepsilon$ and $\ln(\sigma - A)$ after carrying out the linear fitting by substituting the values of stress and strain into Equation (3). The values of n and $\ln B$, represent the slope and the initial value of the fitting curve, respectively. As a result, the coefficient can be calculated as $n = 1.33$ and $B = 1422.26$ MPa.

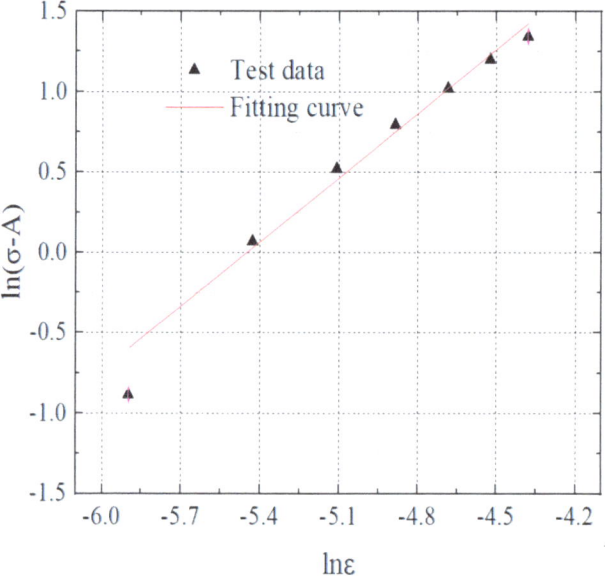

Figure 6. The relationship of $\ln\varepsilon$ and $\ln(\sigma - A)$.

5.1.2. Identification of Parameter C

Under the deformation temperature in this experiment, $T = T_R = 293$ K, Equation (1) was rearranged as:

$$C \ln\left(\frac{\dot{\varepsilon}}{\dot{\varepsilon}_R}\right) + 1 = \frac{\sigma}{(A + B\varepsilon^n)} \qquad (4)$$

Figure 7 depicts the relationship between $\ln(\dot{\varepsilon}/\dot{\varepsilon}_R)$ and $\sigma/(A + B\varepsilon^n)$ after carrying out the linear fitting by substituting 11 strain values and 3 strain rates obtained in this experiment into Equation (4). The value of C represents the slope of the fitting curve. According to the test data of FRP composites, the value of C can be calculated as 0.45.

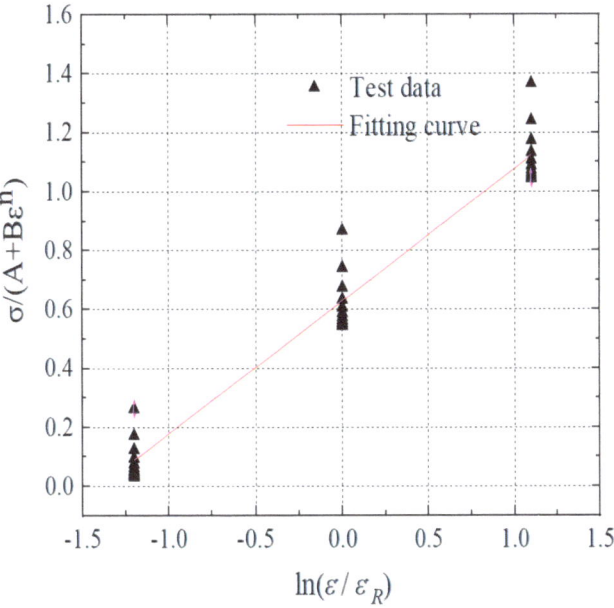

Figure 7. The relationship between $\ln(\dot{\varepsilon}/\dot{\varepsilon}_R)$ and $\sigma/(A + B\varepsilon^n)$.

5.1.3. Identification of Parameter m

Under the deformation rate in this experiment, $\dot{\varepsilon} = \dot{\varepsilon}_R = 0.005$ s^{-1}, Equation (1) was rearranged as:

$$m \ln \frac{T - T_R}{T_m - T_R} = \ln[1 - \frac{\sigma}{(A + B\varepsilon^n)}] \qquad (5)$$

Figure 8 depicts the $\ln[(T - T_R)/(T_m - T_R)]$–$\ln[1 - \sigma/(A + B\varepsilon^n)]$ curve after carrying out the linear fitting by substituting the 11 strain values and 4 deformation temperatures determined in this study into Equation (5). The value of *m* represents the slope of the fitting curves. Based on the experimental data of FRP composites, the value of *m* is 0.45.

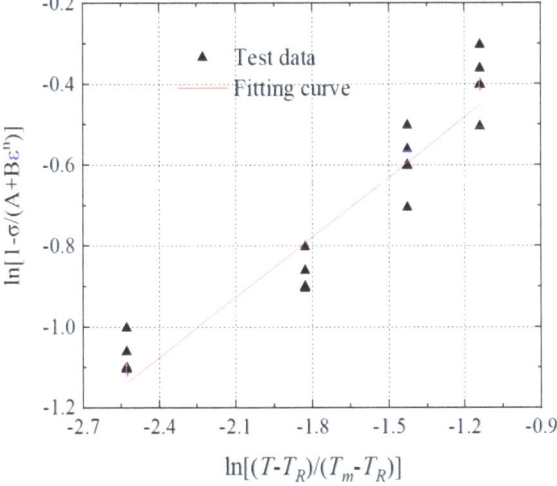

Figure 8. The relationship of $\ln[(T - T_R)/(T_m - T_R)]$ and $\ln[1 - \sigma/(A + B\varepsilon^n)]$.

Finally, the relationship among stress σ, strain ε, deformation temperature T and deformation rate $\dot{\varepsilon}$ was established according to the Johnson–Cook model:

$$\sigma = \left[9.43 + 1422.26 \times \varepsilon^{1.33}\right] \times \left[1 + 0.45 \times \ln\frac{\dot{\varepsilon}}{0.005}\right] \times \left[1 - \left(\frac{T-293}{1007}\right)^{0.78}\right] - T_0 \quad (6)$$

5.2. Verification of the Constitutive Model

The cross-section of the FRP specimens was taken into account as a quantitative parameter throughout the stretching process for the engineering stress–strain curves. This primarily refers to how the cross-section of specimens changed in response to the tensile load. In reality, before tension fracture, the specimen's cross-section steadily declines. The true stress–strain curves of various materials clearly illustrate the impact of the elevated temperatures, where the σ_T and ε_T can be converted by the following expression:

$$\sigma_T = \sigma_E(1 + \varepsilon_E) \quad (7)$$

$$\varepsilon_T = \ln(1 + \varepsilon_E) \quad (8)$$

where σ_E represents the engineering stress and σ_T represents the true stress, and ε_E and ε_T represent the engineering strain and the true strain, respectively.

In this paper, model validation was conducted by comparing true stress–strain curves from experiments with those obtained from computer simulations. The validation was conducted using experimental true stress–strain curves for FRP composites at the deformation rate of 0.005 s^{-1}, which were used to establish model parameters. The Johnson–Cook constitutive model for FRP composites at elevated temperatures was used to finally identify the material properties listed in Table 1. Figure 9 represents the comparison between test data and simulated data by the Johnson–Cook constitutive model at elevated temperatures. As seen in Figure 9, some deviation was seen, and the linear assumption was mostly to account for the inaccuracy. The findings were generally satisfactory, indicating that the linear assumption is appropriate and that this proposed Johnson–Cook constitutive model can accurately depict the true stress–strain behavior of FRP composites in the fire scenario.

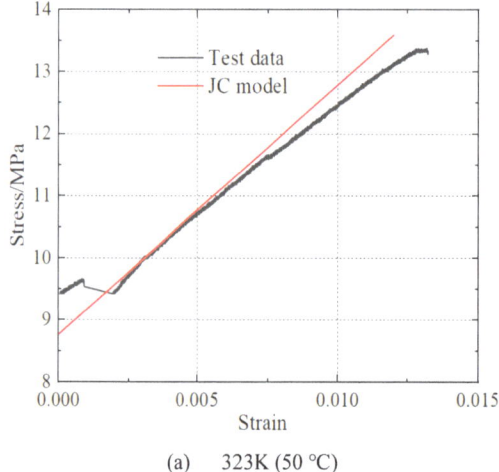
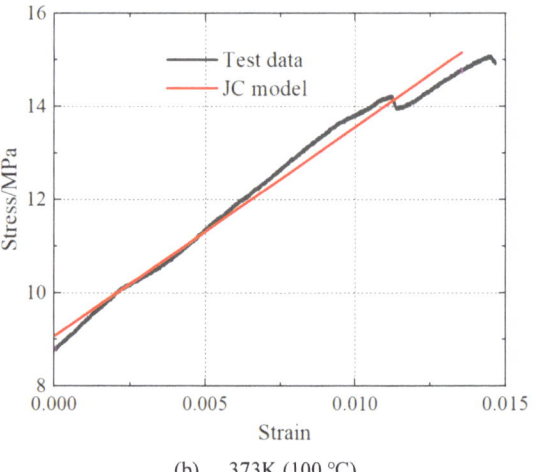

(a) 323K (50 °C)　　　　(b) 373K (100 °C)

Figure 9. Cont.

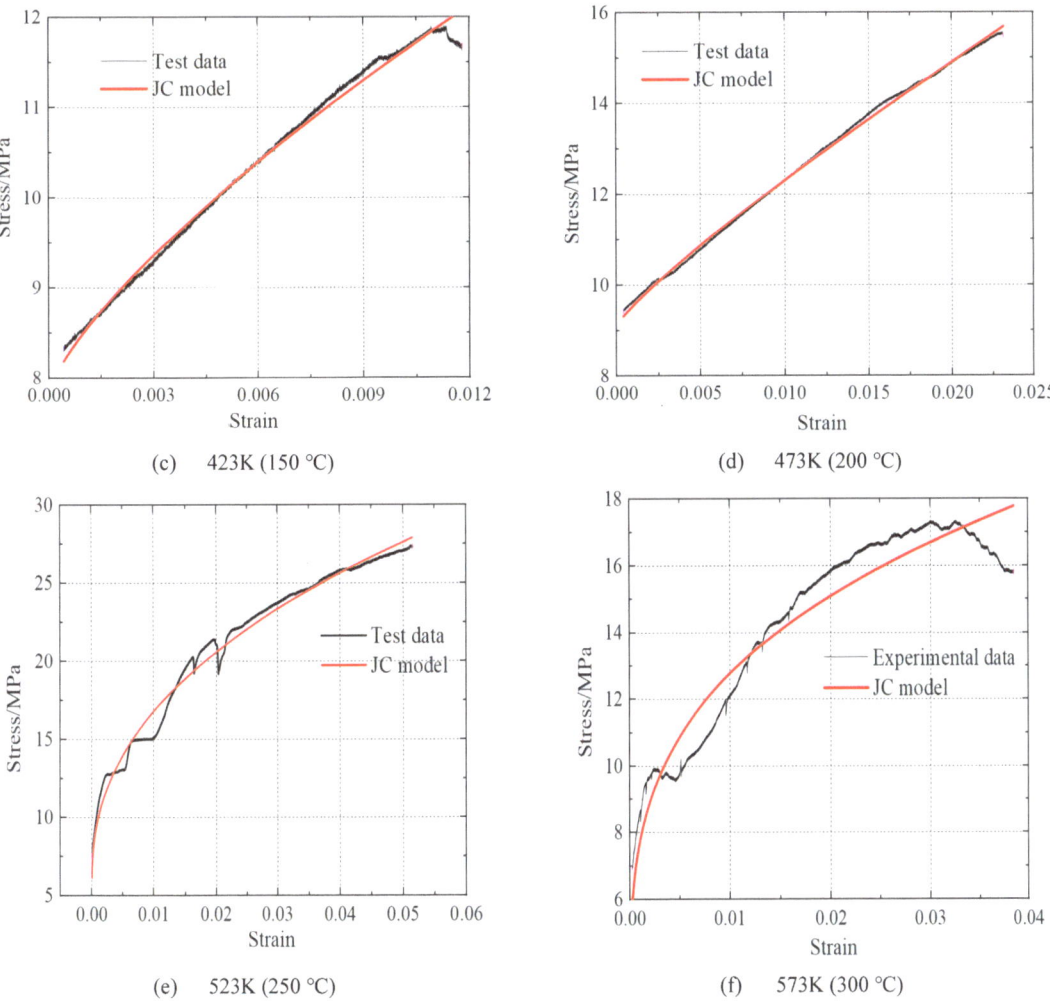

Figure 9. Comparison between test data and simulated data by the Johnson−Cook constitutive model at various high temperatures.

6. Conclusions

To better comprehend how elevated temperatures affect composites made of fiber, an investigation on the mechanical properties of FRP composites exposed to temperatures of 20–350 °C was experimentally researched. Simultaneously, the FRP specimens were axially loaded until fracture to observe the failure visual observations and mechanical properties. Finally, a new Johnson−Cook constitutive model was proposed to predict the behavior of FRP specimens in the fire scenario. The following are the significant conclusions of this experiment:

1. The mechanical properties of FRP composites had a critical temperature of 200 °C. When exposed to temperatures below 200 °C, elevated temperatures had a minor influence on the mechanical properties of FRP composites. When exposed to temperatures above 200 °C, the mechanical properties of FRP composites exhibited significant differences.
2. The ultimate bearing temperature of FRP composites was 300 °C. When exposed to temperatures above 300 °C, the mechanical properties which include ultimate load, fracture load, fracture displacement, and elastic modulus decreased sharply.

3. The elevated temperatures exerted a significant influence on the surface color of the FRP composites. The surface color of FRP composites gradually changed from fully brown to black with increasing temperatures.
4. This proposed Johnson−Cook constitutive model can accurately depict the true stress−strain behavior of FRP composites at elevated temperatures.

Author Contributions: Conceptualization, Methodology, Investigation, Data curation, Writing—reviewing & editing, C.Z.; Investigation, Data curation, Writing—Original draft preparation, Y.L.; Investigation, J.W. All authors have read and agreed to the published version of the manuscript.

Funding: This research was funded by the National Natural Science Foundation of China (grant No. 51508482) and the Natural Science Foundation of Tibet Autonomous (grant No. CGZH2018000014).

Institutional Review Board Statement: Not applicable.

Informed Consent Statement: Not applicable.

Data Availability Statement: The data presented in this study are available on request from the corresponding author. The data are not publicly available due to [Fund Requirements].

Conflicts of Interest: The authors declare no conflict of interest.

References

1. Leone, M.; Matthys, S.; Aiello, M.A. Effect of elevated service temperature on bond between FRP EBR systems and concrete. *Compos. B Eng.* **2009**, *40*, 85–93. [CrossRef]
2. Al-Salloum, Y.A.; Elsanadedy, H.M.; Abadel, A.A. Behavior of FRP-strengthened concrete after high temperature exposure. *Constr. Build. Mater.* **2011**, *25*, 838–850. [CrossRef]
3. Nguyen, P.L.; Vu, X.H.; Ferrier, E. Elevated temperature behaviour of carbon fibre-reinforced polymer applied by hand lay-up (M-CFRP) under simultaneous thermal and mechanical loadings: Experimental and analytical investigation. *Fire Saf. J.* **2018**, *100*, 103–117. [CrossRef]
4. Nguyen, P.L.; Vu, X.H.; Ferrier, E. Characterization of pultruded carbon fibre-reinforced polymer (P-CFRP) under two elevated temperature-mechanical load cases: Residual and thermo-mechanical regimes. *Constr. Build. Mater.* **2018**, *165*, 395–412. [CrossRef]
5. Mouritz, A.; Mathys, Z. Post-fire mechanical properties of marine polymer composites. *Compos. Struct.* **1999**, *47*, 643–653. [CrossRef]
6. Cao, S.; Zhi, W.; Wang, X. Tensile properties of CFRP and hybrid FRP composites at elevated temperatures. *J. Compos. Mater.* **2009**, *43*, 315–330.
7. Bazli, M.; Ashrafi, H.; Oskouei, A.V. Experiments and probabilistic models of bond strength between GFRP bar and different types of concrete under aggressive environments. *Constr. Build. Mater.* **2017**, *148*, 429–443. [CrossRef]
8. Chen, J.; Young, B.; Uy, B. Behavior of high strength structural steel at elevated temperatures. *J. Struct. Eng.* **2006**, *132*, 1948–1954. [CrossRef]
9. Lange, J.; Schneider, R. Constitutive equations of structural steel S460 at high temperatures. *J. Struct. Fire Eng.* **2009**, *2*, 217–230.
10. Wang, W.Y.; Liu, B.; Kodur, V. Effect of temperature on strength and elastic modulus of high-strength steel. *J. Mater. Civ. Eng.* **2013**, *25*, 174–182. [CrossRef]
11. Wang, W.Y.; Wang, K.; Kodur, V.; Wang, B. Mechanical properties of high-strength Q690 steel at elevated temperature. *J. Mater. Civ. Eng.* **2018**, *30*, 04018062. [CrossRef]
12. Wang, W.Y.; Zhang, Y.H.; Xu, L.; Li, X. Mechanical properties of high-strength Q960 steel at elevated temperature. *Fire Saf. J.* **2020**, *114*, 10301. [CrossRef]
13. Qiang, X.H.; Bijlaard, F.; Kolstein, H. Post-fire mechanical properties of high strength structural steels S460 and S690. *Eng. Struct.* **2012**, *35*, 1–10. [CrossRef]
14. Qiang, X.H.; Bijlaard, F.; Kolstein, H. Post-fire performance of very high strength steel S960. *J. Constr. Steel Res.* **2012**, *80*, 235–242. [CrossRef]
15. Gunalan, S.; Mahendran, M. Experimental investigation of post-fire mechanical properties of cold-formed steels. *Thin-Walled Struct.* **2014**, *84*, 241–254. [CrossRef]
16. Chiew, S.P.; Zhao, M.S.; Lee, C.K. Mechanical properties of heat-treated high strength steel under fire/post-fire conditions. *J. Constr. Steel Res.* **2014**, *98*, 12–19. [CrossRef]
17. Chen, Z.; Lu, J.; Liu, H.; Liao, X. Experimental investigation on the post-fire mechanical properties of structural aluminum alloys 6061-T6 and 7075-T73. *Thin-Walled Struct.* **2016**, *106*, 187–200. [CrossRef]
18. Tao, Z.; Wang, X.Q.; Hassan, M.K.; Song, T.Y.; Xie, L.A. Behavior of three types of stainless steel after exposure to elevated temperatures. *J. Constr. Steel Res.* **2019**, *152*, 296–311. [CrossRef]
19. Yu, Y.; Lan, L.; Ding, F.; Wang, L. Mechanical properties of hot-rolled and cold-formed steels after exposure to elevated temperature: A review. *Constr. Build. Mater.* **2019**, *213*, 360–376. [CrossRef]

20. Zhang, C.; Zhu, H.; Zhu, L. Effect of interaction between corrosion and high temperature on mechanical properties of Q355 structural steel. *Constr. Build. Mater.* **2021**, *271*, 121605. [CrossRef]
21. Ren, C.; Dai, L.; Huang, Y.; He, W. Experimental investigation of post-fire mechanical properties of Q235 cold-formed steel. *Thin-Walled Struct.* **2020**, *150*, 106651. [CrossRef]
22. Jiang, J.; Bao, W.; Peng, Z.Y.; Wang, Y.B.; Liu, J.; Dai, X.H. Experimental investigation on mechanical behaviors of TMCP high strength steel. *Constr. Build. Mater.* **2019**, *200*, 664–680. [CrossRef]
23. Zhang, C.; Wang, R.; Song, G. Post-fire mechanical properties of Q460 and Q690 high strength steels after fire-fighting foam cooling. *Thin-Walled Struct.* **2020**, *156*, 106983. [CrossRef]
24. Zhang, C.; Wang, R.; Zhu, L. Mechanical properties of Q345 structural steel after artificial cooling from elevated temperatures. *J. Constr. Steel Res.* **2021**, *176*, 106432. [CrossRef]
25. Li, G.Q.; Liu, H.; Zhang, C. Post-fire mechanical properties of high strength Q690 structural steel. *J. Constr. Steel Res.* **2017**, *132*, 108–116. [CrossRef]
26. Wang, F.; Lui, E.M. Experimental study of the post-fire mechanical properties of Q690 high strength steel. *J. Constr. Steel Res.* **2020**, *167*, 105966. [CrossRef]
27. Zhang, C.; Jia, B.; Wang, J. Influence of artificial cooling methods on post-fire mechanical properties of Q355 structural steel. *Constr. Build. Mater.* **2020**, *252*, 11909. [CrossRef]
28. Zhang, C.; Gong, M.; Zhu, L. Post-fire mechanical behavior of Q345 structural steel after repeated cooling from elevated temperatures with fire-extinguishing foam. *J. Constr. Steel Res.* **2022**, *191*, 107201. [CrossRef]
29. Zhang, C.; Wang, R.; Song, G. Effects of pre-fatigue damage on mechanical properties of Q690 high-strength steel. *Constr. Build. Mater.* **2020**, *252*, 118–845. [CrossRef]
30. *GB/T 228.1-2010, eqv. ISO 6892-1: 2009*; MOD Metallic Materials- Tensile Testing- Part 1: Method of Test at Room Temperature. General Administration of Quality Supervision, and Standardization Administration of the People's Republic of China: Beijing, China, 2010.
31. Johnson, G.R.; Cook, W.H. Fracture characteristics of three metals subjected to various strains, strain rates, temperatures and pressures. *Eng Fract Mech.* **1985**, *21*, 31–48. [CrossRef]
32. Li, H.Y.; Li, Y.H.; Wang, X.F.; Liu, J.J.; Wu, Y. A comparative study on modified Johnson Cook, modified Zerilli–Armstrong and Arrhenius-type constitutive models to predict the hot deformation behavior in 28CrMnMoV steel. *Mater Des.* **2013**, *49*, 493–501. [CrossRef]
33. Samantaray, D.; Mandal, S.; Bhaduri, A.K. A comparative study on Johnson–Cook, modified Zerilli–Armstrong and Arrhenius-type constitutive models to predict elevated temperature flow behavior in modified 9Cr–1Mo steel. *Comput Mater Sci.* **2009**, *47*, 568–576. [CrossRef]
34. Shrot, A.; Bäker, M. Determination of Johnson–Cook parameters from machining simulations. *Comput Mater Sci.* **2012**, *52*, 298–304. [CrossRef]

Disclaimer/Publisher's Note: The statements, opinions and data contained in all publications are solely those of the individual author(s) and contributor(s) and not of MDPI and/or the editor(s). MDPI and/or the editor(s) disclaim responsibility for any injury to people or property resulting from any ideas, methods, instructions or products referred to in the content.

Article

Numerical and Experimental Study on Large-Diameter FRP Cable Anchoring System with Dispersed Tendons

Jingyang Zhou [1,2], Xin Wang [1,2,*], Lining Ding [3], Shui Liu [1,2] and Zhishen Wu [1,2]

1. Key Laboratory of C & PC Structures Ministry of Education, Southeast University, Nanjing 211189, China
2. National and Local Unified Engineering Research Center for Basalt Fiber Production and Application Technology, International Institute for Urban Systems Engineering, Southeast University, Nanjing 211189, China
3. School of Civil Engineering, Nanjing Forestry University, Nanjing 210037, China
* Correspondence: xinwang@seu.edu.cn; Tel.: +86-136-1151-4436

Abstract: Based on a previously designed variable-stiffness load transfer component (LTC), a novel dispersed-tendon cable anchor system (CAS) was developed to increase the anchoring efficiency of large-diameter basalt-fiber-reinforced polymer (BFRP) cables. The static behaviors of the CAS are then numerically evaluated by a simplified three-dimensional finite-element (FE) model and implemented in a full-scale BFRP cable. The FE results indicated that the accuracy of the simplified dispersed-tendon model could be effectively ensured by dividing the revised compensation factor. The anchor behavior of the dispersed-tendon CAS was superior to that of the parallel-tendon CAS when the same cable was applied. The radial stress and tensile stress difference can be reduced by decreasing the tendon spacing. The testing and simulated results agreed well with the load–displacement relationship and axial displacement. All tendons fractured in the testing section, and the LTC suffered minimal damage. The ultimate force of the cable with 127 4-mm-diameter tendons was 2419 kN, and the corresponding anchoring efficiency was 93%. The cable axial tensile strain in the anchoring zone decreased linearly from the loading end to the free end. The cable shear stress concentration at the loading end can be avoided by employing a variable-stiffness anchoring method.

Keywords: basalt-fiber-reinforced polymers (BFRP); larger-diameter cable; dispersed-tendon anchoring method; finite-element (FE) analysis; full-scale experiment

Citation: Zhou, J.; Wang, X.; Ding, L.; Liu, S.; Wu, Z. Numerical and Experimental Study on Large-Diameter FRP Cable Anchoring System with Dispersed Tendons. *Buildings* 2023, *13*, 92. https://doi.org/10.3390/buildings13010092

Academic Editor: Elena Ferretti

Received: 21 November 2022
Revised: 26 December 2022
Accepted: 27 December 2022
Published: 30 December 2022

Copyright: © 2022 by the authors. Licensee MDPI, Basel, Switzerland. This article is an open access article distributed under the terms and conditions of the Creative Commons Attribution (CC BY) license (https:// creativecommons.org/licenses/by/ 4.0/).

1. Introduction

Fiber-reinforced polymers (FRPs) have many outstanding features [1–3], such as light weight, high tensile strength, and corrosion resistance. FRPs are frequently considered ideal materials for solving the bottleneck problems of steel cables [4–6], including easy corrosion and considerable weight. The application of FRP cables in civil engineering has gradually increased [7–9]. Generally, carbon FRP (CFRP) cables are used as partial or full structural members to strengthen engineering structures [10–12]. For example, a network arch bridge with a 124 m span was built across a highway in Stuttgart, Germany, in 2020. All the hangers were created from CFRP. CFRP hangers have many advantages over steel hangers [13], such as a small cross-section, low cost, long service life, and flexible aesthetic arrangement. In another case, a highway cable-stayed bridge having two 100 m main spans, supported by sixty-eight steel cables and four CFRP cables, was successfully built in Shandong, China, in 2022. The CFRP cable was manufactured using 121 7 mm-diameter tendons. The mean ultimate load was 10,440 kN. The mean anchoring efficiency, denoted as the ratio of the measured ultimate to the standard value of the tensile load of the cable, was 107%. Recently, a self-anchored CFRP cable, developed by Feng et al. [14], was applied in steel trusses in Shanghai, China, in 2022. The ultimate load of the cable reached 4550 kN.

In summary, an increasing number of engineering applications indicates that CFRP cables are becoming competitive substitutes for traditional steel cables [15–19].

However, CFRP cables also have some limitations [20], such as high price, brittle fracturing, and low ductility; thus, their application range is limited to some extent. High-performance structural materials should be developed to fill the application blank zone of CFRP cables. Compared with CFRP, basalt FRP (BFRP) is a recently emerging prestressing material [21–23]. BFRP materials are primarily used to enhance new structures and reinforce existing structures owing to their outstanding characteristics, such as high creep fracture stress, high ductility, high performance–price ratio, and eco-friendliness [24]. However, the elastic modulus of BFRPs is significantly lower than those of CFRPs and steel [25]. Consequently, the tensile deformation of BFRP cables is inevitably larger than that of CFRP and steel cables, resulting in excessive deformation of the structures supported by cables. The development of large-diameter BFRP cables is frequently considered an effective solution to solving this problem. However, the anchoring efficiency of BFRP cables also decreases with an increase in cable diameter. Therefore, the premise of utilizing large-diameter BFRP cables is to increase their anchoring efficiency.

Many anchoring methods for FRP cables were proposed by using different inner shape in steel anchorages and designing novel load transfer components (LTCs) [26], but few were adapted to large-diameter BFRP cables. For instance, Wang [27] developed a sectional LTC composed of different continuous fibers and epoxy resin to decrease the stress concentration of BFRP cables with 37 4-mm-diameter tendons. The mean tensile load and anchor efficiency of the BFRP cables were 592 kN and 99%, respectively. Shi [28] developed a composite-wedge anchorage for BFRP tendons involving three variable-stiffness wedges composed of resin, impregnated chopped fiber, silica sand, and fiber sheet. The anchor efficiency and fatigue cycle of the cable anchor system (CAS) were 91% and more than 200 million cycles, respectively. Compared with existing CFRP cables, the proposed methods are most suitable for small-diameter BFRP cables. In summary, there is an urgent need to develop anchoring methods for large-diameter BFRP cables.

In a previous study [29], a sectional variable-stiffness LTC fabricated with epoxy resin and chopped glass fibers with a length of 900 μm and a diameter of 13 μm was proposed (Figure 1). The fiber volume fractions of the segmented LTCs, defined as the volume ratio of fiber to fiber and epoxy resin, were 0%, 8.9%, 22.8%, and 40% [30], respectively. The corresponding elastic moduli were 2.8, 5.17, 7.53, and 9.9 GPa [30], respectively. The anchor-zone cable was arranged in parallel, and the tendon spacing was 1.5 mm. The lengths of the sectional LTCs were 115 (containing a horizontal segment of 15 mm), 100, 100, and 100 mm, respectively. Through full-scale experiments, the mean ultimate tensile load and anchoring efficiency of the three BFRP cables were observed to be 1919 kN and 95% [30], respectively. The proposed method was preliminarily verified to be suitable for anchoring large-diameter cables. However, the anchoring efficiency tends to decline, and the coil radius of the cable increases when the cable diameter exceeds a specific limit. To solve the problems, the cable arrangement in the anchoring zone can be changed from parallel to dispersed. The improvement effect of the method on anchoring efficiency was verified in a previous study on CFRP cables composed of 37 7-mm-diameter tendons [30]. The tensile strength of the CFRP cables exceeded 3000 MPa. The principle of improvement is to use the dispersed tendons to bear part of the axial shear force to avoid shear failure in the LTC. In addition, the coil radius of the cable can be decreased by reducing the tendon spacing and diameter.

As shown in Figure 2, a novel anchoring method was further proposed based on the above optimization strategy. Compared with the previous CAS shown in Figure 1, only the tendon diameter, tendon number, and arrangement of the BFRP cable were changed. The anchorage was derived from a previous experiment, and the sectional LTCs were also unchanged. The BFRP cable was composed of 127 4-mm-diameter tendons in a regular hexagonal arrangement. From the center layer to the outermost layer, the bending angles of the tendons, denoted as the arctangent of the ratio of tendon transverse deviation distance in the free end to the anchorage length [30], were 0°, 0.75°, 1.5°, 2.25°, 3°, 3.75°, and 4.5°, respectively. The tendon spacing of the BFRP cable in the test section was determined using subsequent simulation analysis.

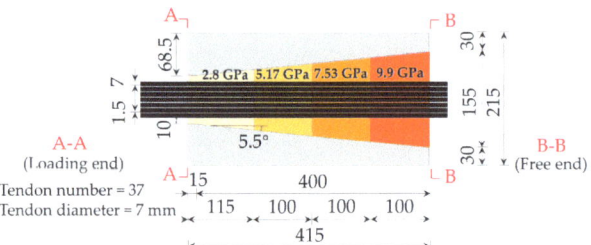

Figure 1. Previous parallel-tendon CAS (dimensions in mm).

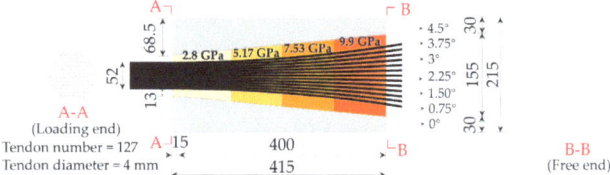

Figure 2. Dispersed-tendon CAS (dimensions in mm).

In this study, the tensile behaviors of the BFRP tendons were evaluated. Based on the previous simplified parallel-tendon modeling method [31], a simplified dispersed-tendon three-dimensional (3D) solid finite-element (FE) model was proposed. Compared with a full 3D FE model, the simplified model aims to improve the modeling speed and computational efficiency while significantly reducing the number of elements. Based on the simplified model, the complex stress states of the anchor-zone cable and LTC were further revealed, and the tendon spacing was determined. A full-scale experiment was carried out to evaluate the static anchoring behavior of the dispersed-tendon BFRP CAS. Compared with published papers on BFRP cables, the BFRP cable with 127 4-mm-diameter tendons may be the most significant specification. The test results will further verify the possibility of BFRP cables used as large-tonnage cables. Thus, the application range of high-performance BFRP cables will be further expanded. The validity of the simplified dispersed-tendon 3D FE model was confirmed by comparing the experimental results. Based on the verified simplified model, optimization design for different FRP CAS can be well undertaken with high calculation accuracy and efficiency. Finally, the optimal FRP CAS can be realized for different application requirements.

2. Tensile Behavior of the BFRP Tendon

2.1. Materials and Preparation

The tensile behavior of the BFRP tendons, a key index for use as a prestressed component, was first evaluated based on 11 identical specimens. BFRP tendons with a shallow ribbed surface, consisting of 21 1200-tex and one 800-tex fiber roving, were manufactured using continuous pultrusion using an epoxy resin matrix. Compared with the BFRP tendons used in a previous study [27], the preparation technology for testing BFRP tendons was improved by decreasing the rib width and height and increasing the fiber volume fraction to obtain a higher tensile strength and reduce the amount of external fiber bending. The mean value of the equivalent diameter of the BFRP tendons, measured using the drainage method [29], was 4 mm. As shown in Figure 3a, the lengths of the anchoring and test sections were 350 and 400 mm, respectively. The two ends of the BFRP tendon were anchored using hollow steel pipes. The outer diameter and wall thickness of the pipes were 14 mm and 3 mm, respectively. The filling material was the epoxy resin that can solidify at room temperature. An extensometer with a gauge length of 50 mm was used to measure the tensile strain of the BFRP tendon.

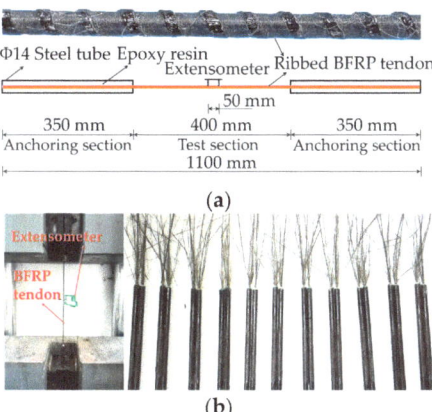

Figure 3. Tensile test of BFRP tendons: (**a**) anchoring scheme and (**b**) loading and failure.

2.2. Calculation Method

The tensile properties of the BFRP tendons can be calculated according to Chinese Standard (GB/T 30022–2013) [32]. The equations are listed in Equations (1)–(4). Specifically, σ_u is the tensile strength (MPa). F_u is the ultimate tensile load (kN). A_{st} is the cross-sectional area (mm^2). E_L is the elastic modulus (GPa). ε_u is the ultimate tensile strain (με). ε_1 and ε_2 are the tensile strains when the loads are F_1 and F_2, respectively. F_1 and F_2 are 20% and 50% of F_u, respectively. f_{fpk} is the standard value of σ_u, which considers the dispersion of BFRP materials and 95% strength assurance rate [30]. f_a is the average σ_u. σ is the standard deviation (SD) of σ_u. The coefficient of variation (CV) is the ratio of σ to f_a.

$$\sigma_u = \frac{F_u}{A_{st}} \quad (1)$$

$$E_L = \frac{F_1 - F_2}{(\varepsilon_1 - \varepsilon_2)A_{st}} \quad (2)$$

$$\varepsilon_u = \frac{F_u}{E_L A_{st}} \quad (3)$$

$$f_{fpk} = f_a(1 - 1.645\text{CV}) = f_a - 1.645\sigma \quad (4)$$

2.3. Results and Discussions

As shown in Figure 3b, a universal machine with a loading capacity of 2000 kN was employed for the BFRP tendons. A tensile speed of 2 mm/min was determined to obtain the elastic modulus of the BFRP tendons according to the standard [32]. In the loading process, the tensile force increased linearly with piston displacement, verifying the linear elastic property of the BFRP tendon. Fiber breakage began to occur gradually when the tensile force exceeded 80% of the ultimate load. The final fracture of the BFRP tendons was abrupt with hardly a warning. All the specimens exhibited a typical multifilament fracture at the test section, indicating that the anchoring scheme was effective and that the material strength was utilized entirely.

The experimental results for the BFRP tendons are displayed in Table 1, in which σ_u, F_u, E_L, ε_u and f_{fpk} were calculated by Equations (1)–(4). The mean values of σ_u, F_u, E_L, and ε_u were 21 kN, 1702 MPa, 55 GPa, and 3.11%, respectively, which were superior to the tensile performance of BFRP tendons in [27]. In addition, the CVs of the tendons were less than 4%, indicating that the tendons with low material dispersion can be employed as prestressing cable members. According to Equation (4), the f_{fpk} of the eleven BFRP tendons was 1638 MPa, which was the basis for calculating the standard value of the ultimate tensile force of the multi-tendon BFRP cable in the following analysis.

Table 1. Testing results for the BFRP tendons.

Specimen	F_u/kN	σ_u/MPa	E_L/MPa	ε_u/%
1	22	1715	54	3.16
2	21	1689	55	3.06
3	21	1707	54	3.14
4	21	1709	54	3.14
5	21	1654	55	2.99
6	22	1754	53	3.28
7	21	1678	55	3.04
8	20	1625	55	2.94
9	22	1726	55	3.12
10	22	1764	54	3.25
11	21	1697	55	3.07
Mean	21	1702	55	3.11
SD	0.48	38.53	0.61	0.10
CV/%	2.26	2.26	1.12	3.16

3. Simulation Analysis of the Dispersed-Tendon CAS

3.1. Proposal of the Simplified Model

3.1.1. Shortcomings of the Full Model

A 3D FE model (called the full model) for parallel-tendon CAS with 37 7-mm-diameter BFRP tendons was created using the ABAQUS/Standard program and C3D8R elements to simulate actual experiments with high accuracy in the previous study [30] (see Figure 4). The full model is characterized by the ability to consider the existence of multiple independent FRP tendons. The accuracy of the full model was verified to be high compared with previous experimental results [29,33]. However, the full model also contributed to an excessive computing burden resulting from the rapidly increasing number of elements and convergence challenges [31]. Additionally, meshing of the full model used for dispersed-tendon CAS was significantly tricky because of the irregular inner geometric shape of the LTC and dispersed tendons. Therefore, a simplified 3D FE model is required to make the calculation and meshing more feasible.

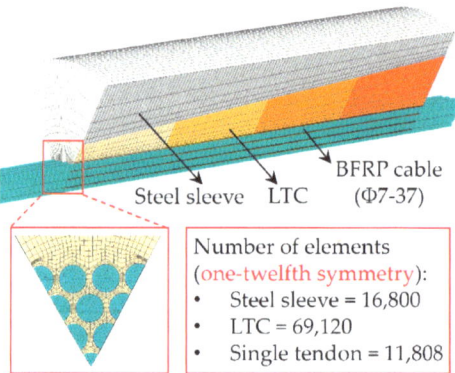

Figure 4. Full FE model.

3.1.2. Theoretical Analysis for the Simplified Model

In a previous study [31], a simplified model for parallel-tendon CAS was proposed using Ansys 15.0 and Solid 186 elements (Figure 5). The simplified model was characterized by the ability to replace multiple tendons with several rings. However, the disadvantages of the simplified model deserve attention. The equivalent ring model has a ring hoop effect. The ring cable will generate circumferential resistance when subjected to radial extrusion. Thus, the deformation of the cable was limited. This deformation characteristic is completely different from that of the multiple-tendon cable. In this model, the tendons in each layer are equivalent to the corresponding concentric rings based on the principle of equal cross-sectional areas. The center tendon is defined as the first ring. The radius and thickness t of the ith ring are expressed as Equations (5) and (6), where D_t is the tendon diameter (mm), δ is the tendon spacing (mm), and $N(i)$ is the number of tendons in the ith layer. For a regular hexagonal arrangement, $N(i) = 1, 6, 12, 18 \ldots$, for $i = 1, 2, 3, 4 \ldots$, respectively [31].

$$R^{center}(i) = (i-1)(D_t + \delta) \tag{5}$$

$$t(i) = R^{outer}(i) - R^{inner}(i) = \frac{N(i)D_t^2}{8R^{center}(i)} \tag{6}$$

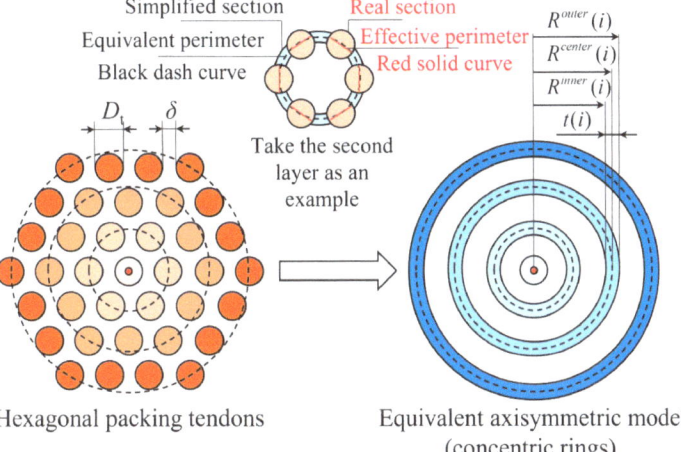

Figure 5. Simplification of the parallel-tendon cable.

Based on this assumption, the radial and shear stress of BFRP cables were consistently underestimated owing to the absence of tendon spacing. Therefore, a compensation factor φ, defined as the ratio of the effective perimeter of the real section to that of the simplified section [31] (see Figure 5), was proposed. When $i \geq 2$, φ can be expressed as Equation (7), where φ is a constant when $i \geq 2$. This phenomenon can be explained by the constant wall thickness of the rings in the axial direction. The radial and shear stresses of the BFRP cable, obtained from the simplified FE model with parallel concentric rings along the axial direction, can be further adjusted by dividing φ. However, for the dispersed-tendon CAS, the derived φ cannot be employed directly because of the varying wall thickness of the rings along the axial direction. Thus, φ must be further revised to be consistent with the full dispersed-tendon model.

$$\varphi(i) = \frac{N(i)D_t}{2\pi R^{center}(i)} = \frac{N(i)D_t}{2\pi(i-1)(D_t+\delta)} \tag{7}$$

As shown in Figure 6, the geometric shape of the dispersed tendons in the ith layer can be regarded an arc. According to the geometrical relationship, $y(i)$ and $R^{center}(i)$ can be expressed as Equations (8)–(10). Specifically, L is the anchoring length. $\theta(i)$ is the bending angle of the dispersed tendons in the ith layer. $R(i)$ is the bending radius of the dispersed tendons in the ith layer. $x(i)$ is the horizontal distance between the point in the center position of the ith layer and the loading end. $y(i)$ is the vertical distance between the point in the center position of the *i*th layer and the center line of the first layer. Based on Equations (8)–(10), Equation (7) can be revised as Equation (11). Based on the given dispersed-tendon CAS in Figure 2, the following parameters can be determined: D_t = 4 mm, L = 415 mm, and $\theta(i)$ = 0°, 0.75°, 1.5°, 2.25°, 3°, 3.75°, 4.5°, for i = 1, 2, 3, 4, 5, 6, 7, respectively.

$$R(i) = \frac{L}{2\sin\theta(i)\cos\theta(i)} \quad (8)$$

$$y(i) = R(i) - \sqrt{R(i)^2 - x(i)^2} \quad (9)$$

$$R^{center}(i) = (i-1)(D_t + \delta) + y(i) \quad (10)$$

$$\varphi = \frac{N(i)D_t}{2\pi\left[(i-1)(Dt+\delta) + \frac{L}{2\sin\theta(i)\cos\theta(i)} - \sqrt{\left(\frac{L}{2\sin\theta(i)\cos\theta(i)}\right)^2 - x(i)^2}\right]} \quad (11)$$

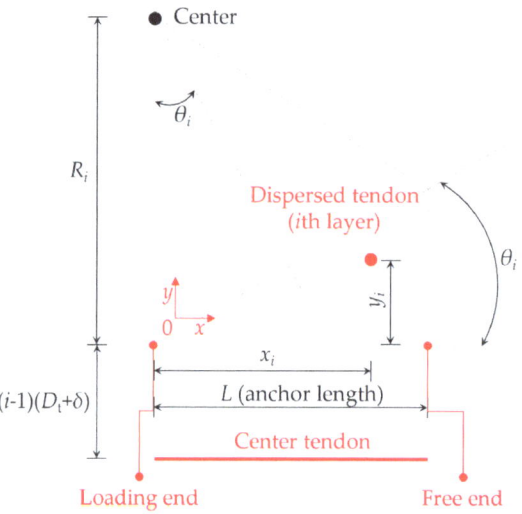

Figure 6. Geometric relation of dispersed tendons.

3.2. FE Modeling

For the proposed CAS shown in Figure 2, the tendon spacing was first determined based on a parallel-tendon model. The advantage of using a parallel-tendon model is that it can fully consider the influence of the change in tendon spacing on cables in different positions. Subsequently, the superiority of the dispersed tendon was verified by comparing it with the parallel-tendon CAS based on the full and simplified models. According to the symmetry of the CAS having 127 4-mm-diameter tendons, three types of FE models, namely full parallel-tendon, simplified parallel-tendon, and simplified dispersed-tendon models, were established based on ABAQUS/Standard (Figure 7). The number of elements along the part thickness direction is not less than three. Based on the same meshing principles, the 1/12 full FE model includes 121,968 elements for cable and 30,840 elements for LTC. In comparison, the 1/12 simplified FE model includes only 44,604 elements for cable and 24,840 elements for LTC. Thus, the total number of elements can be dramatically reduced through simplifying. The detailed dimensions of the anchorage and the LTC are shown in Figure 2.

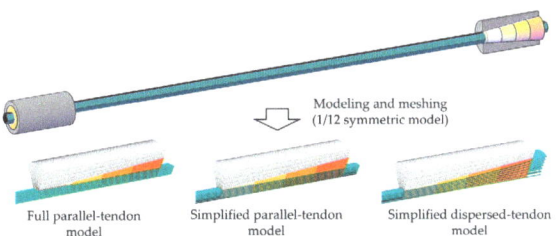

Figure 7. Full and simplified FE models.

A friction coefficient of 0.15 at the interface of the LTC and the anchorage was set based on previous simulations [26,29]. The interaction between the outer surfaces of the cable and the inner surfaces of the LTC was considered a tie because of sufficient cohesive action. The symmetric model was implemented by applying cyclic symmetry constraints to two symmetric planes and axial symmetry constraints to the cross-section of the cable at the middle portion. The eight-node solid element C3D8R was selected for the entire model to obtain high simulation accuracy [30]. A concentrated load was applied to the reference point, which was coupled to the surface of the anchorage perpendicular to the tensile direction of the cable. Based on f_{fpk} and the nominal diameter of the tendons, the load for the 1/12 model was 435,346 N. The material properties of the FE model critically affect the accuracy of the simulation and the convergence of the model. Only linear elasticity was employed in the simulation process to reveal the stress distributions and deformations of the CAS. The mechanical properties of the BFRP CAS that were obtained from Table 1 and previous studies are listed in Table 2, where some data derived from Wang et al. [34] and Zhou et al. [30] were used to create the table.

Table 2. Model parameters for the BFRP CAS.

Property		Cable	LTC	Sleeve
Poisson's ratio	Major	0.30	0.3	0.3
	Minor	0.02	0.3	0.3
Elasticity modulus (GPa)	Longitudinal	55	2.8–9.9	200
	Transverse	8	2.8–9.9	200
Shear modulus (GPa)	Longitudinal	6.0	/	77
	Transverse	6.0	/	77
Tensile strength (MPa)	Longitudinal	1702	/	560
	Transverse	/	/	560
Compressive strength (MPa)	Transverse	143	85–125	540
Elongation (%)	Longitudinal	3.11	/	>4

3.3. Simulated Results and Discussions

3.3.1. Determination of Tendon Spacing

Four tendon spacings (0, 0.5, 1, and 1.5 mm) were studied using a full parallel-tendon FE model. In a previous study, an apparent stress concentration in the outermost tendons at the sharp corners was observed. Thus, tendon-127 (T-127) was selected for further analysis. As shown in Figure 8a, the tendon spacing had a slight effect on the axial tensile stress of T-127, primarily attributed to the higher tensile elastic modulus of the BFRP tendons than that of the LTC. As shown in Figure 8b, the tensile stress difference between the inside and outside ($\Delta \sigma$) of T-127 and the radial compressive stress reached a maximum value near the loading end. The $\Delta \sigma$ of T-127 near the loading end also increased with an increase in tendon spacing, primarily caused by the tension-bending deformation of the tendon. The tensile stress difference between the inside and outside of the tendons decreased by 38%. The radial compressive stress in Figure 8c exhibited variation characteristics similar to $\Delta \sigma$. The loading-end radial stress decreased by 12% when the tendon spacing decreased from 1.5 mm to 0. In summary, the overall stress distributions of the cable in the anchoring zone (except for the loading end) were slightly influenced by the change in tendon spacing. A tendon spacing of 0 mm was selected to minimize the diameter and bending radius of the BFRP cable.

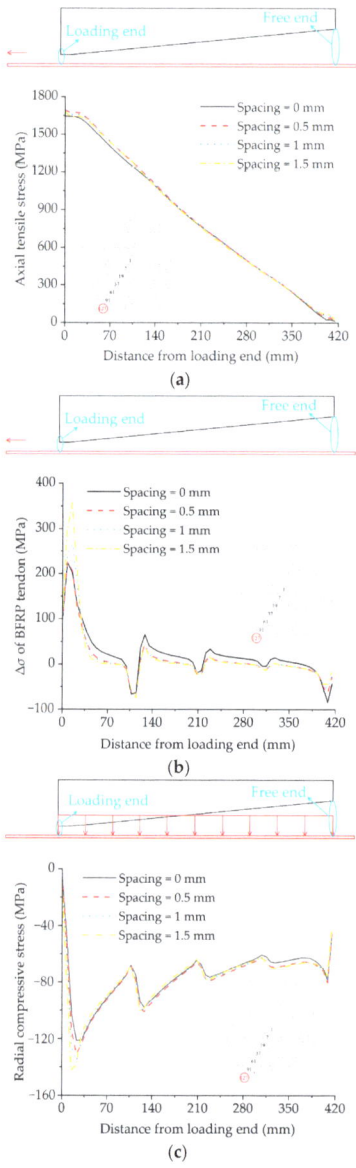

Figure 8. Simulated results of full model with tendon spacing: (**a**) axial tensile stress, (**b**) Δσ of BFRP tendon, and (**c**) radial compressive stress.

3.3.2. Comparison of Three Simulation Methods

A previous study indicated that the anchoring efficiency of the CFRP cables' high-strength CFRP tendons could be significantly improved by turning parallel tendons into dispersed tendons in the anchoring zone [30]. Thus, parallel-tendon and dispersed-tendon CASs were implemented to anchor the BFRP cable with a tendon spacing of 0. Full and simplified models to simulate the parallel-tendon CAS were employed to illustrate the effectiveness of the simplified method. Although the computing time of the model is greatly affected by the computer performance, the proportion of computing time of different models on the same computer should be relatively fixed. For this simulation, the calculation time of 263 min for the full parallel-tendon model was larger than that of 47 min for the simplified one. The calculation efficiency of the simplified model was significantly higher than that of the full model.

As shown in Figure 9a, the axial tensile stress of T-1 in the centralized position was studied using different anchoring methods. From the loading end to the free end, the axial tensile stress decreased with a nearly linear trend. It was slightly influenced by the anchoring and modeling methods, which was consistent with the simplified concept. As shown in Figure 9b, the radial compressive stress of T-127 in the simplified parallel-tendon model was close to that of T-127 in the full parallel-tendon model. This phenomenon indicated that the accuracy of the radial compressive stress of the simplified parallel-tendon model can be increased by dividing the compensation factor φ. When the distance from the loading end exceeded 120 mm, the radial compressive stress in the simplified dispersed-tendon model was lower than that in the simplified parallel-tendon model. This phenomenon can be explained by the gradual increase in tendon spacing from the loading end to the free end because of the bending arrangement, thereby resulting in reduced extrusion action between the tendons. The large longitudinal elastic modulus difference between the cable and LTC also contributed to this result.

As shown in Figure 9c, the shear stress difference between the full and the simplified parallel-tendon models gradually decreased from the loading end to the free end. It can be explained by stress concentration and magnitude. The shear stress in the simplified parallel-tendon model was the mean value because of the symmetric circular section. However, T-127 in the full parallel-tendon model was located at the stress concentration position, resulting in excessive shear stress. Additionally, the shear stress decreased with fluctuations. Consequently, the shear stress difference between the two models decreased as the shear stress decreased. The shear stress in the simplified dispersed-tendon model was also lower than that in the simplified parallel-tendon model. This result can be explained by the fact that the farther the shear stress is from the center of the cable, the smaller the shear stress. A similar conclusion was obtained in a previous study on high-strength CFRP CAS [30].

In summary, the dispersed-tendon CAS was superior to the parallel-tendon CAS when the same cable was anchored. Meanwhile, it is feasible to use a simplified dispersed-tendon model to simulate dispersed-tendon CAS. Further simulation verification of the simplified dispersed-tendon model using a full-scale experiment will be implemented in the following study.

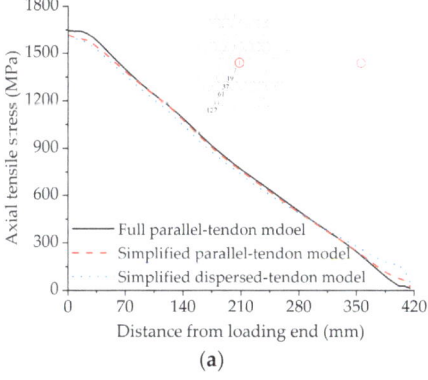

(a)

Figure 9. *Cont.*

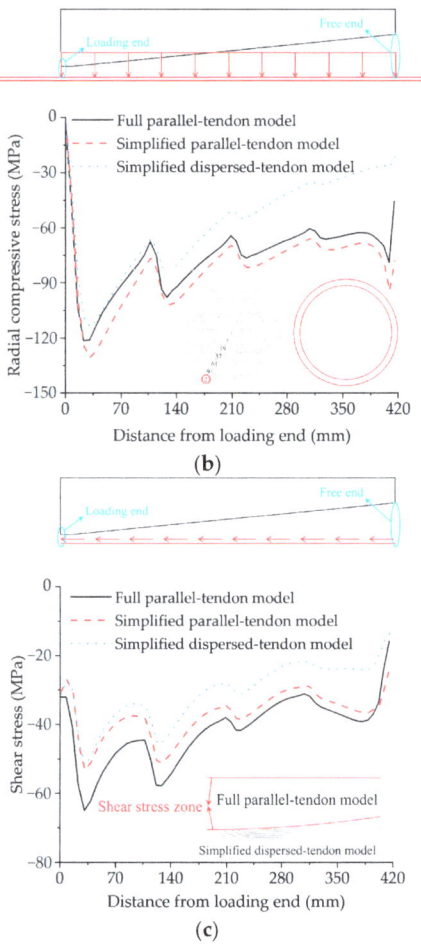

Figure 9. Simulated results of three FE models: (**a**) axial tensile stress of T-1, (**b**) radial compressive stress of T-127, and (**c**) shear stress of the LTC.

4. Experimental Verification in Dispersed-Tendon CAS
4.1. Manufacturing

In the manufacturing process, the combination of the cable in the test section and the dispersion of the cable in the anchoring zone are two major challenges. Only one BFRP cable was fabricated owing to the high cost and difficulty of cable preparation. As shown in Figure 10, the cable, characterized by a tendon spacing of 0 and a regular hexagon section, was first combined by hand with two positioning plates having a hole spacing of 2 mm [33]. The assembled parts of the cable were then fixed by wrapping them with medical tape that had the advantages of toughness and viscosity at specific intervals. The test section of the cable was further compressed by wrapping it with polyethylene (PE) tape with a width of 50 mm. Subsequently, cable dispersion was realized using separate and integral polyvinyl chloride (PVC) positioning plates. Separate PVC positioning plates with a hole diameter of 4.2 mm and thickness of 2 mm were first employed to disperse the anchoring-zone cable layer by layer. The combined cable was then placed into the prepared anchorages, which were connected using two channel steel with a length of 3150 mm. Notably, the inner surfaces of the anchorages were coated with silicon grease to make removing the LTC easier after loading. The separate positioning plates were removed after all tendons were threaded into the holes of the integrated positioning plate. The segmented LTCs were cast along the vertical direction, segment by segment. The casting interval was determined to be approximately 2 h to ensure reliable bonding between the interfaces of the adjacent LTCs. The casted LTCs were set to rest for 7 d to cure the epoxy resin adequately.

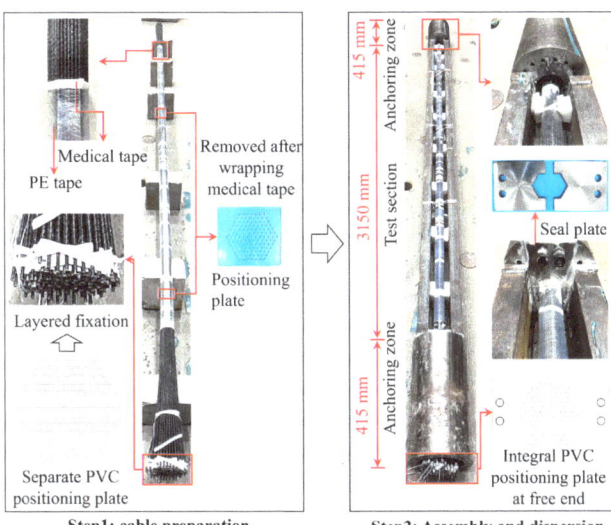

Figure 10. Preparation of dispersed-tendon CAS.

4.2. Loading and Measuring

A horizontal stretching device with a maximum loading capacity of 6500 kN was used for tensile testing of the cable (Figure 11a). The active anchorage was connected to the load cell by a nut and 20 high-strength bolts. A steel base plate was placed between the die anchorage and bearing plate. The maximum stroke of the piston rod was 400 mm. To eliminate the assembly clearance of the loading device, a tensile force of about 5 kN to the cable to produce a pretension was applied. A graded loading method including four stages was implemented at 100 MPa/min loading speed according to the Chinese standard [35]. The load was maintained for no less than 10 min when it reached 20%, 40%, and 60% of the standard value of the ultimate tensile force of the cable.

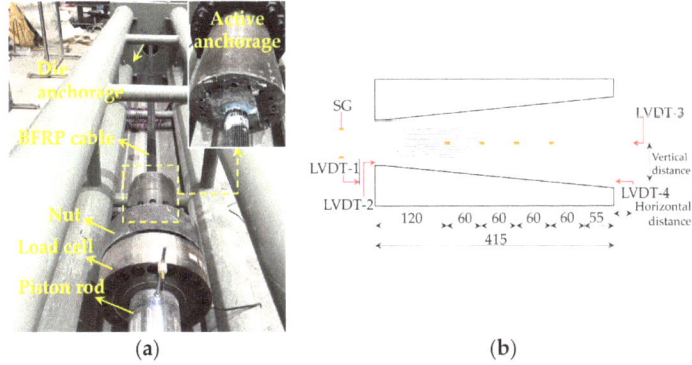

Figure 11. Loading and measurement schemes: (**a**) loading device and (**b**) measurement arrangement.

The measurement arrangement of the displacement and strain is shown in Figure 11b. The cable axial tensile strains at the test section and anchoring zone were collected using eighteen and four strain gauges (SGs), respectively. The ultimate measurement range of the SGs was 20,000 με. The distance between SGs in the anchoring zone was set to 60 mm. The cable and LTC axial displacements at the loading and free ends were measured by four linear variable differential transformers (LVDTs) with an ultimate measuring range of 100 mm. The data acquisition frequency was 1 Hz.

4.3. Experimental Results and Discussions

4.3.1. Failure Mode

During the loading process, slippage of the LTC along the axial direction was observed, resulting from the smooth inner surface of the anchorage. When the load exceeded approximately 80% of the standard value of the ultimate tensile force of the cable, the external fibers of the BFRP tendons began to rupture at random. When the ultimate load was reached, several BFRP tendons in the test section began to rupture one after another. Subsequently, the remaining tendons ruptured within a short time. The uneven stress of the tendons caused this phenomenon at the test section. The final failure mode of the cable is shown in Figure 12. The LTCs in the die and active anchorages suffered slight damage, indicating that the shear and compressive strengths of the LTCs were sufficient to anchor the cable. The cable cross-sections at the loading end also indicated that the bonding between the interface of the LTC and the cable was reliable. Almost no radial extrusion damage occurred to the loading-end cable, indicating that the variable-stiffness LTC could effectively reduce the cable radial stress concentration.

Figure 12. Failure mode of the CAS.

4.3.2. Load–Displacement Curve

As shown in Figure 13, the experimental load–displacement curve increased linearly with several fluctuations resulting from the slippage of the LTC. A horizontal section was then clearly observed at the end of the curve because of the rupture of a few tendons, demonstrating that the BFRP CAS had certain ductility characteristics. This phenomenon is mainly caused by the relatively low elastic modulus and high fracture elongation of the cable. Finally, the curve decreased rapidly after reaching an ultimate load of 2419 kN, which was significantly larger than the mean ultimate load of 1919 kN tested in a previous study [29]. Additionally, the experimental curve was followed closely by the simulated curve, indicating that the load–displacement relationships can be adequately simulated using the simplified dispersed-tendon model with high accuracy.

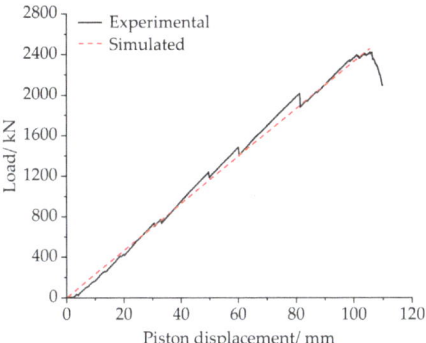

Figure 13. Load–displacement curve of the cable.

4.3.3. Anchoring Efficiency

The anchoring efficiency defined in Equation (12) [30] is typically employed to evaluate the bearing capacity of CASs. η_a is the anchoring efficiency, F_{fTu} is the measured ultimate tensile load of the cable, F_{fpk} is the standard value of the ultimate tensile force of the cable, and N is the number of tendons. According to Table 1 and the cable configuration, f_{fpk}, A_{st}, and N of the BFRP cable were 1638 MPa, 12.56 mm², and 127, respectively. F_{fpk} was calculated to be 2613 kN. η_a of the CAS was calculated to be 93%, slightly lower than the 95% specified by the standard [35]. This result can be explained by several potential reasons, such as the anchorage (no redesign), assembly errors, and material dispersion.

$$\eta_a = \frac{F_{fTu}}{F_{fpk}} = \frac{F_{fTu}}{f_{fpk} A_{st} N} \tag{12}$$

4.3.4. Axial Displacement

As shown in Figure 14, the experimental displacement at the loading end was noticeably larger than that at the free end, which was caused by the tensile deformations of the cable and LTC. Meanwhile, the experimental displacements of the cable were always larger than those of the LTC, which can be explained by the vertical distance difference between the LVDTs and the axial deformations of the cable and LTC. Additionally, the maximum difference between the simulated and experimental displacements was only 0.4 mm, indicating that the axial displacement can be reasonably simulated using the simplified dispersed-tendon model.

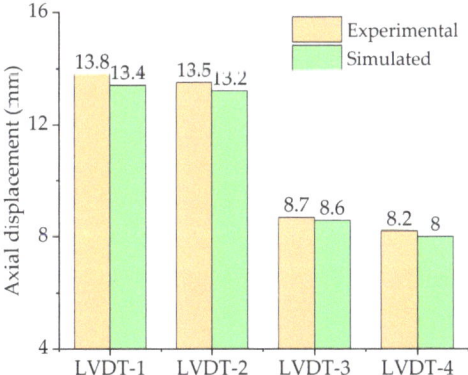

Figure 14. Axial displacement of the LTC and cable.

4.3.5. Axial Tensile Strain

The axial tensile strains of the BFRP tendons in the outermost layer arranged at intervals were measured (Figure 15). Generally, the tensile strain of the tendons increased linearly with an increase in the load, which was in good agreement with the linear elastic characteristic of BFRP materials. The tensile strain differences between the tendons increased with increasing load, and the maximum value reached 2551 $\mu\varepsilon$ when the load was 1756 kN. This phenomenon may be explained from three aspects: the length difference of tendons caused by cutting error and different initial bending shapes of tendons, the unparallel measuring direction of SGs, and the varying thicknesses of glue between tendons and SGs.

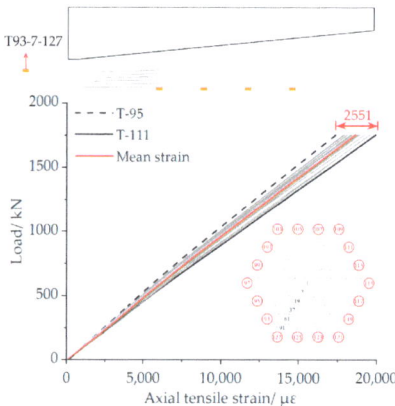

Figure 15. Axial tensile strains of the BFRP cable in the test section.

The changing laws of the axial tensile strains of T-1 at different load grades are shown in Figure 16, where D represents the distance from the loading end. Linear decreasing trends in the axial tensile strains of T-1 were observed at different load grades owing to the high longitudinal elastic modulus ratio of the BFRP cable to the LTC. The ratio of the maximum to minimum tensile strains generally increased with an increase in the load, primarily resulting from the axial wedge action of the LTC that restricted the cable deformation. When the load was 2000 kN, the plotting points were one less than the other cases. The main reason is that the strain gauge at the loading end was damaged when the load increased from 1600 to 2000 kN. Thus, the strain gauge could not collect valid data.

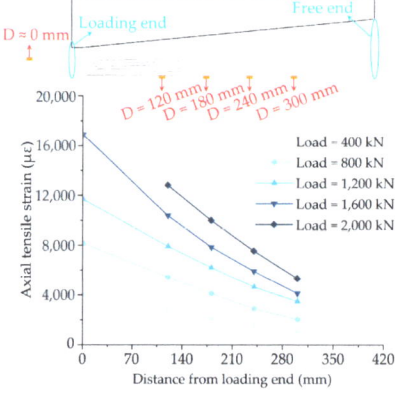

Figure 16. Axial tensile strains of T-1 at different load grades.

Based on the measured tensile strains in Figure 16, the mean shear stress of T-1 was calculated using the formula proposed in a previous study [27]. In this formula, the mean axial tensile strain, distance between two adjacent SGs, and elastic modulus of the BFRP tendons are considered. As shown in Figure 17, when the load was less than 1200 kN, the mean shear stress decreased gradually from the loading end to the free end. This phenomenon can be attributed to the varying elastic modulus of the LTC and the relatively small wedge displacement. Additionally, the mean shear stress near the loading end decreased drastically when the load was greater than 1200 kN. This phenomenon is possibly caused by the excessive wedge displacement (radial extrusion) of the LTC that limits its axial tensile deformation and the measuring error of the SGs. In general, no shear stress concentration was observed in the anchoring zone, illustrating that the variable-stiffness design for the LTC was beneficial for reducing the stress concentration caused by the excessive stiffness in the loading-end LTC.

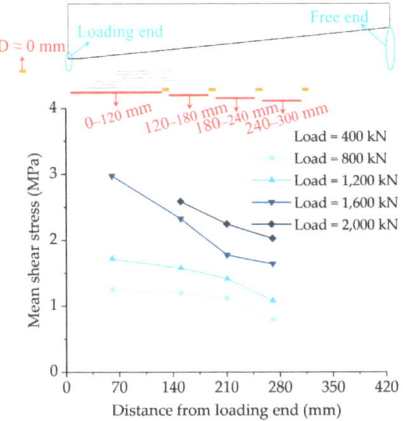

Figure 17. Mean shear stress of T-1 at different load grades.

5. Contribution of the Research to Practical Implementation

The research contribution mainly proposes a simplified model for anchoring design, cable manufacturing, and performance verification. Generally, the optimization design for anchoring cable is the premise of cable preparation. For different FRP cables, a simplified simulation method with high accuracy and efficiency can shorten the development cycle of FRP CAS. Based on the designed FRP CAS, the developed manufacturing method in the laboratory can provide helpful preparation strategies and details for actual factory production. The large-diameter BFRP cable in this paper was verified to be high anchoring efficiency. Furthermore, the anchoring efficiency of the cable may be further improved by optimizing the anchoring system. The design and testing results show that the BFRP cable is capable of engineering application. The cost-effective BFRP cables will also be a solid complement to the CFRP cables in the future. Thus, it is necessary to carry out in-depth scientific research on BFRP cables.

6. Conclusions

The main conclusions are as follows.

Compared with the full model, the simplified model has apparent advantages in terms of the decrease in element number, increase in computing efficiency, and convergence. The simulated results of the simplified dispersed-tendon model established using the concentric ring simplification method can be corrected by dividing the revised compensation factor from the geometric relations among the tendons.

The FE results showed that the loading-end radial stress decreased by 12% when the tendon spacing decreased from 1.5 mm to 0. Meanwhile, the tensile stress difference between the inside and outside of the tendons decreased by 38%. The radial compressive stress of the cable and the shear stress of the LTC in the dispersed-tendon CAS were lower than those in the parallel-tendon CAS when the same cable was anchored. In addition, the loading ends of the cable and LTC were verified as the maximum stress positions.

The experimental load–displacement relationship and axial displacement agreed well with the simulated results. All tendons were completely ruptured in the testing section, and the LTC was nearly unchanged. The ultimate tensile force of the cable was 2419 kN, and the corresponding anchoring efficiency was 93%. The axial tensile strain of the anchoring-zone cable generally decreased at a rate of 10–43 $\mu\varepsilon$/m when the load increased from 400 to 2000 kN. The shear stress concentration of the cable in the anchoring zone can be eliminated by applying a variable-stiffness LTC.

In general, a novel dispersed-tendon CAS for large-diameter BFRP cables was developed based on a previously developed LTC. The static behaviors of the CAS were then numerically evaluated using a simplified 3D FE model and implemented in a full-scale experiment. The shortcomings of the simplified model will be further overcome by releasing the circumferential stress of the anchor-zone cable. The developed large-diameter BFRP cable CAS is the foundation for further promotion and application.

Author Contributions: Conceptualization, X.W.; methodology, J.Z., L.D. and S.L.; software, J.Z.; validation, J.Z. and X.W.; formal analysis, J.Z. and L.D.; investigation, J.Z., L.D. and S.L; resources, X.W.; data curation, J.Z.; writing—original draft preparation, J.Z.; writing—review and editing, J.Z. and X.W.; supervision, Z.W.; project administration, J.Z. and X.W.; funding acquisition, X.W. All authors have read and agreed to the published version of the manuscript.

Funding: The research was funded by the National Natural Science Foundation of China (Grant 52208233 & 52278244), the Natural Science Foundation of Jiangsu Province (Grant No. BK20220855), the China Postdoctoral Science Foundation (Grant No. 2022M720726), the Excellent Postdoctoral Program of Jiangsu Province (Grant No. 2022ZB132), and the National Key Research and Development Program of China (No. 2022YFB3706503).

Data Availability Statement: The data presented in this study are available on request from the corresponding author.

Acknowledgments: The authors also acknowledge Jiangsu Green Materials Valley New Material T&D Co., Ltd. for providing BFRP tendons.

Conflicts of Interest: The authors declare no conflict of interest.

References

1. Wang, Q.; Zhu, H.; Zhang, B.; Tong, Y.; Teng, F.; Su, W. Anchorage systems for reinforced concrete structures strengthened with fiber-reinforced polymer composites: State-of-the-art review. *J. Reinf. Plast. Comp.* **2020**, *39*, 327–344. [CrossRef]
2. Feng, B.; Wang, X.; Wu, Z.; Yang, Y.; Pan, Z. Performance of anchorage assemblies for CFRP cables under fatigue loads. *Structures* **2021**, *29*, 947–953. [CrossRef]
3. Xie, G.H.; Tao, Z.; Wang, C.; Yan, P.; Liu, Y. Prediction of Elastic-Softening-Debonding behavior for CFRP Tendon-Adhesively bonded anchors. *Structures* **2022**, *40*, 659–666. [CrossRef]
4. Meier, U.; Farshad, M. Connecting high-performance carbon-fiber-reinforced polymer cables of suspension and cable-stayed bridges through the use of gradient materials. *J. Comput.-Aided Mater. Des.* **1996**, *3*, 379–384. [CrossRef]
5. Wang, X.; Wu, Z.; Wu, G.; Zeng, F. Enhancement of basalt FRP by hybridization for long-span cable-stayed bridge. *Compos. Part B Eng.* **2013**, *44*, 184–192. [CrossRef]
6. Shi, J.; Wang, X.; Wu, Z.; Zhu, Z. Optimization of anchorage and deviator for concrete beams prestressed with external fiber-reinforced polymer tendons. *Compos. Struct.* **2022**, *297*, 115970. [CrossRef]
7. Liu, Y.; Zwingmann, B.; Schlaich, M. Carbon fiber reinforced polymer for cable structures—A review. *Polymers* **2015**, *7*, 2078–2099. [CrossRef]
8. Wang, L.; Zhang, J.; Xu, J.; Han, Q. Anchorage systems of CFRP cables in cable structures—A review. *Constr. Build. Mater.* **2018**, *160*, 82–99. [CrossRef]
9. Zhao, J.; Mei, K.; Wu, J. Long-term mechanical properties of FRP tendon-anchor systems—A review. *Constr. Build. Mater.* **2020**, *230*, 117017. [CrossRef]
10. Schlaich, M.; Liu, Y.; Zwingmann, B. Carbon Fiber Reinforced Polymer for Orthogonally Loaded Cable Net Structures. *Struct. Eng. Int.* **2015**, *25*, 34–42. [CrossRef]

11. Mei, K.; Seracino, R.; Lv, Z. An experimental study on bond-type anchorages for carbon fiber-reinforced polymer cables. *Constr. Build. Mater.* **2016**, *106*, 584–591. [CrossRef]
12. Xie, G.; Yin, J.; Liu, R.; Chen, B.; Cai, D. Experimental and numerical investigation on the static and dynamic behaviors of cable-stayed bridges with CFRP cables. *Compos. Part B Eng.* **2017**, *111*, 235–242. [CrossRef]
13. Meier, U.; Winistorfer, A.; Haspel, L. Reinforced polymer hangers world's first large bridge fully relying on carbon fiber reinforced polymer hangers. *SAMPE J.* **2021**, *57*, 22–30.
14. Ai, P.; Feng, P.; Lin, H.; Zhu, P.; Ding, G. Novel self-anchored CFRP cable system: Concept and anchorage behavior. *Compos. Struct.* **2021**, *263*, 113736. [CrossRef]
15. Fang, Z.; Wang, C.; Zhang, H.; Zhang, K.; Xiang, Y. Experimental study on anchoring performance of CFRP strand in reactive power concrete. *China J. Highw. Transport.* **2016**, *29*, 198–206.
16. Yang, D.; Zhang, J.; Song, S.; Zhou, F.; Wang, C. Experimental investigation on the creep property of carbon fiber reinforced polymer tendons under high stress levels. *Materials* **2018**, *11*, 2273. [CrossRef]
17. Feng, B.; Wang, X.; Wu, Z. Evaluation and prediction of carbon fiber–reinforced polymer cable anchorage for large capacity. *Adv. Struct. Eng.* **2019**, *22*, 1952–1964. [CrossRef]
18. Song, S.; Zang, H.; Duan, N.; Jiang, J. Experimental research and analysis on fatigue life of carbon fiber reinforced polymer (CFRP) tendons. *Materials* **2019**, *12*, 3383. [CrossRef]
19. Kim, T.; Jung, W. Improvement of anchorage performance of carbon fiber-reinforced polymer cables. *Polymers* **2022**, *14*, 1239. [CrossRef]
20. Yang, Y.; Fahmy, M.; Guan, S.; Pan, Z.; Zhan, Y.; Zhao, T. Properties and applications of FRP cable on long-span cable-supported bridges: A review. *Compos. Part B Eng.* **2020**, *190*, 107934. [CrossRef]
21. Shi, J.; Wang, X.; Wu, Z.; Zhu, Z. Creep behavior enhancement of a basalt fiber-reinforced polymer tendon. *Constr. Build. Mater.* **2015**, *94*, 750–757. [CrossRef]
22. Peng, Z.; Wang, X.; Wu, Z. A bundle-based shear-lag model for tensile failure prediction of unidirectional fiber-reinforced polymer composites. *Mater. Des.* **2020**, *196*, 109103. [CrossRef]
23. Shi, J.; Wang, X.; Wu, Z.; Wei, X.; Ma, X. Long-term mechanical behaviors of uncracked concrete beams prestressed with external basalt fiber-reinforced polymer tendons. *Eng. Struct.* **2022**, *262*, 114309. [CrossRef]
24. Wang, X.; Shi, J.; Liu, J.; Yang, L.; Wu, Z. Creep behavior of basalt fiber reinforced polymer tendons for prestressing application. *Mater. Des.* **2014**, *59*, 558–564. [CrossRef]
25. Yang, Y.; Wang, X.; Wu, Z. Long-span cable-stayed bridge with hybrid arrangement of FRP cables. *Compos. Struct.* **2020**, *237*, 111966. [CrossRef]
26. Zhou, J.; Wang, X.; Peng, Z.; Wu, Z.; Zhu, Z. Failure mechanism and optimization of fiber-reinforced polymer cable-anchor system based on 3D finite element model. *Eng. Struct.* **2021**, *243*, 112664. [CrossRef]
27. Wang, X.; Xu, P.; Wu, Z.; Shi, J. A novel anchor method for multitendon FRP cable: Manufacturing and experimental study. *J. Compos. Constr.* **2015**, *19*, 04015010. [CrossRef]
28. Shi, J.; Wang, X.; Zhang, L.; Wu, Z.; Zhu, Z. Composite-wedge anchorage for fiber-reinforced polymer tendons. *J. Compos. Constr.* **2022**, *26*, 04022005. [CrossRef]
29. Zhou, J.; Wang, X.; Peng, Z.; Wu, Z.; Zhu, Z. Evaluation of a large-tonnage FRP cable anchor system: Anchorage design and full-scale experiment. *Eng. Struct.* **2022**, *251*, 113551. [CrossRef]
30. Zhou, J.; Wang, X.; Wu, Z.; Zhu, Z. A Large-tonnage high-strength CFRP cable-anchor system: Experimental investigation and FE study. *J. Compos. Constr.* **2022**, *26*, 04022053. [CrossRef]
31. Peng, Z.; Wang, X.; Zhou, J.; Wu, Z. Reliability assessment of fiber-reinforced polymer cable-anchorage system. *Compos. Struct.* **2021**, *273*, 114308. [CrossRef]
32. *Chinese Standard GB/T 3002-2013*; Test Method for Basic Mechanical Properties of fiber Reinforced Polymer Bar. Chinese Standard Press: Beijing, China, 2013.
33. Zhou, J.; Wang, X.; Peng, Z.; Wu, Z.; Wei, X. Enhancement of FRP cable anchor system: Optimization of load transfer component and full-scale cable experiment. *J. Compos. Constr.* **2022**, *26*, 04022008. [CrossRef]
34. Wang, X.; Xu, P.; Wu, Z.; Shi, J. A novel anchor method for multi-tendon FRP cable: Concept and FE study. *Compos Struct.* **2015**, *120*, 552–564. [CrossRef]
35. *Chinese Standard T/CECS 10112-2020*; Anchorage and Grip for Prestressing Fiber-Reinforced Polymer. Chinese Standard Press: Beijing, China, 2020.

Disclaimer/Publisher's Note: The statements, opinions and data contained in all publications are solely those of the individual author(s) and contributor(s) and not of MDPI and/or the editor(s). MDPI and/or the editor(s) disclaim responsibility for any injury to people or property resulting from any ideas, methods, instructions or products referred to in the content.

Article

Phenomenological 2D and 3D Models of Ductile Fracture for Girth Weld of X80 Pipeline

Naixian Li [1], Bin Jia [1,2,*], Junhong Chen [3], Ying Sheng [1] and Songwen Deng [1]

[1] School of Civil Engineering and Architecture, Southwest University of Science and Technology, Mianyang 621010, China
[2] Shanghai Key Laboratory of Engineering Structure Safety, Shanghai 200032, China
[3] Gas Transmission Management Division Southwest Oil, Gasfield Company PetroChina, Chengdu 610041, China

Citation: Li, N.; Jia, B.; Chen, J.; Sheng, Y.; Deng, S. Phenomenological 2D and 3D Models of Ductile Fracture for Girth Weld of X80 Pipeline. *Buildings* 2023, *13*, 283. https://doi.org/10.3390/buildings13020283

Academic Editor: Oldrich Sucharda

Received: 23 November 2022
Revised: 10 January 2023
Accepted: 16 January 2023
Published: 18 January 2023

Copyright: © 2023 by the author. Licensee MDPI, Basel, Switzerland. This article is an open access article distributed under the terms and conditions of the Creative Commons Attribution (CC BY) license (https://creativecommons.org/licenses/by/4.0/).

Abstract: Welding is the main method for oil/gas steel pipeline connection, and a large number of girth welds are a weak part of the pipeline. Under extremely complex loads, a steel pipeline undergoes significant plastic deformations and eventually leads to pipeline fracture. A damage mechanics model is a promising approach, capable of describing material fracture problems according to the stress states of the materials. In this study, an uncoupled fracture 2D model with a function of fracture strain and stress triaxiality, two uncoupled 3D fracture models, a consider the effect of Lode parameter stress-modified critical strain (LSMCS) model, and an extended Rice–Tracey (ERT) criterion were applied to X80 pipeline girth welds. Comprehensive experimental research was conducted on different notched specimens, covering a wide range of stress states, and the corresponding finite element models were established. A phenomenon-based hybrid numerical–experimental calibration method was also applied to determine the fracture parameter for these three models, and the stress triaxiality of the influence law of the tensile strength was analyzed. The results showed that the proposed fracture criterion could better characterize the ductile fracture behaviors of the girth welds of the X80 pipeline; however, the prediction accuracy of the 3D fracture model was higher than that of the 2D fracture model. The functional relationship between the tensile strength and stress triaxiality of the X80 pipeline girth welds satisfied the distribution form of the quadratic function and increased monotonically. The research results can be used to predict the fracture of X80 pipeline girth welds under various complex loads.

Keywords: X80 pipeline girth weld; ductile fracture; damage model; tensile strength

1. Introduction

Pipelines made of high-strength steel are the most commonly used means of transporting oil and gas (O&G) for onshore and offshore installations. X80 pipeline steel is widely used for oil and gas pipelines because of its excellent mechanical properties, and its laying quantity is also increasing [1,2]. Given the fixed length of pipes, long-distance natural gas transmission pipelines have to be connected by manual girth welding every 12 m. Owing to several limitations, girth welds inevitably produce defects during the welding process, making them a vulnerable point in pipeline structures. In actual projects, pipelines are laid in a variety of environments, such as mountainous areas, deserts, glaciers, and oceans. When these environments change or a third party carries out construction activities in the vicinity, geological disasters, such as settlement and landslides, can occur in the location of the pipeline. As weak points of the pipelines, girth welds are most likely to be destroyed during these events. Recently, a strain-based method has been developed, which has a better resolution to deal with global plastic deformations, compared with the traditional stress-based method. This method has been applied to standards and recommended practices to evaluate defects in pipes [3–6]. However, it is difficult to capture the actual failure physics involved in the localized plastic deformation and material tearing in this method [7].

Fracture and damage mechanics theories are commonly employed to study ductile fractures. Both theories require a full understanding of the mechanical properties of materials. Based on 104 experimental results, Khalaj [8,9] realized the prediction of mechanical properties of metal materials through artificial neural networks and successfully predicted the tensile strength of X70 pipeline steel by chemical composition elements. Pouraliakbar [10] used an adaptive neuro-fuzzy interface system (ANFIS) to simulate the Charpy impact energy of Al6061-SiCp laminated nanocomposites prepared by mechanical alloying in two configurations of splitting and crack resistance, which provided a new idea for obtaining the mechanical properties of materials. In traditional fracture mechanics, material fracture toughness data are required, and fracture toughness is considered in terms of the resistance curve. Traditional fracture mechanics generally have certain limitations. First, they must be studied based on material cracks, therefore, prediction of object fractures without prefabricated cracks is impossible. Second, the crack propagation direction, including the crack path deviations, cannot be predicted without a user-defined criterion [11].

However, the damage mechanics theory can describe the initiation and propagation of cracks by allowing damage evolution to include the influences of both local stress and strain variables [11]. Thus, the limitations of traditional fracture mechanics can be effectively removed. The Gurson–Tvergaard–Needleman (GTN) model, based on micromechanics, is a classical coupled damage mechanics model. This model was developed from the porous plastic model developed by Gurson [12] and was used to describe the effect of holes on the plastic behavior of materials. The model was then modified by Tvergaard and Needleman, and the failure process of the materials was expressed by describing the evolution behaviors of the holes [13,14]. However, the original GTN model could not describe the material failure under shear stress. The modified GTN model [15–17] considering shear failure resulted in an increase in the free parameters of the model, which make the calibration difficult. The other model is an uncoupled fracture model that excludes the effect of damage on the mechanical properties of materials, and is the common method used in industry. The uncoupled fracture model typically consists of a mathematical expression based on the relationship between the fracture strain and stress state, forming a fracture criterion. Many uncoupled fracture models, such as Johnson–Cook criterion [18], stress-modified critical strain model [19], Xue–Wierzbicki model [20], and extended Rice–Tracey (ERT) criterion [21], have been developed.

Damage-mechanics models have been widely used for metal fractures [22–24]. Several mechanical damage models have been applied to oil/gas pipelines. Oh [25–28] formulated a series of notch tensile specimens for X65 pipeline steel, developed GTN and uncoupled fracture models, and verified the accuracy of the model through blasting experiments with full-scale pipelines having defects. Shinohara [29] integrated the GTN and Thomason models to develop a fracture model capable of describing the anisotropic fracture behavior of X100 pipeline steel. Paredes [30] calibrated the MMC (Mohr–Coulomb) parameters for X70 pipeline steel. Han [31] calibrated the MMC material parameters versus ERT for X80 pipeline steel, based on experiments. Regarding the research on the ductile fracture of girth welds in oil and gas pipelines, Sarzosa [32] developed the fracture track between the Lode parameter, stress triaxiality, and fracture strain for the girth weld on the X65 pipeline, and the simulated fracture prediction effect was in good agreement with the test results. Although damage models have been frequently applied by many researchers in simulating ductile fractures in pipe steels, the feasibility of these models remains subject to verification, and the calibration of parameters has been controversial. In addition, ductile fracture models for pipeline girth welds are lacking.

In this study, a series of welded tensile specimens, under different stress states, was designed for X80 pipeline girth welds, and a corresponding finite element model was established. The girth-weld fracture parameters of the X80 pipeline under different stress states were determined and obtained through experimental tests and finite element simulations.

A 2D fracture model of the stress triaxial degree and fracture strain was established, and the uncoupled 3D fracture model parameters of an extended Rice–Tracey (ERT) model and the consider the effect of Lode parameter stress-modified critical strain (LSMCS) model were fitted. The established constitutive and fracture models can be used for the safety risk assessment of the girth welds of X80 pipelines.

2. Overview of the Damage Model

2.1. Characterization of Stress States

In previous studies on ductile fracture, stress triaxiality and the Lode parameter has been extensively applied to characterize the stress states of materials and were the two main parameters that affect fracture strain. The stress triaxial with the Lode parameter is expressed as

$$\eta = \frac{\sigma_m}{\bar{\sigma}} \quad (1)$$

$$\bar{\theta} = 1 - \frac{2}{\pi}\arccos\frac{27 J_3}{2\bar{\sigma}^3} \quad (2)$$

where η is the stress triaxiality, $\bar{\theta}$ is the dimensionless Lode angle parameter, σ_m is the mean stress, $\bar{\sigma}$ is the equivalent stress, and J_3 is the third invariant of the deviatoric stress.

$$\bar{\sigma} = \sqrt{3 J_2} = \sqrt{\frac{1}{2}\left[(\sigma_1 - \sigma_2)^2 + (\sigma_2 - \sigma_3)^2 + (\sigma_3 - \sigma_1)^2\right]} \quad (3)$$

$$J_3 = (\sigma_1 - \sigma_m)(\sigma_2 - \sigma_m)(\sigma_2 - \sigma_m) \quad (4)$$

$$\sigma_m = \frac{1}{3}(\sigma_1 + \sigma_2 + \sigma_3) \quad (5)$$

where J_2 is the second invariant of the deviatoric stress and $\sigma_1, \sigma_2, \sigma_3$ denotes the three principal stresses in the stress tensor.

2.2. Uncoupled Fracture Models

2.2.1. Extended Rice–Tracey (ERT) Model

Rice and Tracey studied the growth of isolated spherical voids in an infinite ideal rigid-plastic matrix and developed the well-known Rice–Tracey (RT) fracture criterion [33]. Gruben [21] introduced the influence of the Lode parameters in the RT model using the correction proposed by Nahshon [17] and obtained the ERT model as follows:

$$\bar{\varepsilon}_f^{pl} = \frac{1}{R_1 \exp(R_2 \eta) + R_3 \sin^2[\frac{\pi}{2}(1 - \bar{\theta})]} \quad (6)$$

where R_1, R_2, R_3 are three material parameters to be determined, and $\bar{\varepsilon}_f^{pl}$ is fracture strain.

2.2.2. Consider the Effect of Lode Parameter Stress-Modified Critical Strain (LSMCS) Model

Based on the stress-modified critical strain model (SMCS), which only considers stress triaxiality on fracture strain, Huang [34] added the influence term of the Lode parameters and developed the LSMCS model as follows:

$$\bar{\varepsilon}_f^{pl} = \alpha \exp(-1.5\eta)[\gamma + (1 - \gamma)\bar{\theta}^2] \quad (7)$$

where α, γ are the two material parameters to be determined.

3. Experimental Methods

In this study, the ductile fracture of X80 pipe girth weld is studied by combining test and simulation, and the flow chart is shown in Figure 1. Given that fabricating small-sized tensile specimens for girth welds is difficult, a study was conducted to design heteromorphic specimens with welded joints. To obtain the fracture parameters of X80 pipeline girth welds, the constitutive model of X80 pipeline steel must be determined. For X80 pipeline steel, standard tensile specimens (smooth round bar: SRB) were designed based on the requirements of GB/T +228.1−2010 [35], as shown in Figure 2. For weld specimens, based on HB 5214−1996 [36], GB/T 7314−2017 [37], and HB 6736−1993 [38], notched round bar tension specimens (radius: 1 mm, 3 mm, 5 mm, NRB1, NRB3, NRB5) and compression and flat shear (FS) specimens were designed. Following a previous study [31], a central-hole tensile specimen (CH6) with a central round hole diameter of 6 mm was designed, as shown in Figure 3. The stress state of high-stress triaxiality was determined using round bar notch weld specimens; low-stress triaxiality was determined using shear weld specimens; negative stress triaxiality was obtained using compression specimens; and the stress state of stress triaxiality near 0.4 was determined using central-hole tensile specimens.

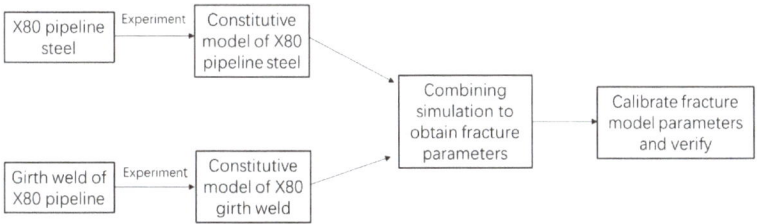

Figure 1. Research flow chart.

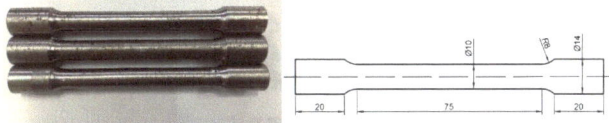

Figure 2. Standard tensile specimen (dimensions in mm).

The experiment was performed using a CMT5150 universal test machine, with a maximum load of 100 kN, manufactured by Meister Industrial System (China) Co., Ltd., in June 2010. The distance of the extensometer was 50 mm, which has a grade 1 accuracy. All the experiments were conducted with a loading rate of 0.45 mm/min. The material used in this research was an X80 girth-weld pipe, with a diameter of 1219 mm and a pipe thickness of 18.4 mm, The welding material was 91T8, and a steel grade can be additionally qualified according to TR/ISO 15608 [39]. The average weld parameters used for preparing the test weld using the FCAW process are (1) welding current 170 Amps; (2) welding voltage 20 V; (3) wire feed speed 5 m/min, and front arc energy 0.72 kJ/mm, while the V Groove and welding were performed in accordance with GB/T 31032-2014 [40], and the welding method according to ISO 4063 [41]. The size of the weld joint is shown in Figure 4. To ensure the quality of the welding, non-destructive testing of the welds was performed; non-destructive testing includes radiographic testing and time of flight diffraction, ensuring that the source of specimen welds without defects. All specimens were obtained from the longitudinal direction of the pipe, and the girth weld was ensured to be located at the notch of the special-shaped specimen, as shown in Figure 5.

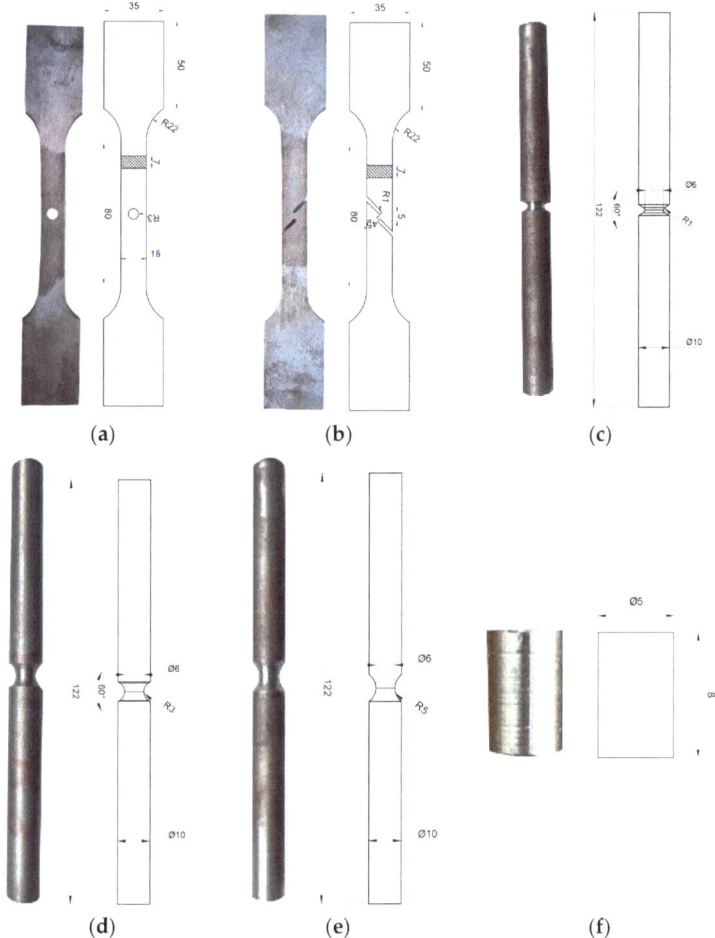

Figure 3. Weld special-shaped specimens (dimensions in mm): (**a**) CH6, (**b**) FS, (**c**) NRB1, (**d**) NRB3, (**e**) NRB5, and (**f**) compression.

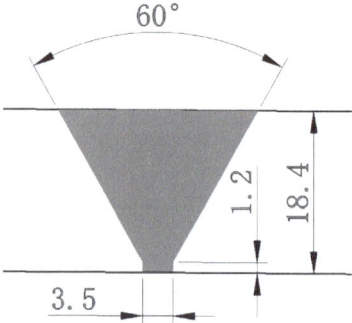

Figure 4. Weld joint (dimensions in mm).

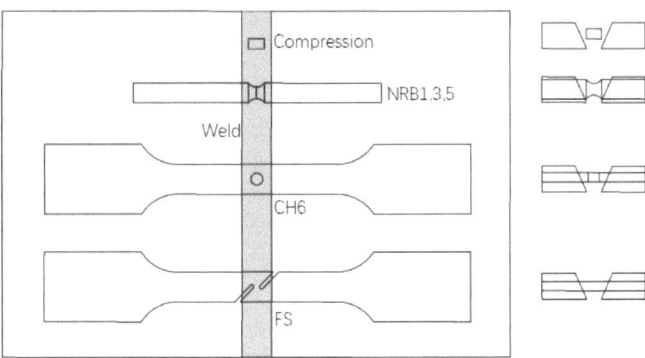

Figure 5. Schematic of weld specimen pick-up.

4. Constitutive Equations

4.1. Constitutive Model of X80 Pipeline Steel

Figure 5 shows the true stress-strain curve of the X80 pipeline steel drawn based on the results of the experiment. It was observed that X80 pipeline steel had no obvious yield plateau; therefore, the stress when 0.2% plastic strain occurred was adopted as the yield strength (Rp0.2). The mechanical parameters of X80 pipeline steel are summarized in Table 1.

Table 1. Mechanical parameters of X80 pipeline steel.

Young's Modulus (MPa)	Poisson's Ratio	Yield Strength/Rp0.2 (MPa)	Tensile Strength (MPa)
206,000	0.3	638	739

Under large strains, the difference in the stress anisotropy of X80 pipeline steel could be ignored [31]; therefore, X80 pipeline steel could be regarded as an isotropic material in this study. For invalid data of specimens after necking, stress-strain data before necking were adopted, and the Johnson–Cook constitutive model was applied to fit the hardening curve which describes the hardening behavior after necking. This parameter is expressed as [42]

$$\bar{\sigma} = \left[A + B(\bar{\varepsilon}^{pl})^n\right]\left[1 + C \ln\left(\frac{\dot{\bar{\varepsilon}}^{pl}}{\dot{\varepsilon}_0}\right)\right](1 - \hat{\theta}^m) \quad (8)$$

where $\bar{\varepsilon}^{pl}$ is the equivalent plastic strain, $\frac{\dot{\bar{\varepsilon}}^{pl}}{\dot{\varepsilon}_0}$ is the dimensionless plastic strain rate, $1 - \hat{\theta}^m$ is a temperature-related term, and A, B, C, n, m are parameters to be determined experimentally.

In this study, quasi-static loading was adopted, and the temperature changes of the specimens during loading were ignored. Therefore, the temperature and strain rate terms in Equation (8) were equated to 1, simplifying the equation into

$$\bar{\sigma} = A + B\left(\bar{\varepsilon}^{pl}\right)^n \quad (9)$$

The hardening curve data before neck shrinkage were selected for the preliminary parameter fitting. The load-displacement curve of the standard tensile specimen was drawn using a finite element software and was compared with the experimental results. The slope of the hardening curve after the neck shrinkage was continuously adjusted until the simulation (FEM) results were consistent with the experimental results. Finally, parameters A, B, and n were determined to be 506.94 MPa, 398.467 MPa, and 0.17402, respectively. The hardening curve is shown in Figure 6b, and a comparison of the experimental and finite element results is shown in Figure 6c. The established constitutive model can characterize the mechanical behavior of X80 pipeline steel.

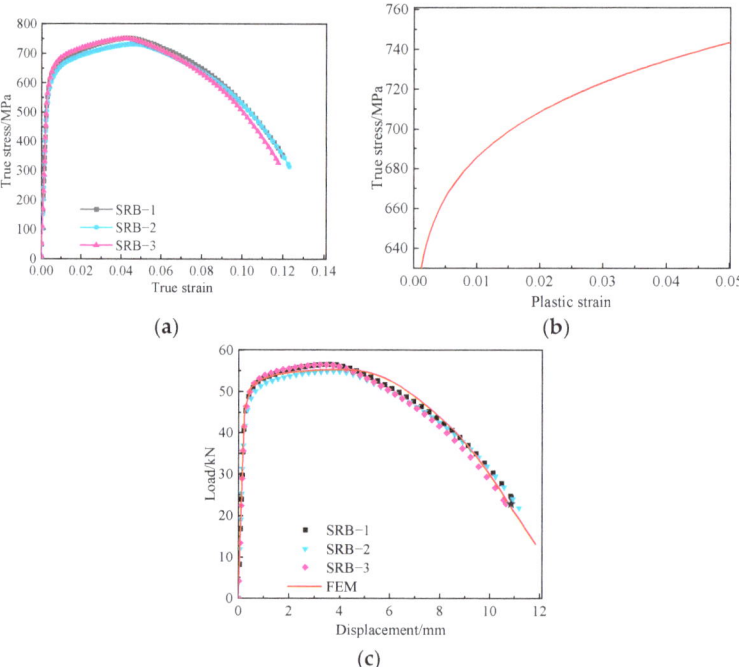

Figure 6. X80 pipeline steel stress-strain curve: (**a**) true stress-strain curve, (**b**) Johnson–Cook hardening curve, and (**c**) comparison of finite element and experimental results.

4.2. Constitutive Model of Girth Weld

According to numerous studies [43–46], the mechanical properties of the elastic stage of a girth weld are consistent with those of the base metal, which was also considered in this work. In this study, the constitutive relationship of the girth weld was determined using compression specimens. Owing to the influence of the friction force at the loading end of the compression machine, the compressed specimens formed obvious "drum shapes". This shape led to the inaccuracy of the constitutive relationship obtained directly through experiments and the correct hardening constitutive curve of the weld material established with the corresponding finite element model. Finite element software uses ABAQUS2022, and Figure 7 shows the finite element model of the compression specimens. Two rigid plates were set up in which the bottom plate was completely fixed, the upper plate was loaded with a downward displacement, the friction coefficient (μ) between the rigid body presses and the specimens was set to 0.15, and the mesh size was 0.2, totaling into 25,600 elements. The quasi-static calculation was carried out using the C3D8R solid element and explicit dynamics (the kinetic energy proportion was less than 5% of the internal energy). The constitutive curve was constantly corrected using the finite element software, verifying until it was consistent with the experimental conditions. The final plastic hardening constitutive curve of the X80 girth weld is shown in Figure 8a, and Figure 8b shows the comparison results between test and finite element model (FEM). The hardening constitutive model finally determined can well reflect the mechanics of materials and uses a piecewise function description of

$$\begin{cases} \bar{\sigma} = 490.2 + 478.1(\bar{\varepsilon}^{pl})^{0.1918}(0 < \bar{\varepsilon}^{pl} < 0.272) \\ \bar{\sigma} = 120\bar{\varepsilon}^{pl} + 829(0.272 < \bar{\varepsilon}^{pl}) \end{cases} \quad (10)$$

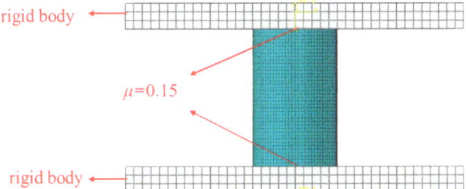

Figure 7. Compressed specimen loading model.

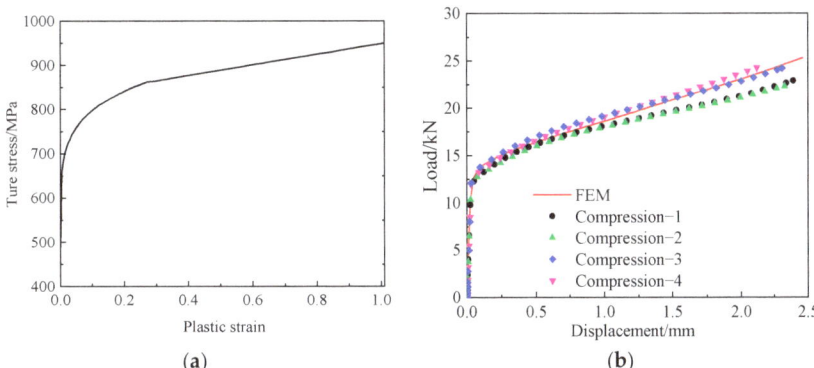

Figure 8. X80 girth weld constitutive model: (**a**) hardening curve; (**b**) comparison of test results with FEM.

5. Calibration Procedure
5.1. Finite Element Model

The fracture strain, stress triaxiality, and Lode parameters of the test specimens were analyzed and calibrated according to the elastic-plastic finite element model. ABAQUS was used to establish a three-dimensional finite element model of the special-shaped notch tensile specimen to obtain the simulated load displacement curves. A large deformation was observed at the weld gap, and its heat-affected zone was either still in the elastic stage or the plastic stage was small and could be neglected, therefore, the material softening of the heat-affected zone could be assumed negligible. The notch position was endowed with the attributes of the weld material, and the weld was considered to be an isotropic uniform material [47]. One end of the specimen was articulated, and the other end was loaded with displacement, as shown in Figure 9. C3D8R solid elements were used for all specimens. The global element size was 1mm, the notch part was encrypted, the CH6 and FS middle densified element size was 0.2, while the element of the notch part of the round bar specimen was densified along the longitudinal direction and the element size was 0.2 mm. Calculate using explicit dynamics (the kinetic energy proportion is less than 5% of the internal energy) and reduce integral control.

It was seen from the load displacement curves drawn based on the finite element and experimental results shown in Figure 10 that the established base material and girth weld constitutive models could reflect the mechanical behaviors of the weld tensile specimens under different stress states. The initial fracture points of the specimens are marked with "★" on their respective curves in Figure 10. The experimental results revealed notched round bar cracks at the specimen center and plate-pattern cracks on the notched surface. The critical element determined by each specimen is shown in Figure 10. The equivalent plastic strain of the critical element under the state of fracture displacement is the fracture strain. The maximum load simulation value of NRB3 was slightly higher than the test value, which may be caused by the mechanical non-uniformity of the weld. However, according to the simulation of all test specimens, it is reasonable to regard the weld as isotropic. The element body at the fracture position was selected to evaluate the evolution processes of the stress triaxiality, Lode parameters, and equivalent plastic strain, as illustrated in Figure 11. The Lode parameter of the round bar specimen was constant at 1. It was found that the round-hole plate in the process of tensile change was not near 0.5, and the shear specimen remained near 0. In addition, the stress triaxiality of the sheet metal remained unchanged during the tensile process, whereas those of the notched round bar tensile specimens all increased to different levels, and the stress triaxiality of the specimens with 1 mm radius was increased significantly.

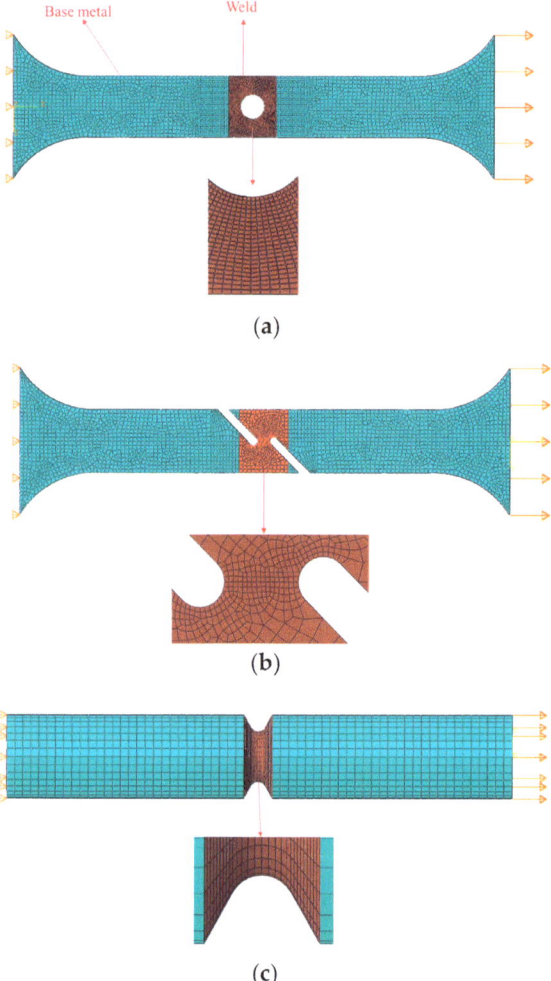

Figure 9. *Cont.*

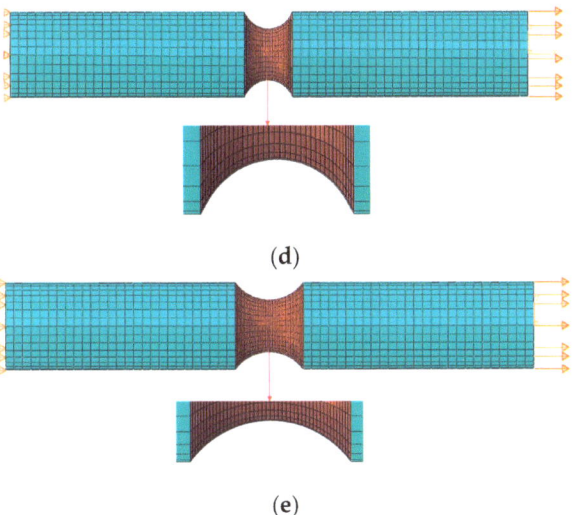

Figure 9. Finite element mesh models: (**a**) CH6; (**b**) FS; (**c**) NRB1; (**d**) NRB3; (**e**) NRB5.

5.2. Fitting of Experiment Data

The stress states of the specimens constantly changed during the entire tensile process; therefore, the stress triaxiality of the element had to be averaged with the Lode parameters as follows [48,49]:

$$(\eta)_{av} = \frac{1}{\bar{\varepsilon}_f^{pl}} \int_0^{\bar{\varepsilon}_f^{pl}} \eta \, d\bar{\varepsilon}^{pl} \tag{11}$$

$$(\bar{\theta})_{av} = \frac{1}{\bar{\varepsilon}_f^{pl}} \int_0^{\bar{\varepsilon}_f^{pl}} \bar{\theta} \, d\bar{\varepsilon}^{pl} \tag{12}$$

where $(\eta)_{av}$ denotes the average stress triaxiality, and $(\bar{\theta})_{av}$ denotes the average Lode parameter.

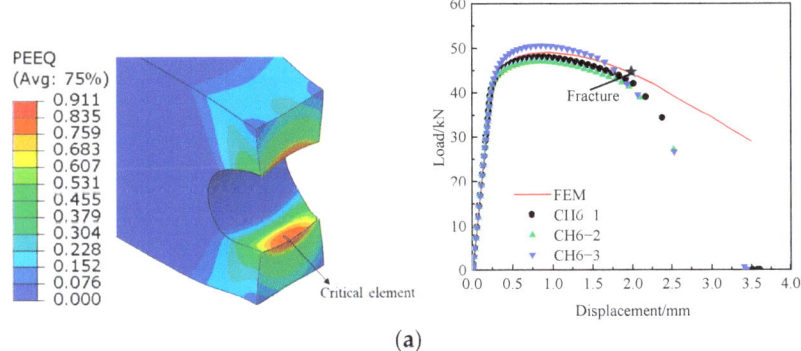

Figure 10. *Cont.*

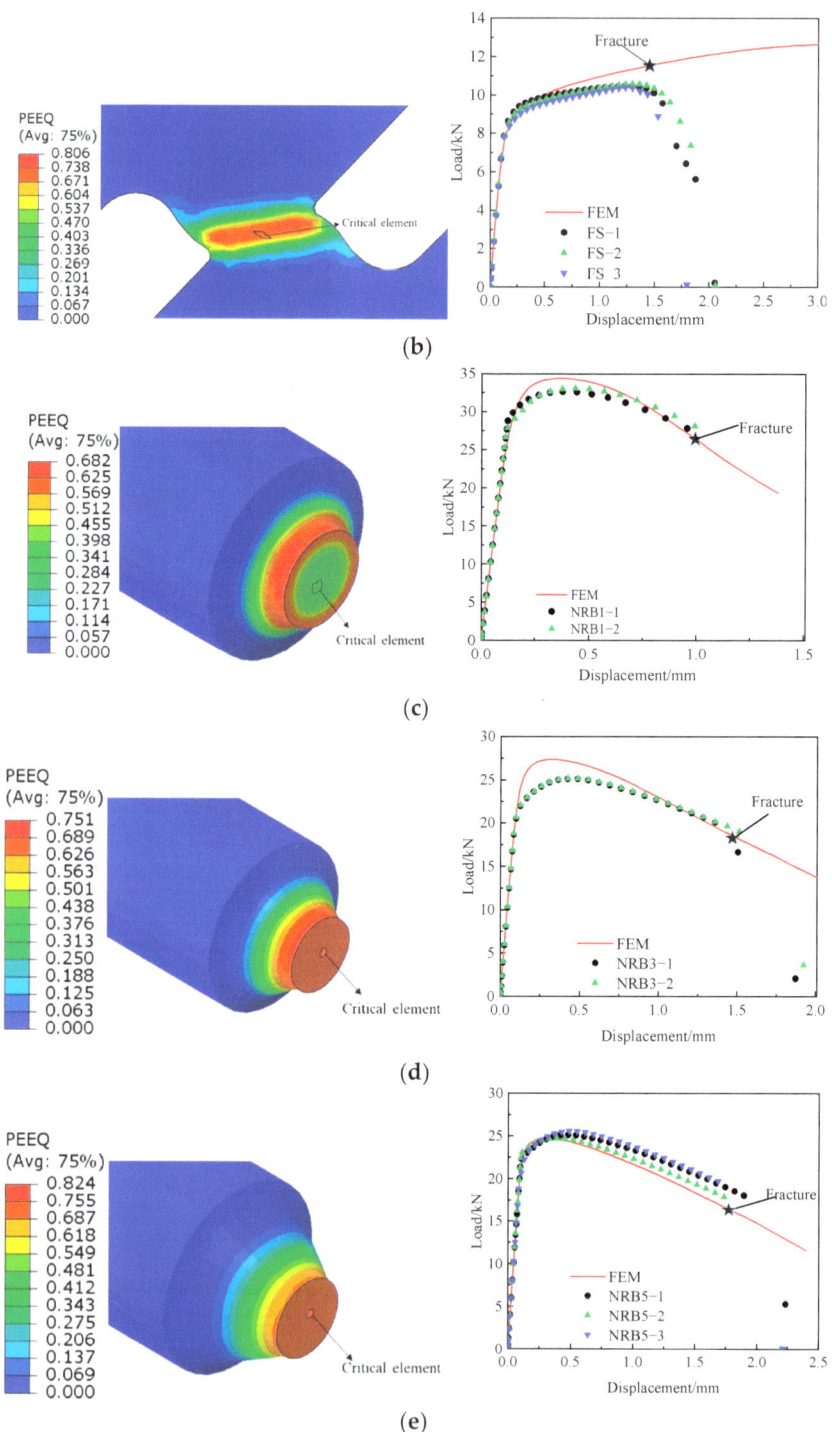

Figure 10. Determination of critical displacement: (**a**) CH6; (**b**) FS; (**c**) NRB1; (**d**) NRB3; (**e**) NRB5.

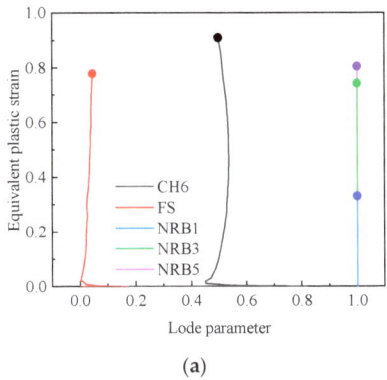
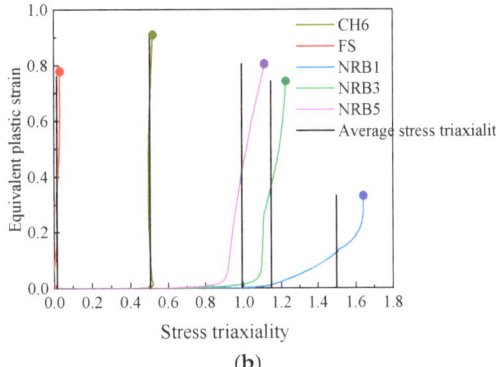

Figure 11. Evolution of equivalent plastic strain and stress states: (**a**) evolution of equivalent plastic strain and Lode parameters; and (**b**) evolution of equivalent plastic strain and stress triaxiality.

5.2.1. Fitting of 2D Fracture Criteria

By neglecting the effects of the Lode parameters on the fracture strain, the fracture criterion was expressed as a function of the stress triaxiality and fracture strain. The Johnson–Cook failure model presented a good fracture prediction model for materials under high-stress triaxiality. In this study, the Johnson–Cook failure model was adopted to determine the relationship between fracture strain and stress triaxiality under high stress triaxiality, as follows [18]:

$$\bar{\varepsilon}_f^{pl} = [d_1 + d_2 \exp(d_3 \eta)]\left[1 + d_4 \ln\left(\frac{\dot{\bar{\varepsilon}}^{pl}}{\dot{\varepsilon}_0}\right)\right](1 + d_5 \hat{\theta}^m) \qquad (13)$$

where $d_1 - d_5$ are parameters to be determined by experiment, and $(1 + d_5 \hat{\theta}^m)$ is a temperature-related term.

Quasi-static loading was applied in this study, and the temperature changes in the specimens during loading were ignored. Therefore, the temperature and strain rate terms in Equation (13) were eliminated, which simplifies it to

$$\bar{\varepsilon}_f^{pl} = d_1 + d_2 \exp(d_3 \eta) \qquad (14)$$

The average stress triaxiality and fracture strain values determined above are summarized in Table 2. Based on the presented data, a failure model function of the girth weld of the X80 pipeline was developed. The Johnson–Cook failure model was applied to fit the round-hole tensile and notch specimens. The fracture strain at a stress triaxiality of 0.333 was determined by extension, which was described by a linear function for the shear specimen, and the fracture strain when the stress triaxiality was 0 was determined by linear function extension.

Table 2. Average stress triaxiality, Lode parameter, and fracture strain of each specimen.

	NRB1	NRB3	NRB5	CH6	FS
Average stress triaxiality	1.498	1.146	0.992	0.503	0.019
Average Lode parameter	1	1	1	0.521	0.031
Fracture strain	0.331	0.743	0.806	0.910	0.780

Bao [50] obtained the fracture cutoff effect of materials; that is, when the stress triaxiality of the material is $-1/3$, the fracture strain tends to approach infinite at this point. No cracks were observed in the specimens during the compression experiments. Therefore, a fracture strain equal to 5 was used to characterize the uncracked specimens when the stress triaxiality was $-1/3$. The power function was applied to describe the fracture strain when the stress triaxiality was $-1/3$ to 0, as stated in Equation (15). The established failure model is illustrated in Figure 12.

$$\begin{cases} \bar{\varepsilon}_f^{pl} = -0.6686 + 2.016\exp(-0.4175\eta)(0.333 < \eta < 1.5) \\ \bar{\varepsilon}_f^{pl} = 1.052\eta + 0.7346(0 < \eta < 0.333) \\ \bar{\varepsilon}_f^{pl} = \frac{1}{0.4918+\eta} - 1.299(-0.333 < \eta < 0) \end{cases} \quad (15)$$

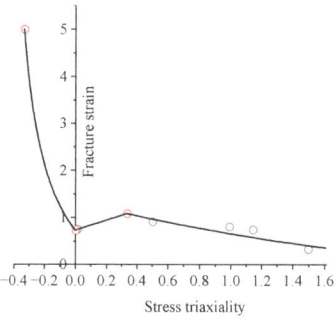

Figure 12. 2D fracture model.

It can be seen from Figure 12 that the monotonic and functional types of the fracture strain curve, at different stress triaxiality intervals, are different. In the stress triaxiality range of $-1/3$–0, the fracture strain function presented a power function distribution and monotonically decreased. From 0 to 0.333, the function monotonically increases into a linear function. In the range 0.333–1.5, the function distribution conformed to the Johnson–Cook failure model and decreased monotonically.

5.2.2. Determination of 3D Fracture Model

While considering the Lode parameters in the fracture model, a 3D model is developed in which the stress triaxiality and Lode parameters affect the fracture strain. Equations (6) and (7) were fitted based on the data in Table 2 to determine the unknown parameters in these equations. The resulting fracture model is illustrated in Figure 13 and the determined parameters are summarized in Table 3.

As shown in Figure 13, the ERT and LSMCS models exhibit the same trends in the overall distribution. Under certain Lode parameters, the fracture strain decreased along with an increase in the stress triaxiality. However, slight numerical differences were observed between the two models, mainly in the peak distribution of the fracture strain.

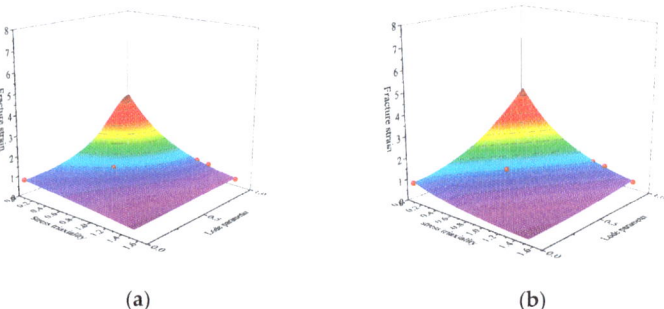

Figure 13. 3D fracture model: (**a**) ERT and (**b**) LSMCS.

Table 3. Determined model parameters.

	ERT			LSMCS	
R_1	R_2	R_3		α	γ
0.279	1.477	1.030		3.883	0.218

6. Comparison with Notched Tensile Specimen Experiments

To verify the accuracy of the established girth weld failure model for X80 pipelines, a failure model was introduced using ABAQUS to conduct a fracture simulation of the experimental processes mentioned above. Several points on the function of the 2D model were selected using the built-in ductile damage of ABAQUS. The 3D model defined the functional relationship between stress triaxiality, Lode parameters, and fracture strain through the keywords in the INPUT file provided by ABAQUS, which defined the correlation between the Lode angle function and fracture strain. The conversion relationship between the Lode angle function and Lode angle parameters is as follows:

$$\bar{\theta} = 1 - \frac{2}{\pi}\arccos(\zeta) \tag{16}$$

where ζ is the Lode angle function.

The simulation results are shown in Figure 14. It was observed that the fracture model that considered the influences of the Lode parameters was significantly superior to the 2D model. The maximum error between the simulated crack initiation displacement and the average experimental crack initiation displacement of the 2D fracture model was 18%, and that of the 3D fracture model was less than 10%. Figure 15 is a comparison of experimental and simulated fracture morphology, and the fracture morphology is basically consistent.

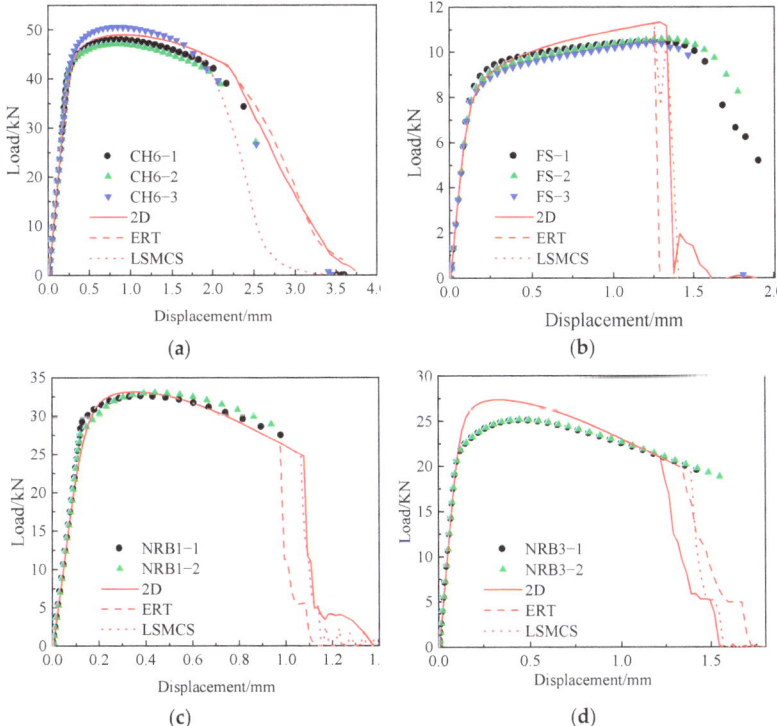

Figure 14. *Cont.*

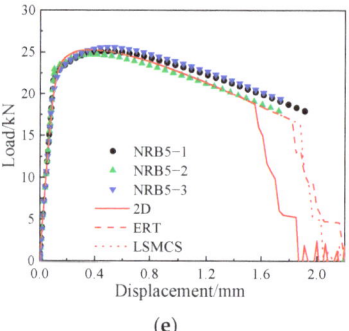

(e)

Figure 14. Comparison between finite element and experimental fracture: (**a**) CH6, (**b**) FS, (**c**) NRB1, (**d**) NRB3, and (**e**) NRB5.

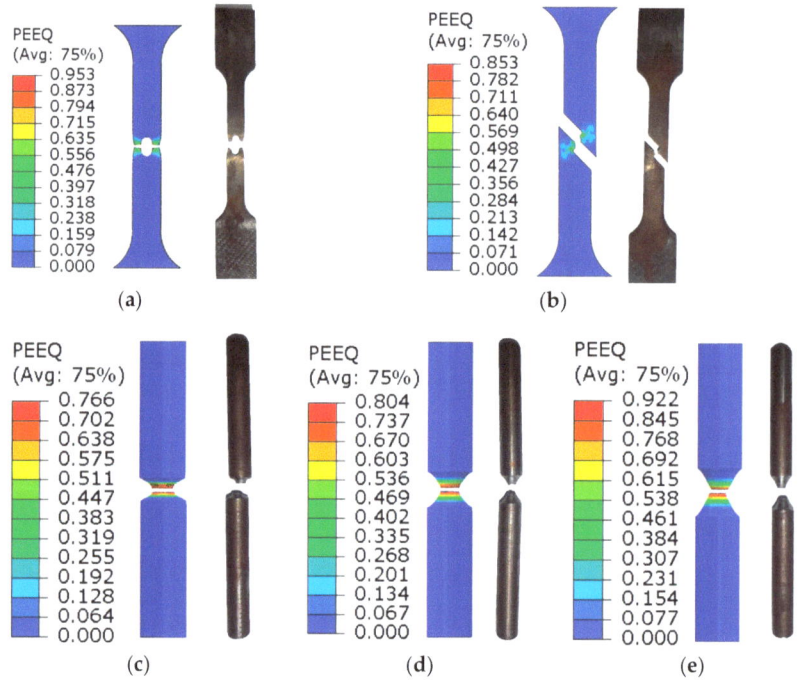

Figure 15. Comparison of experimental and simulated fracture morphology: (**a**) CH6, (**b**) FS, (**c**) NRB1, (**d**) NRB3, and (**e**) NRB5.

7. Influence of Stress Triaxiality on Tensile Strength

As shown in Figure 11, the stress triaxiality of each tensile specimen exhibited a constantly increasing trend during the tensile process; thus, the average stress triaxiality was applied to describe the relationship between the stress triaxiality and tensile strength. HB 5214-96 defines notch tensile strength as the maximum load divided by the minimum initial cross-sectional area. The fracture displacement decreased, indicating that ductility decreased with an increase in stress triaxiality. The average value of the tensile strength of the specimen was equated to the tensile strength under each working condition. The determined tensile strength is listed in Table 4, and it can be seen from Figure 16 that the tensile strength increased with an increase in stress triaxiality [51–53]. The relationship between the average stress triaxiality and ultimate tensile strength is expressed as

$$\sigma_{\text{tensile}} = 506.023(\eta)_{\text{av}}^2 - 622.018(\eta)_{\text{av}} + 956.467 \tag{17}$$

Table 4. Average stress triaxiality and tensile strength of notched specimens.

	NRB1	NRB3	NRB5	CH6
Average stress triaxiality	1.498	1.146	0.992	0.503
Average tensile strength (MPa)	1162	891.1	889.8	770.9
Fracture displacement (mm)	1.000	1.460	1.760	2.070

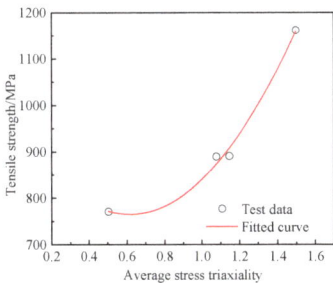

Figure 16. Function relationship between tensile strength and average stress triaxiality.

8. Conclusions

In this study, the ductile fracture of X80 pipeline girth weld was studied experimentally and numerically. The mechanical properties of base metal and weld were obtained through experiments, and the fracture parameters of weld under different stress states were obtained through special-shaped specimens. An uncoupled 2D fracture model and two uncoupled 3D fracture models of the X80 pipeline girth welds were introduced through a phenomenological approach. The 2D model is a function of fracture strain and stress triaxiality. The 3D models were a consider the effect of Lode parameter stress-modified critical strain (LSMCS) model and the extended Rice–Tracey (ERT) criterion.

Tensile failure tests under five different stress states were carried out and corresponding finite element models were established. Fracture parameters from high stress triaxiality to low stress triaxiality were obtained by combining test and numerical simulation. Fracture strain and stress triaxiality were obtained to develop a 2D fracture model. Based on the 2D model, the related parameters of ERT and LSMCS are calibrated considering the influence of the Lode parameters. In addition, a simulation proved the validity of the developed fracture model.

Compared with the 2D fracture models, the 3D fracture models yielded more accurate prediction results. In the fracture prediction of the CH6 and FS specimens, the 2D and 3D fracture models provided good prediction results for the fracture displacement. However, in the fracture prediction of the round bar notch specimens, the 2D fracture model generated results in which the fracture displacement was larger or smaller, but the 3D fracture model generated consistently good test results.

The influence of stress triaxiality of weld materials on tensile strength is similar to that of metal materials previously studied, the functional relationship between the tensile strength and stress triaxiality of the X80 pipeline girth welds satisfied the distribution form of the quadratic function and increased monotonically. Ductility decreases with an increase in stress triaxiality.

The research results can be used to predict the fracture of X80 pipeline girth weld under various complex loads, and can be used to evaluate the safety of pipeline girth weld in practical projects.

Author Contributions: N.L.: software, data curation, methodology, writing—original draft preparation. B.J.: conceptualization, resources, supervision. J.C.: writing—review & editing, investigation. Y.S.: formal analysis. S.D.: software. All authors have read and agreed to the published version of the manuscript.

Funding: This research was funded by Natural Science Foundation of Sichuan Province (2022NSFSC0317), the Open Fund of Shanghai Key Laboratory of Engineering Structure Safety (No. 2019-KF04), Sichuan Provincial Science and Technology Department Project (2020JDTD0021, 2023NSFSC0887) and National Natural Science Foundation of China (51908476). And the APC was funded by Bin Jia.

Data Availability Statement: The data used in the study is available with the authors and can be shared upon reasonable request.

Conflicts of Interest: The authors declare that they have no known competing financial interests or personal relationships that could have appeared to influence the work reported in this paper.

References

1. Witek, M. Possibilities of Using X80, X100, X120 High-Strength Steels for Onshore Gas Transmission Pipelines. *J. Nat. Gas Sci. Eng.* **2015**, *27*, 374–384. [CrossRef]
2. American Petroleum Institute. *Specification for Line Pipe, API Specification 5L*; American Petroleum Institute: Washington, DC, USA, 2018; Volume 46.
3. *British Standard BS 7910*; Guide to Methods for Assessing the Acceptability of Flaws in Metallic Structures. BSI Standards Publication: London, UK, 2013.
4. *API 579-2000*; API Recommended Practice 579. Fitness for Service. API Standards. American Petroleum Institute: Washington, DC, USA, 2000.
5. Kocak, M.; Webster, S.; Janosch, J.J.; Ainsworth, R.A.; Koers, R. *FITNET Fitness-for-Service (FFS) Procedure*; GKSS Research Centre Geesthacht: Geesthacht, Germany, 2008.
6. Milne, I.; Ainsworth, R.A.; Dowling, A.R.; Stewart, A.T. Assessment of the Integrity of Structures Containing Defects. *Int. J. Press. Vessel. Pip.* **1988**, *32*, 3–104. [CrossRef]
7. Fairchild, D.P.; Crapps, J.M.; Cheng, W.; Tang, H.; Shafrova, S. Full-Scale Pipe Strain Test Quality and Safety Factor Determination for Strain-Based Engineering Critical Assessment. In Proceedings of the 2016 11th International Pipeline Conference, Calgary, AB, Canada, 26–30 September 2016; Volume 2.
8. Khalaj, G.; Nazari, A.; Pouraliakbar, H. Prediction of Martensite Fraction of Microalloyed Steel by Artificial Neural Networks. *Neural Netw. World* **2013**, *13*, 117–130.
9. Khalaj, G.; Azimzadegan, T.; Khoeini, M.; Etaat, M. Artificial Neural Networks Application to Predict the Ultimate Tensile Strength of X70 Pipeline Steels. *Neural Comput. Appl.* **2013**, *23*, 2301–2308. [CrossRef]
10. Pouraliakbar, H.; Nazari, A.; Fataei, P.; Livary, A.K.; Jandaghi, M. Predicting Charpy Impact Energy of Al6061/SiCp Laminated Nanocomposites in Crack Divider and Crack Arrester Forms. *Ceram. Int.* **2013**, *39*, 6099–6106. [CrossRef]
11. Kofiani, K.; Nonn, A.; Wierzbicki, T. New Calibration Method for High and Low Triaxiality and Validation OnSENT Specimens of API X70. *Int. J. Press. Vessel. Pip.* **2013**, *111*, 187–201. [CrossRef]
12. Gurson, A.L. Continuum Theory of Ductile Rupture by Void Nucleation and Growth: Part 1—Yield Criteria and Flow Rules for Porous Ductile Media. *J. Eng. Mater. Technol. Trans. ASME* **1977**, *99*, 2–15. [CrossRef]
13. Tvergaard, V. On Localization in Ductile Materials Containing Spherical Voids. *Int. J. Fract.* **1982**, *18*, 237–252. [CrossRef]
14. Tvergaard, V.; Needleman, A. Analysis of the Cup-Cone Fracture in a Round Tensile Bar. *Acta Metall.* **1984**, *32*, 157–169. [CrossRef]
15. Nielsen, K.L.; Tvergaard, V. Failure by Void Coalescence in Metallic Materials Containing Primary and Secondary Voids Subject to Intense Shearing. *Int. J. Solids Struct.* **2011**, *48*, 1255–1267. [CrossRef]
16. Jackiewicz, J. Use of a Modified Gurson Model Approach for the Simulation of Ductile Fracture by Growth and Coalescence of Microvoids under Low, Medium and High Stress Triaxiality Loadings. *Eng. Fract. Mech.* **2011**, *78*, 487–502. [CrossRef]
17. Nahshon, K.; Hutchinson, J.W. Modification of the Gurson Model for Shear Failure. *Eur. J. Mech. A/Solids* **2008**, *27*, 1–17. [CrossRef]
18. Johnson, G.R.; Cook, W.H. Fracture Characteristics of Three Metals Subjected to Various Strains, Strain Rates, Temperatures and Pressures. *Eng. Fract. Mech.* **1985**, *21*, 31–48. [CrossRef]
19. Chi, W.-M.; Kanvinde, A.M.; Deierlein, G.G. Prediction of Ductile Fracture in Steel Connections Using SMCS Criterion. *J. Struct. Eng.* **2006**, *132*, 171–181. [CrossRef]
20. Wierzbicki, T.; Bao, Y.; Lee, Y.W.; Bai, Y. Calibration and Evaluation of Seven Fracture Models. *Int. J. Mech. Sci.* **2005**, *47*, 719–743. [CrossRef]
21. Gruben, G.; Hopperstad, O.S.; Borvik, T. Evaluation of Uncoupled Ductile Fracture Criteria for the Dual-Phase Steel Docol 600DL. *Int. J. Mech. Sci.* **2012**, *62*, 133–146. [CrossRef]

22. Zhang, Y.; Liu, Y.; Yang, F. Ductile Fracture Modelling of Steel Plates under Tensile and Shear Dominated States. *J. Constr. Steel Res.* **2022**, *197*, 107469. [CrossRef]
23. Kacem, A.; Laurent, H.; Thuillier, S. Experimental and Numerical Investigation of Ductile Fracture for AA6061-T6 Sheets at Room and Elevated Temperatures. *Int. J. Mech. Sci.* **2022**, *222*, 107201. [CrossRef]
24. Zhang, Y.; Shen, F.; Zheng, J.; Münstermann, S.; Li, T.; Han, W.; Huang, S. Ductility Prediction of HPDC Aluminum Alloy Using a Probabilistic Ductile Fracture Model. *Theor. Appl. Fract. Mech.* **2022**, *119*, 103381. [CrossRef]
25. Oh, C.K.; Kim, Y.J.; Baek, J.H.; Kim, Y.P.; Kim, W.S. Ductile Failure Analysis of API X65 Pipes with Notch-Type Defects Using a Local Fracture Criterion. *Int. J. Press. Vessel. Pip.* **2007**, *84*, 512–525. [CrossRef]
26. Oh, C.K.; Kim, Y.J.; Baek, J.H.; Kim, W.S. Development of Stress-Modified Fracture Strain for Ductile Failure of API X65 Steel. *Int. J. Fract.* **2007**, *143*, 119–133. [CrossRef]
27. Oh, C.K.; Kim, Y.J.; Baek, J.H.; Kim, Y.P.; Kim, W. A Phenomenological Model of Ductile Fracture for API X65 Steel. *Int. J. Mech. Sci.* **2007**, *49*, 1399–1412. [CrossRef]
28. Oh, C.S.; Kim, N.H.; Kim, Y.J.; Baek, J.H.; Kim, Y.P.; Kim, W.S. A Finite Element Ductile Failure Simulation Method Using Stress-Modified Fracture Strain Model. *Eng. Fract. Mech.* **2011**, *78*, 124–137. [CrossRef]
29. Shinohara, Y.; Madi, Y.; Besson, J. Anisotropic Ductile Failure of a High-Strength Line Pipe Steel. *Int. J. Fract.* **2016**, *197*, 127–145. [CrossRef]
30. Paredes, M.; Wierzbicki, T.; Zelenak, P. Prediction of Crack Initiation and Propagation in X70 Pipeline Steels. *Eng. Fract. Mech.* **2016**, *168*, 92–111. [CrossRef]
31. Han, P.; Cheng, P.; Yuan, S.; Bai, Y. Characterization of Ductile Fracture Criterion for API X80 Pipeline Steel Based on a Phenomenological Approach. *Thin-Walled Struct.* **2021**, *164*, 107254. [CrossRef]
32. Sarzosa, D.F.; Paredes, M.; Savioli, R.; Ruggieri, C.; Leite, L.G.; da Silva, N.S.; Garmbis, A.G. Experimental and Numerical Study on the Ductile Fracture Response of X65 Girth-Welded Joint Made of Inconel 625 Alloy. *Theor. Appl. Fract. Mech.* **2022**, *121*, 103533. [CrossRef]
33. Rice, J.R.; Tracey, D.M. On the Ductile Enlargement of Voids in Triaxial Stress Fields. *J. Mech. Phys. Solids* **1969**, *17*, 201–217. [CrossRef]
34. Huang, X.; Zhao, W.; Zhao, J.; Wang, Z. Fracture Model of Q235B Steel Considering the Influence of Stress Triaxiality and Lode Parameter. *J. Basic Sci. Eng.* **2019**, *27*, 1172–1187. (In Chinese) [CrossRef]
35. GB/T 228.1-2010; Metallic Materials-Tensile Testing-Part 1: Method of Test at Room Temperature. Standardization Administration of the People's Republic of China (SAC): Beijing, China, 2010.
36. HB 5214-1996; Method of Notch Tensile Test for Metals at Room Temperature. Aviation Industry Corporation of China: Beijing, China, 1996.
37. GB/T 7314-2017; Metallic Materials-Compression Test Method at Room Temperature. Standardization Administration of the People's Republic of China (SAC): Beijing, China, 2017.
38. HB 6736-1993; Shear Test Method for Sheet Metal. Aviation Industry Corporation of China: Beijing, China, 1993.
39. ISO/TR 15608; Welding—Guidelines for a metallic materials grouping system. European Committee for Standardization (CEN): Brussels, Belgium,, 2006.
40. GB/T 31032-2014; Welding and Acceptance Standard for Steel Pipings and Pipelines. Standardization Administration of the People's Republic of China (SAC): Beijing, China, 2014.
41. SS-EN ISO 4063; 2010 Welding and Allied Processes—Nomenclature of Processes and Reference Numbers. European Committee for Standardization (CEN): Brussels, Belgium, 2012.
42. Johnson, G.R.; Cook, W.H. A Constitutive Model and Data for Metals Subjected to Large Strains, High Strain Rates and High Temperatures. In Proceedings of the 7th International Symposium on Ballistics, The Hague, The Netherlands, 19–21 April 1983; Volume 547.
43. Bastola, A.; Wang, J.; Shitamoto, H.; Mirzaee-Sisan, A.; Hamada, M.; Hisamune, N. Full- and Small-Scale Tests on Strain Capacity of X80 Seamless Pipes. *Procedia Struct. Integr.* **2016**, *2*, 1894–1903. [CrossRef]
44. Nie, H.; Ma, W.; Xue, K.; Ren, J.; Dang, W.; Wang, K.; Cao, J.; Yao, T.; Liang, X. A Novel Test Method for Mechanical Properties of Characteristic Zones of Girth Welds. *Int. J. Press. Vessel. Pip.* **2021**, *194*, 104533. [CrossRef]
45. Wu, X.; Shuai, J.; Xu, K.; Lv, Z.; Shan, K. Determination of Local True Stress-Strain Response of X80 and Q235 Girth-Welded Joints Based on Digital Image Correlation and Numerical Simulation. *Int. J. Press. Vessel. Pip.* **2020**, *188*, 104232. [CrossRef]
46. Zhang, Y.; Shuai, J.; Ren, W.; Lv, Z. Investigation of the Tensile Strain Response of the Girth Weld of High-Strength Steel Pipeline. *J. Constr. Steel Res.* **2022**, *188*, 107047. [CrossRef]
47. Li, Y.J.; Li, Q.; Wu, A.P.; Ma, N.X.; Wang, G.Q.; Murakawa, H.; Yan, D.Y.; Wu, H.Q. Determination of Local Constitutive Behavior and Simulation on Tensile Test of 2219-T87 Aluminum Alloy GTAW Joints. *Trans. Nonferrous Met. Soc. China* **2015**, *25*, 3072–3079. [CrossRef]
48. Bai, Y.; Wierzbicki, T. Application of Extended Mohr-Coulomb Criterion to Ductile Fracture. *Int. J. Fract.* **2010**, *161*, 1–20. [CrossRef]
49. Bao, Y.; Wierzbicki, T. On Fracture Locus in the Equivalent Strain and Stress Triaxiality Space. *Int. J. Mech. Sci.* **2004**, *46*, 81–98. [CrossRef]
50. Bao, Y.; Wierzbicki, T. On the Cut-off Value of Negative Triaxiality for Fracture. *Eng. Fract. Mech.* **2005**, *72*, 1049–1069. [CrossRef]

51. Jenkins, W.D.; Willard, W.A. Effect of Temperature and Notch Geometry on the Tensile Behavior of a Titanium Alloy. *J. Res. Natl. Bur. Stand. Sect. C Eng. Instrum.* **1966**, *70*, 5–11. [CrossRef]
52. Kumar, J.G.; Nandagopal, M.; Parameswaran, P.; Laha, K.; Mathew, M.D. Effect of Notch Root Radius on Tensile Behaviour of 316L(N) Stainless Steel. *Mater. High Temp.* **2014**, *31*, 239–248. [CrossRef]
53. Lei, X.; Li, C.; Shi, X.; Xu, X.; Wei, Y. Notch Strengthening or Weakening Governed by Transition of Shear Failure to Normal Mode Fracture. *Sci. Rep.* **2015**, *5*, 10537. [CrossRef]

Disclaimer/Publisher's Note: The statements, opinions and data contained in all publications are solely those of the individual author(s) and contributor(s) and not of MDPI and/or the editor(s). MDPI and/or the editor(s) disclaim responsibility for any injury to people or property resulting from any ideas, methods, instructions or products referred to in the content.

Article

Flexural Response of Concrete Beams Reinforced with Steel and Fiber Reinforced Polymers

Noura Khaled Shawki Ali *, Sameh Youssef Mahfouz and Nabil Hassan Amer

Construction and Building Engineering Department, College of Engineering and Technology, Arab Academy for Science, Technology and Maritime Transport (AASTMT), B 2401 Smart Village, Giza 12577, Egypt
* Correspondence: nora@aast.edu

Abstract: This paper numerically investigates the flexural response of concrete beams reinforced with steel and four types of Fiber-Reinforced Polymers (FRP), i.e., Carbon FRP (CFRP), Glass FRP (GFRP), Aramid FRP (AFRP), and Basalt FRP (BFRP). The flexural responses of forty beams with two boundary conditions (simply supported and over-hanging beams) were determined using ABAQUS. Subsequently, the finite element models were validated using experimental results. Eventually, the impact of the reinforcement ratios ranging between 0.15% and 0.60% on the flexural capacity, crack pattern, and fracture energy were investigated for all beams. The results revealed that, for the low reinforcement ratios, the flexural performance of CFRP significantly surpassed that of steel and other FRP types. As the reinforcement ratio reached 0.60%, the steel bars exhibited the best flexural performance.

Keywords: ABAQUS; finite element analysis (FEA); fiber reinforcement polymer (FRP); concrete damage plasticity (CDP)

1. Introduction

Although steel is the most popular material for reinforcing concrete elements due to its cost and enhanced mechanical features [1–3], the corrosion of steel bars reduces the structure's lifespan and raises maintenance expenses [4,5]. Fiber-Reinforced Polymer (FRP) bars are an appealing option in the reinforcement of concrete elements because they provide excellent levels of durability, corrosion resistance, and fatigue resistance, in addition to having a high strength-to-weight ratio [6–13]. FRP may be used for various purposes, such as reinforcing structures internally or externally by embedding discrete FRP fibers into the concrete, utilizing the near-surface-mounted technique (NSM) for FRP plates and sheets attached to the structure with vinyl-ester glue or epoxy [9,10].

To simulate the behavior of beams, many constitutive numerical models were provided in the existing literature [14–17]. Tejaswini and Raju [14] studied the flexural response for reinforced concrete beam sections with different failure modes using ABAQUS. The aim was to numerically compare the Finite Element Analysis (FEA) experimental findings. Salih et al. [15] executed seventeen models using ANSYS software of concrete beams reinforced with CFRP and GFRP bars. Various parameters were studied, including the number of bars, size, types, and longitudinal configuration for FRP bars. The outcomes were described in the form of a load–deflection diagram. Al Hasani et al. [16] investigated the crack propagation of RC beams reinforced with steel bars using ABAQUS software to simulate crack propagation. A comparison of numerical and experimental data findings has been established. The results indicated that the cracks were initiated from the tension side at the bottom of the beam. Shirmardi and Mohammadizadeh [17] simulated twenty concrete beams reinforced with GFRP in ABAQUS. The study focused on the span/depth ratio and the reinforcement ratio. The results showed that the rigidity of the beam decreases as the span/depth ration increases.

Many experimental tests were also conducted to compare the flexural behavior of the beams reinforced with GFRP and steel bars [18–23]. Shanour et al. [18] investigated

Citation: Shawki Ali, N.K.; Mahfouz, S.Y.; Amer, N.H. Flexural Response of Concrete Beams Reinforced with Steel and Fiber Reinforced Polymers. *Buildings* **2023**, *13*, 394. https://doi.org/10.3390/buildings13020374

Academic Editors: Rui Guo, Bo Wang, Muye Yang, Weidong He and Chuntao Zhang

Received: 29 December 2022
Revised: 17 January 2023
Accepted: 26 January 2023
Published: 29 January 2023

Copyright: © 2023 by the authors. Licensee MDPI, Basel, Switzerland. This article is an open access article distributed under the terms and conditions of the Creative Commons Attribution (CC BY) license (https://creativecommons.org/licenses/by/4.0/).

seven GFRP and steel beams under four-point loading. The primary studied factors were a concrete compressive strength and reinforcement ratio. The fracture breadth and GFRP reinforcing stresses were measured for the tested beams mid-span deflection. The results showed that increasing the reinforcement ratio reduces the crack widths and mid-span deflection. Krasniqi et al. [19] examined GFRP and steel-reinforced beams under four-point loading. The results showed early crack initiation due to the low elastic modulus of GFRP bars. Sirimontree et al. [20] investigated the flexural behavior of concrete reinforced with GFRP and steel bars. The beam was subjected to four-point loading. The stiffness, flexural capacity, and mode of failure were investigated. The results relieved that the stiffness of GFRP-reinforced beams decreased in comparison to steel-reinforced beams.

Arivalagan [21] examined the beams' flexural behavior reinforced with GFRP bars and stainless-steel bars (SSRB). The results showed that the beams reinforced with GFRP experience larger deflections, lower stiffness, and lower ultimate loads than the control beam. This was due to a slip between the rebar and the concrete. Saraswathi and Dhanalakshmi [22] studied the behavior of concrete beams reinforced with GFRP and steel bars. Various factors, including the load capability, load deflection, and mechanism of failure were examined. The GFRP bars exhibit higher deflections due to their low elastic modulus. The GFRP bars failed due to the slip between the rebar and the concrete. To improve the FRP bars' bond behaviors. Murugan and Kumaran [23] studied the flexural tensile behavior of five rectangular concrete beams reinforced with surface-treated GFRP under two-point static loading. The sand-sprinkled and grooved reinforcing bars were used to improve the bond behavior. The investigated parameters were the ultimate load capacity, fracture widths, crack propagation, and beam failure modes. The results showed that the sand-coated GFRP reinforcements had poorer performance than the grooved GFRP beams concerning ultimate load capacity and deflections. To compare the behavior of the beams reinforced with BFRP and steel bars.

Some studies focused on the studying behavior of BFRP and CFRP. Hamdy and Arafa [24] examined six concrete beams reinforced with BFRP bars, dispersed steel fibers and steel tested in four-point bending until failure. The moment carrying capacity and failure loads were calculated and compared with the experimental data. The results showed that the BFRP bar-reinforced beams experienced greater deflection values than steel beams as it has low stiffness of FRP bars. Zhang et al. [25] experimentally studied the flexural deflection of six concrete beams reinforced with BFRP bars and one beam reinforced with steel bars. Additionally, the numerical simulation was performed using FEM. The findings demonstrated that all the BFRP-reinforced concrete beams had either concrete crushing or rupture. Ashour and Habeeb [26] investigated the tensile behavior of the CFRP-reinforced simply supported and continuous beams. It was found that the CFRP beams failed due to the rupture of bars. The use of CFRP bars enhanced the load carrying capacity. Many studies focused on the usage of CFRP in the strengthening of rc beams [27–30]. All the studies indicated that CFRP increased the load carrying capacity in flexural and shear strengthening.

Despite the efforts of the previously mentioned studies to investigate the flexural behavior of concrete beams reinforced with FRP, the majority of these studies focused on the utilization of GFRP bars, disregarding the other FRP types. Furthermore, most studies focused on the application of CFRP bars to strengthen the concrete beams rather than reinforcing them. Thus, the flexural behavior of most FRP types lacks further investigation. Moreover, the fracture energies of the beams reinforced with FRP were not discussed. Therefore, the current study addresses the literature gap by examining the flexural behavior of four FRP types (CFRP, AFRP, BFRP, and GFRP) to help the structural designers to find an effective alternative to steel bars. To this end, the influence of different reinforcement ratios (0.15%, 0.27%, 0.42%, and 0.60%) on the flexural capacity of forty concrete beams. Moreover, the load–deflection relationships and crack patterns of these beams were discussed. In order to verify the models, the results of FEM were compared with the experimental work performed by Issa and Elzeiny [31]. Section 2 of this study provides an overview of the experimental work, specifics of developing numerical FE models, the behavior of materials, and FE model verification. The parametric study and findings are discussed in Section 3. Finally, a brief discussion of the results, conclusions, and recommended work are presented in Section 4.

2. Model Evolution

2.1. General

The experimental investigation performed by Isa and Elzeiny [31] consisted of six overhanging concrete beams that were subjected to a three-point load. The beams' dimensions were 150 × 250 mm; the total length was 2000 mm, including a 600 mm long cantilever. There were three groups, each with varying ratios of GFRP rebars, concrete strength, and rebar types (steel or GFRP). The details of the examined beams are shown in Figure 1.

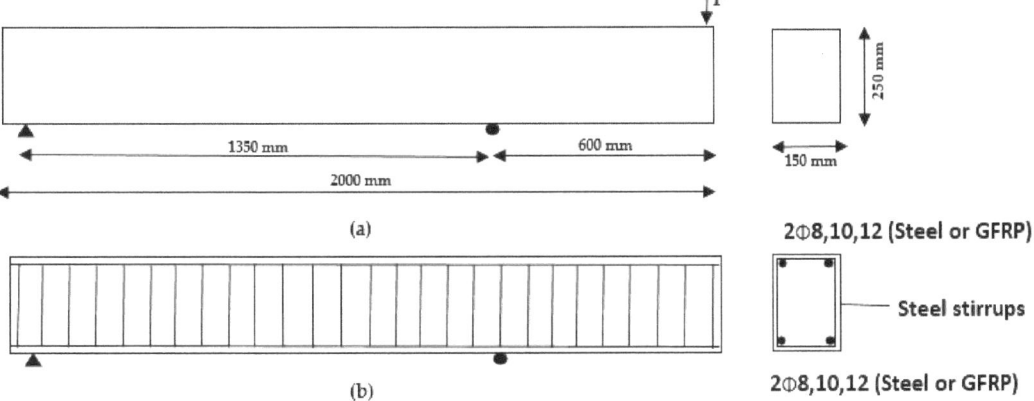

Figure 1. Details of concrete beams: (**a**) concrete dimensions of tested beam and (**b**) longitudinal reinforcement configuration.

2.2. Model Loading Boundary Condition and Meshing

The overhanging beams presented by Isa and Elzeiny [31] were simulated in ABAQUS. In order to simulate the concrete, 3D solid C3D8R element have been used. T3D2 is a 2D truss element used in the modeling of steel and FRP reinforcements. The concrete material was defined according to the damage plasticity models and details of defining steel and FRP materials are discussed in detail in the following section. A static condition of loading was considered. The embedded bar option was used to simulate the bond between FRP and steel reinforcement in concrete. Ties were utilized to simulate the bond between the beams and the plates. A loading point (reference point) at the tip of the cantilever was added at the top of the plate as shown in Figure 2 and the displacement was computed at the node located at the bottom of the beam. The fine mesh size of 20 mm and increment size from 0.01 to 1 to obtain more accurate results. The details of meshing are shown in Figure 2.

2.3. Materials

2.3.1. Damaged Plasticity Model

Three crack models: (i) brittle crack model, (ii) smeared crack model, and (iii) concrete-damaged plasticity (CDP) model in ABAQUS were used to simulate the concrete damage [32]. The CDP model was used to represent the inelastic responses of concrete compression and tension damage characteristics. The models take into account two failure mechanisms, namely tensile cracking and compressive crushing.

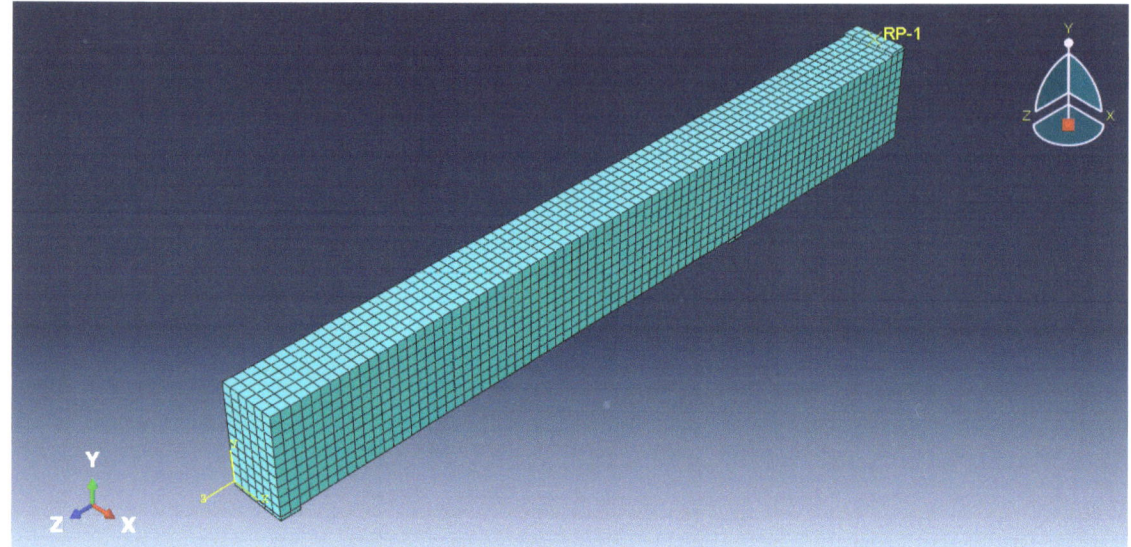

Figure 2. Reference point and finite element meshing.

2.3.2. Tensile Behavior of Concrete

The tensile strain and cracking stress relationship was used to describe the behavior of concrete under tension. The cracking strain was computed using Equation (1):

$$\tilde{\varepsilon}_t^{ck} = \varepsilon_t - \varepsilon_{ot}^{el} \tag{1}$$

where $\tilde{\varepsilon}_t^{ck}$ is the tensile cracking strain, ε_t tensile strain, and ε_{ot}^{el} is the elastic strain of the unaffected material. The model was developed based on the Nayal and Rasheed [33] model of tension-stiffening. Gilbert and Warner [34] created a homogenized stress and strain relationship. The accuracy of the plastic strain values was checked using ABAQUS and are calculated as shown in Equation (2). The inaccurate damage curves result from the tensile plastic strain values that are negative or decreasing.

$$\tilde{\varepsilon}_t^P = \tilde{\varepsilon}_t^{ck} - \frac{d_t}{(1-d_t)} \frac{\sigma_t}{E_0} \tag{2}$$

where $\tilde{\varepsilon}_t^P$ is the tensile plastic strain, d_t refers to tensile damage parameter values, σ_t is the concrete tensile stress, and E_0 is the elastic modulus.

2.3.3. Compressive Behavior of Concrete

The relationship between compressive stress and cracking strain was used in defining the nonlinear compression behavior of concrete. Equation (3) was used to convert compressive strain to inelastic strain.

$$\tilde{\varepsilon}_c^{in} = \varepsilon_c - \varepsilon_{oc}^{el} \tag{3}$$

where $\tilde{\varepsilon}_c^{in}$ is the compressive inelastic strain, ε_c is compressive strain, and ε_{oc}^{el} is elastic strain corresponding to the unaffected material. The accuracy of the plastic strain values was checked using Equation (4) to ensure there were no negative or decreasing values.

$$\tilde{\varepsilon}_c^P = \tilde{\varepsilon}_c^{in} - \frac{d_c}{(1-d_c)} \frac{\sigma_c}{E_0} \tag{4}$$

where the symbol $\tilde{\varepsilon}_c^{\,p}$ is the compressive plastic strain, d_c is the damage parameter values, and σ_c is the compressive strength. In compression, the stress–strain relationship was obtained using a computational model created by Hsu et al. [34]. The concrete materials with compressive strengths of up to 62 MPa can be utilized with this model. The stress–strain of concrete exhibits linear behavior up to 50% of compressive strength in the hardening part. The material was described until the σ_c of 30 MPa in the softening part. The model equations are presented in Equations (5)–(8).

$$\sigma_c = \left(\frac{\beta \left(\frac{\varepsilon_c}{\varepsilon_0} \right)}{\beta - 1 + \left(\frac{\varepsilon_c}{\varepsilon_0} \right)^\beta} \right) \sigma_{cu} \tag{5}$$

$$\beta = \frac{1}{1 - (\sigma_{cu}/(\varepsilon_0 E_0))} \tag{6}$$

$$\varepsilon_0 = 0.000089\, \sigma_{cu} + 0.002114 \tag{7}$$

$$E_0 = 124.3\, \sigma_{cu} + 32831.2 \tag{8}$$

where ε_0 refers to strain at peak stress and β refers to a variable that depends on the shape of stress–strain diagram.

2.3.4. Concrete Damage Parameters (CDP)

The elastic modulus following tensile and compressive failure can be calculated using Equations (9) and (10). The tensile and compressive failure has a range between 0 and 1, where 0 indicates that the material is in its initial state but 1 indicates that there is a loss in material strength. The compression and tensile damage can be calculated using Equations (11) and (12).

$$E_t = E_0(1 - d_t) \tag{9}$$

$$E_c = E_0(1 - d_c) \tag{10}$$

$$d_t = 1 - \left(\frac{\sigma_t}{\sigma_t'} \right) \tag{11}$$

$$d_c = 1 - \left(\frac{\sigma_c}{\sigma_c'} \right) \tag{12}$$

where E_t refer to the tensile damage elastic modulus, E_c is elastic modulus of compressive damage, d_t is tensile damage, d_c is compressive damage, σ_c' is effective compressive strength, and σ_t' is effective tensile strength. The CDP model in ABAQUS is defined by five parameters [35]. Table 1 summarizes the CDP values needed to define the model in ABAQUS [36].

Table 1. Recommended values for CDP model parameters.

CDP Parameters	Symbol	Recommended Values
Dilation angle	Ψ	From 30° to 45°
Eccentricity	ϵ	0.1
Viscosity Parameter	μ	0.0001 to 0.008
Shape Factor	K_c	From 0.667 to 1
Biaxial stress ratio	$\frac{f_{bo}}{f_{co}}$	From 1 to 1.16

2.3.5. FRP Bars Behaviors

The behavior of the FRP bars was assumed to be isotropic linear elastic up to failure without any damage criteria [37,38]. FRPs were defined by their linearly elastic response in the absence of any visible yield point. The stress–strain equation is presented in Equation (13).

$$f_f = E_f\, \varepsilon_f \quad \varepsilon_f \leq \varepsilon_{fu} \tag{13}$$

where f_f refer to the fiber strength, E_f is the elastic modulus, ε_f refer to the strain, and ε_{fu} refer to the ultimate strain of FRP bars.

2.3.6. Behavior of Steel

Steel bars used for reinforcement have linearly elastic behavior under low strains. The yield point of steel is used to explain its plastic behavior. When a material's stress–strain curve reaches a yield point, the elastic behavior changes to a plastic one. The only strains that are produced by the steel's deformation before it reaches the yield point are elastic strains, which are entirely returned when the load is withdrawn. When the steel reaches its yield stress, permanent (plastic) deformation starts to occur [39]. The stress–strain equation is expressed in Equation (14). Kobraei et al. [37] and Abbood et al. [38] suggested proper values for the yield stress, elastic modulus of steel, and elastic modulus of FRP; these are summarized in Table 2.

$$f_s = \begin{cases} E_s \, \varepsilon_s & \varepsilon_s \leq \varepsilon_{sy} \\ f_{sy} & \varepsilon_s > \varepsilon_{sy} \end{cases} \quad (14)$$

where f_s is stress, E_s is modulus of elasticity, ε_s is strain in the steel bars, f_{sy} is yielding strength, and ε_{sy} is yielding strain.

Table 2. Steel and FRP properties.

Bar Type	Yield Stress (N/mm^2)		Tensile Strength (N/mm^2)		Modulus of Elasticity (kN/mm^2)	
	Allowable Range	Chosen Value	Allowable Range	Chosen Value	Allowable Range	Chosen Value
Steel	276–517	450	483–690	500	200	200
GFRP	–	–	483–1600	1045	35–51	40
CFRP	–	–	600–3690	2900	120–580	300
AFRP	–	–	1720–2540	2500	41–125	100
BFRP	–	–	600–1500	1200	50–65	55

2.4. Experimental Program Specimens

The specimens were classified into three groups A, B, and C. Group A consists of three beams reinforced with steel bars (SN8-8, SN10-10, and SN 12-12). The first part of the symbol (SN) reflects the steel reinforcement and the second part 8-8, 10-10, and 12-12 reflects the bar diameter. Group B consists of two beams reinforced with GFRP bars (GN8-8 and GN12-10). Group C represents the beam reinforced with GFRP bars (GM10-10). The σ_c of concrete for group A and B was 42.25 MPa. The σ_c of concrete for group C was 59.26 MPa. The compressive strength of the concrete and the details of reinforcement configuration are summarized in Table 3.

Table 3. Compressive strength of concrete and reinforcement configuration.

Group	Beam	Type of Longitudinal Reinforcement	Top Longitudinal Reinforcement	Bottom Longitudinal Reinforcement	Compressive Strength (MPa)
A	SN 8-8	Steel	2Ø8	2Ø8	42.25
	SN 10-10	Steel	2Ø10	2Ø10	
	SN 12-12	Steel	2Ø12	2Ø12	
B	GN 8-8	GFRP	2Ø8	2Ø8	42.25
	GN 12-10	GFRP	2Ø12	2Ø10	
C	GM 10-10	GFRP	2Ø10	2Ø10	59.26

2.5. FE Models Verification and Discussion

Four statistical indicators were utilized to verify the model [35]:

(i) Nash—Sutcliffe efficiency (NSE) is a statistic indicator that determines the proportionate difference between the observed data variance and residual variance.
(ii) Coefficient of determination (R2) is the percentage of variance or difference that can be statistically explained by one or more independent variables for a dependent variable.
(iii) Modified index of agreement (md) calculates the proportional and additive differences between the experimental and numerical in the means and variances.
(iv) Kling—Gupta efficiency (KGE) evaluates the bias, correlation, and variability between the numerical and experimental data. These indicators are calculated for Equations (15)–(18).

$$\text{NSE} = 1 - \left[\frac{\sum_{a=1}^{N}(\hat{x}_a - x_a)^2}{\sum_{a=1}^{N}(x_a - x^{mean})^2}\right] \quad (15)$$

$$md = 1 - \frac{\sum_{a=1}^{N}|x_a - \hat{x}_a|}{\sum_{a=1}^{N}(|\hat{x}_a - x^{mean}| + |x_a - x^{mean}|)} \quad (16)$$

$$R^2 = \left(\frac{\sum_{a=1}^{N}[(x_a - x^{mean})(\hat{x}_a - \hat{x}^{mean})]}{\sqrt{\sum_{a=1}^{N}[\hat{x}_a - \hat{x}^{mean}]^2}\sqrt{\sum_{a=1}^{N}[x_a - x^{mean}]^2}}\right)^2 \quad (17)$$

$$KGE = 1 - \sqrt{(P_c - 1)^2 + \left(\frac{\hat{x}^{mean}}{x^{mean}} - 1\right)^2 + \left(\frac{\hat{S.D}/\hat{x}^{mean}}{S.D/x^{mean}} - 1\right)^2} \quad (18)$$

where $\hat{x}_a$ denotes a numerical value, $S.D$ is the experimental data standard deviation, N denotes the quantity of data values, x_a denotes the experimentally acquired data value, P_c denotes the Pearson's correlation coefficient, x^{mean} denotes the experimental mean value, and $\hat{S.D}$ is the standard deviation of numerical data. ABAQUS was used to validate six FEM models for the beam features illustrated in Figure 1. The deflection was measured at the tip of the cantilever. Figure 3 illustrates the outcomes of load deflection curves. The results indicated that steel had a greater load capacity than GFRP except for the diameter size (8 mm), which was attributed to the better bond strength of GFRP bars in small diameters when compared to steel. As the diameter increases the bond strength decreases [40]. Moreover, steel exhibits a greater increase than GFRP as the load increases due to the low stiffness and modulus of elasticity of GFRP. The results of R2, KGE, md, and NSE for all beams are presented in Table 4. The outcomes revealed good concordance between experimental and numerical findings.

Table 4. Results of statistical indicator.

Statistical Indicators	SN8-8	SN10-10	SN12-12	GN8-8	GN10-10	GN12-10	Optimal Value
NSE	0.909	0.833	0.743	0.949	0.59	0.874	1
md	0.974	0.963	0.937	0.986	0.971	0.973	1
R	0.944	0.988	0.782	0.971	0.895	0.918	1
KGE	0.867	0.828	0.862	0.885	0.737	0.831	1

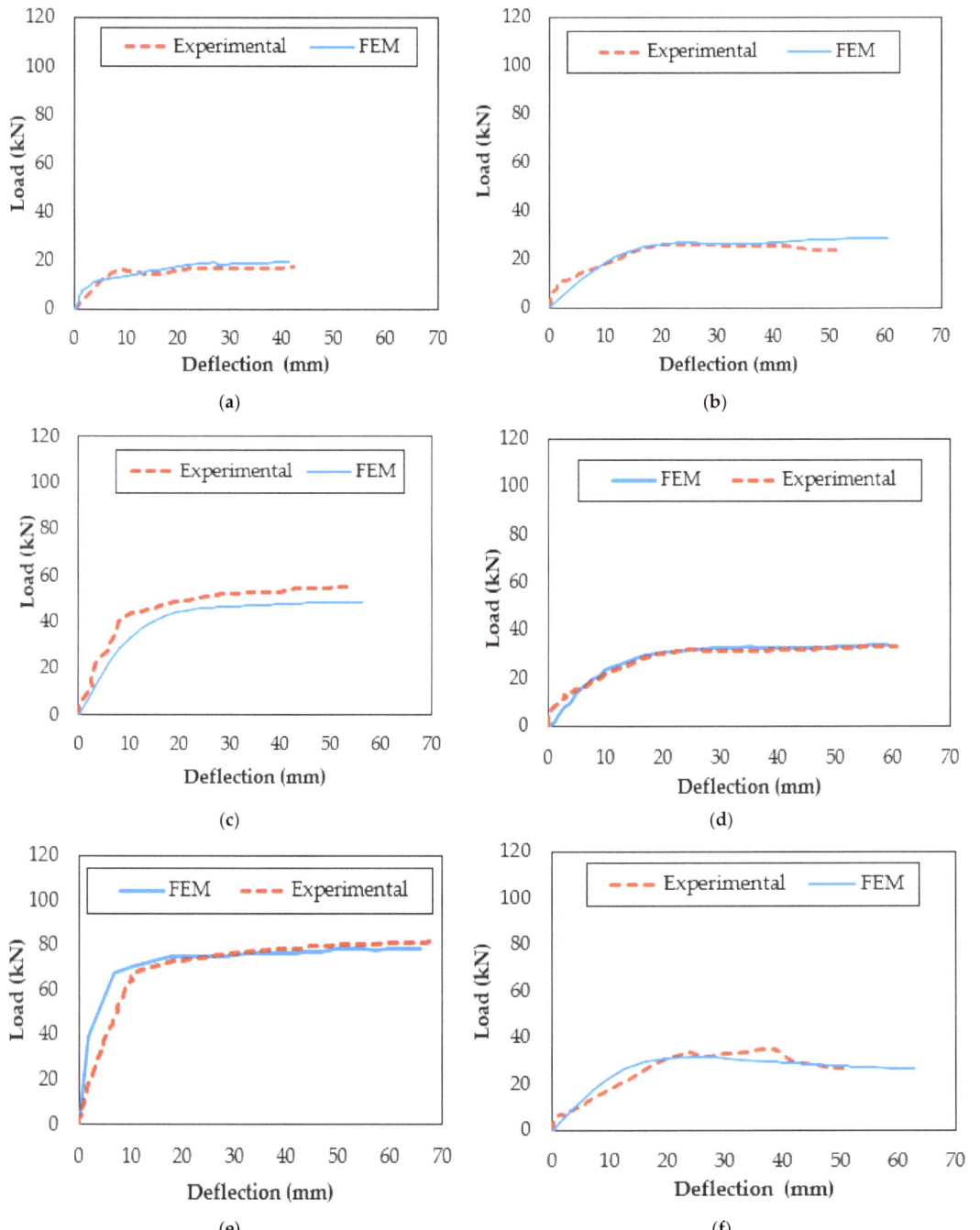

Figure 3. The load–deflection curves for overhanging beams: (**a**) SN8-8; (**b**) GN8-8; (**c**) SN10-10; (**d**) GM10-10; (**e**) SN12-12; and (**f**) GN12-10.

3. Parametric Study and Results

The study includes examining the flexural behavior and fracture propagation of beams using various FRP types and comparing them with those of steel. Two parameters were studied: the bar type and the reinforcement ratio. The reinforcement ratio 0.15% refers to bar diameter 6 mm, 0.27% refers to 8 mm, 0.42% refers to 10 mm, and 0.60% refers to 12 mm. These parameters were evaluated for simply supported and over hanging beams. The beam details are shown in Figure 4. The σ_c of concrete was 42.5 MPa while E_s for steel reinforcement was 200 GPa. The investigation includes various FRP materials such as glass, aramid, carbon, and basalt. The reinforcement configurations for both groups are shown in Table 5. The displacement was measured at the beams' midpoint for simple beams and at the cantilever tip for the overhanging beams.

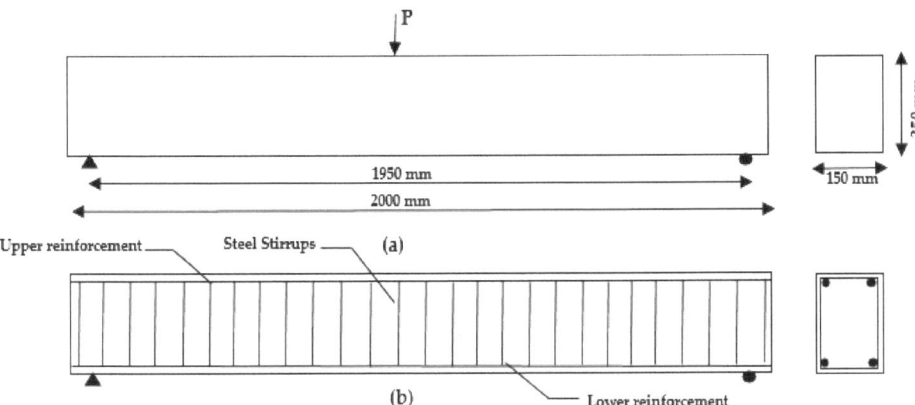

Figure 4. Details of concrete beams: (**a**) concrete dimensions of simple beam and (**b**) longitudinal reinforcement configuration.

Table 5. Reinforcement configuration.

Type of Longitudinal Bars	Beam No.	Top Longitudinal Reinforcement	Bottom Longitudinal Reinforcement	Reinforcement Ratio (%)	Stirrups
CFRP	CFRP 6-6	2ø6	2ø6	0.15	
	CFRP 8-8	2ø8	2ø8	0.27	
	CFRP 10-10	2ø10	2ø10	0.42	
	CFRP 12-12	2ø12	2ø12	0.60	
BFRF	BFRP 6-6	2ø6	2ø6	0.15	
	BFRP 8-8	2ø8	2ø8	0.27	
	BFRP10-10	2ø10	2ø10	0.42	
	BFRP 12-12	2ø12	2ø12	0.60	ø 8 @ 140 mm (Steel)
AFRP	AFRP 6-6	2ø6	2ø6	0.15	
	AFRP 8-8	2ø8	2ø8	0.27	
	AFRP 10-10	2ø10	2ø10	0.42	
	AFRP12-12	2ø12	2ø12	0.60	
GFRP	GFRP 6-6	2ø6	2ø6	0.15	
	GFRP 8-8	2ø8	2ø8	0.27	
	GFRP10-10	2ø10	2ø10	0.42	
	GFRP 12-12	2ø12	2ø12	0.60	
Steel	Steel 6-6	2ø6	2ø6	0.15	
	Steel 8-8	2ø8	2ø8	0.27	
	Steel 10-10	2ø10	2ø10	0.42	
	Steel 12-12	2ø12	2ø12	0.60	

3.1. Results of Simple Beam

The FE models for simple beams were executed. The maximum deflection was measured at the beams' mid span. The results of the load–deflection curves are depicted in Figure 5. The results show that increasing the reinforcement ratios for FRP bars reduce the load capacity due to the loss of bond strength. This can be attributed to the type of bar surface, the surface treatments, the cross section, position and diameter of the rebar, and the concrete strength. For the steel bars, increasing the reinforcement ratios increases the loading carrying capacity. The relationship between ultimate loads and reinforcement ratio is illustrated in Figure 6. It was observed that CFRP had a higher modulus of elasticity leading to lower deflections and better stiffness of beams. The beams reinforced with AFRP showed better ductility than GFRP and BFRP.

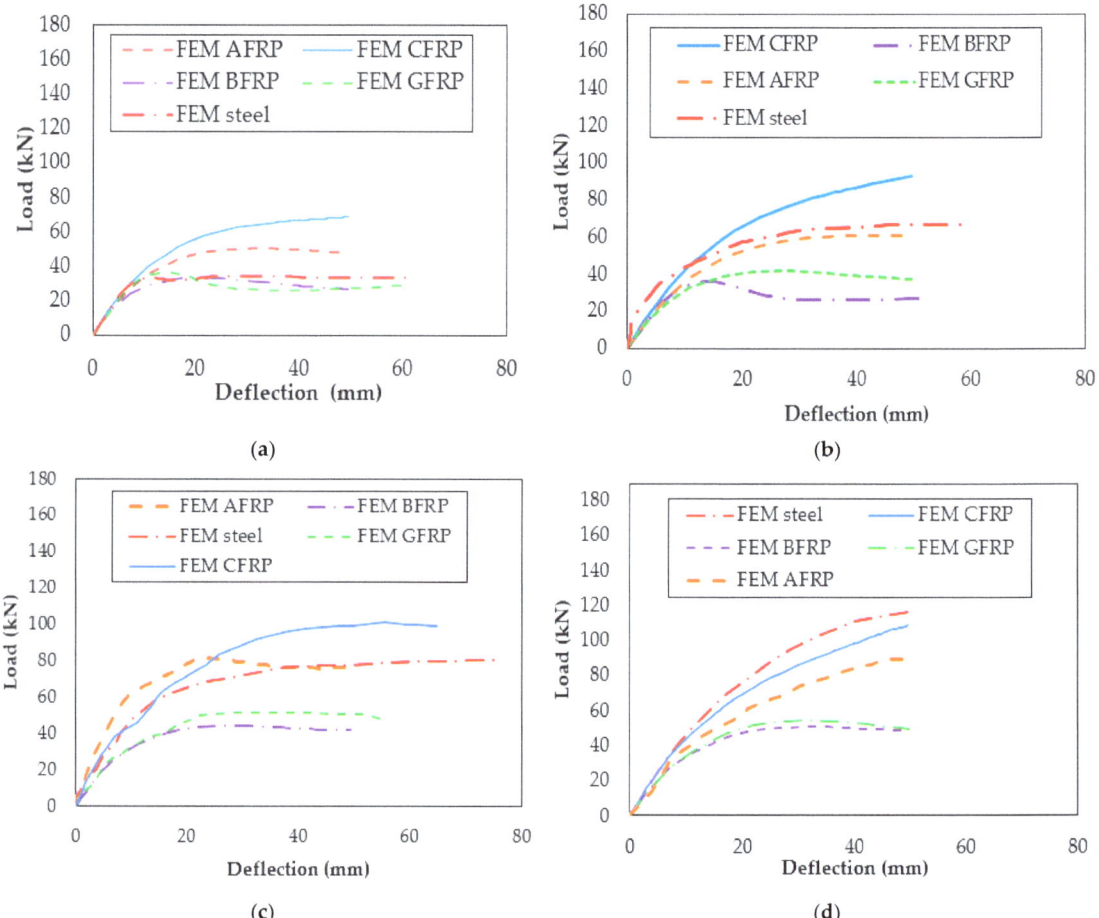

Figure 5. Load Deflection curves for simple beam with different reinforcement ratios: (**a**) 0.15%, (**b**) 0.27%, (**c**) 0.42%, and (**d**) 0.60%.

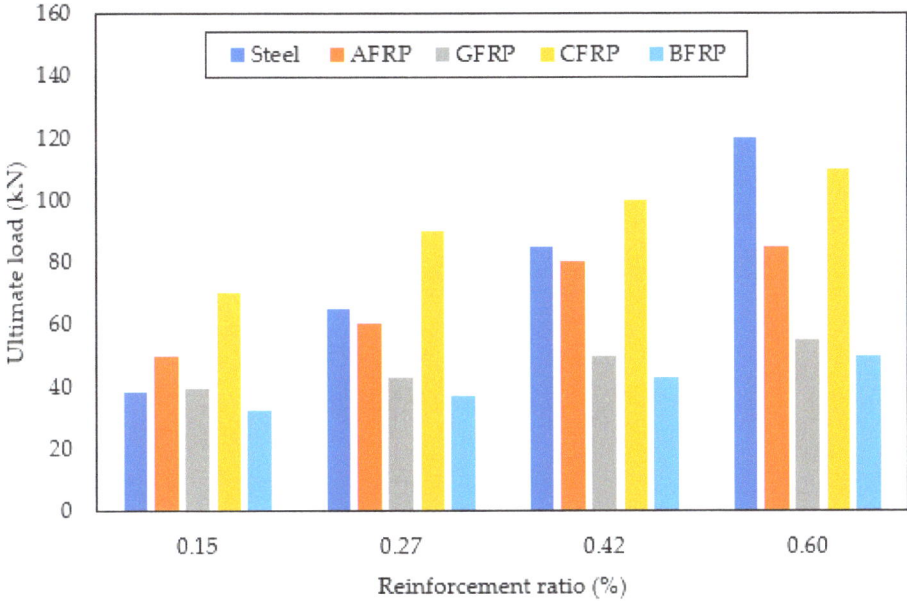

Figure 6. Ultimate load vs. reinforcement ratio.

All the beams showed the same trend for fracture behavior. The failure mode was tension failure as shown in Tables S1–S4 in the Supplementary Materials. Figure 7 shows the crack pattern for CFRP and Steel beams at 0.15% reinforcement ratio as a sample. It can be observed that the crack started at the bottom of the beams near to the maximum tension zone. The fracture energy (G_f) was calculated as follows in Equation (19).

$$G_f = \frac{W_o}{A_L} \tag{19}$$

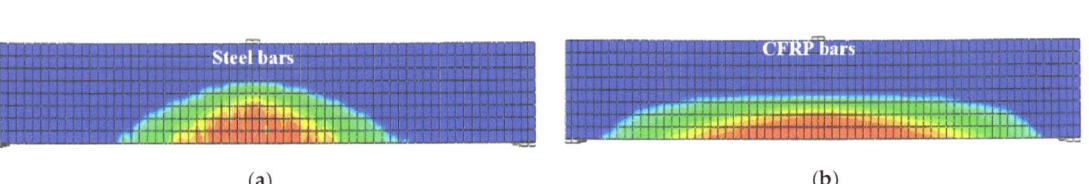

(a)   (b)

Figure 7. Crack Pattern at 0.15% reinforcement ratio (**a**) Steel and (**b**) CFRP beams.

The region under the entire load deflection diagram represents the energy that the beam will absorb during failure [35]. The results of the fracture energy (G_f) are shown in Tables S1–S4 in the Supplementary Materials. At 0.15%, the results show G_f of (70,686.7 N/m) for CFRP and (40,646.7 N/m) for steel and, at 0.60%, G_f was (100,713.3 N/m) for CFRP and (137,260 N/m) for steel as shown in Figure 8. This shows a better ductility of steel bars in higher reinforcement ratios due to the bond strength enhancement.

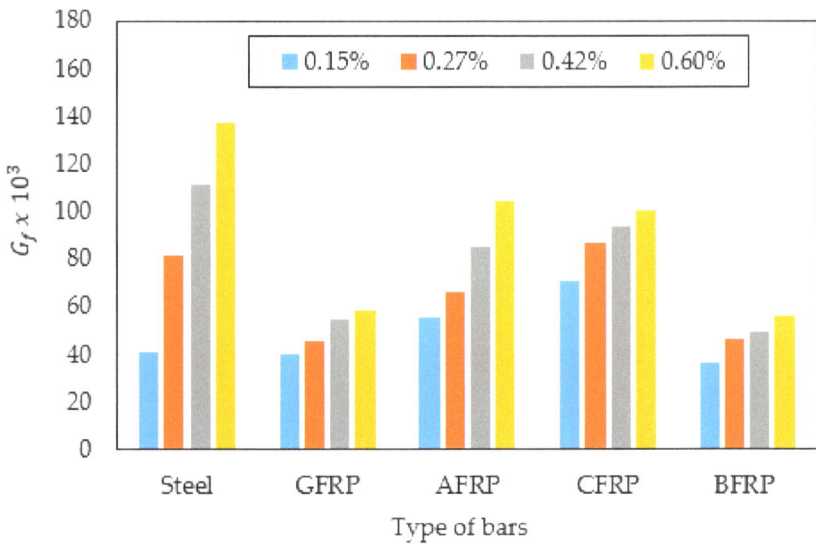

Figure 8. Fracture energy vs. Type of bars.

3.2. Results of Overhanging Beams

The models were executed and the maximum deflection was measured at the cantilever tip. The results of load deflection are illustrated in Figure 9. The reinforcement ratios had an effect on the stiffness of the beam. As the reinforcement ratio increases, the stiffness of the beam increases. As expected, greater deflections were obtained for the beams with lower reinforcement ratios. Furthermore, the deflections in the concrete beams reinforced with GFRP and BFRP in all the models were greater than those reinforced with steel, CFRP, and AFRP at the same load. This is due to the low stiffness and modulus of elasticity of the BFRP and GFRP bars. As a result, increasing the reinforcement ratio increases the ultimate load capacity and decreased the deflection. The relationship between the ultimate load and reinforcement ratio is illustrated in Figure 10. The figure shows that steel reaches a higher ultimate load in the larger reinforcement ratios than the FRP bar types. Moreover, CFRP shows better ductility than other FRP materials due to higher tensile strength. The beams reinforced with BFRP and GFRP had a low stiffness due to the low modulus of elasticity and, consequently, high deformations were obtained.

For the crack patterns, all the beams showed the same trend for fracture behavior. A tension failure was observed as shown in Table S5–S8 in the Supplementary Materials. At 0.15%, for steel and CFRP beams, the crack appeared at the top of the cantilever as shown in Figure 11 as a sample. The results of the fracture energy are shown in Table S5–S8 in the Supplementary Materials. It was found that the maximum load capacity is directly proportional to G_f. The results of the fracture energy were as shown in Figure 12. At 0.15%, the results of G_f were (39,133.3 N/m) for CFRP and (14,666 N/m) for steel and, at 0.60% G_f, were (77,740 N/m) for CFRP and (89,253.3 N/m) for steel. It was observed that, as the reinforcement ratio increases, the fracture energy increases.

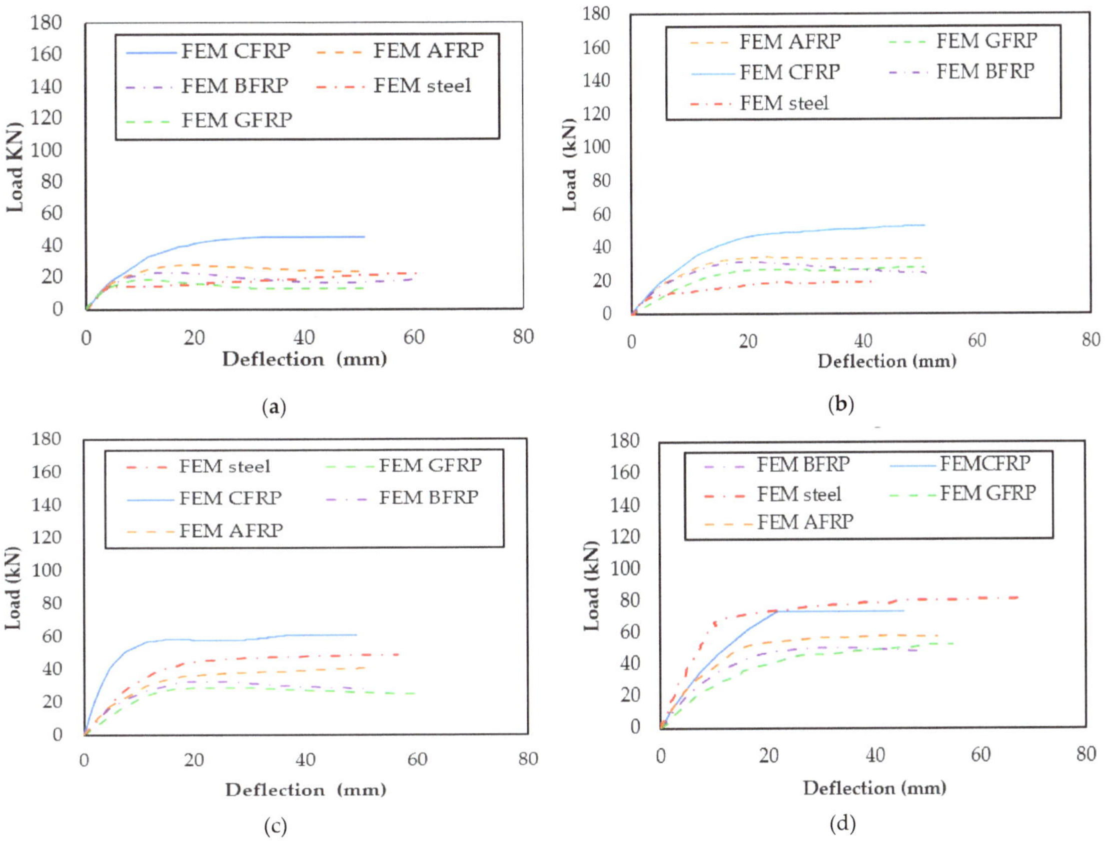

Figure 9. Load–deflection relationship for different reinforcement ratios: (**a**) 0.15%, (**b**) 0.27%, (**c**) 0.42%, and (**d**) 0.60%.

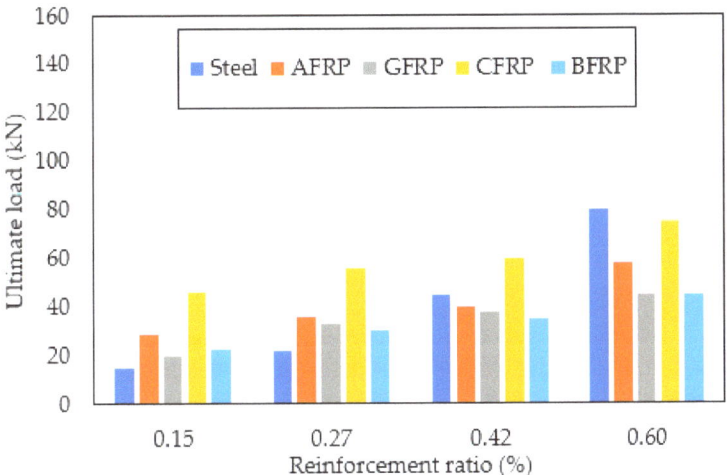

Figure 10. Load vs. Reinforcement ratio.

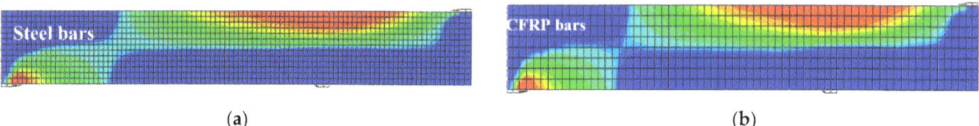

Figure 11. Crack pattern at 0.15% reinforcement ratio: (**a**) Steel and (**b**) CFRP beams.

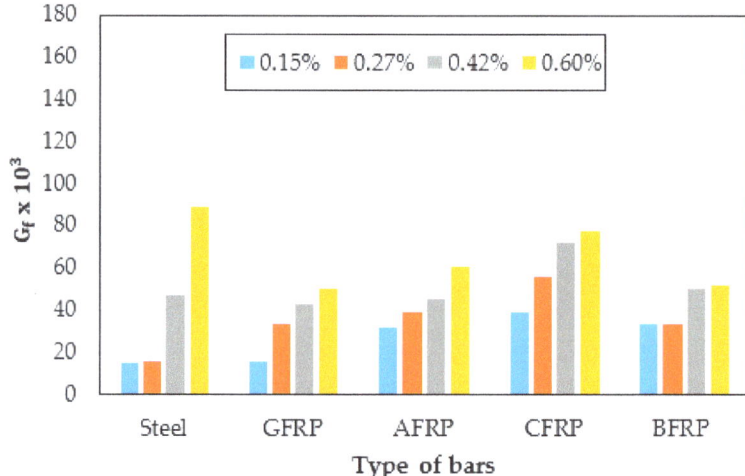

Figure 12. Fracture energy vs. Type of bars.

4. Conclusions

In this paper, FE models were developed to investigate the flexural behavior of steel- and FRP-reinforced concrete beams. Forty simply supported and overhanging beams were simulated using ABAQUS. CDP was employed to express the inelastic responses of concrete. The load–deflection curves, crack propagations, and fracture energy were attained for each beam. The findings showed that:

- The FEM results of overhanging beams were validated using four statistical indicators and they showed good agreement with the experimental results in the literature.
- The CFRP bars could withstand higher load than steel bars by 29% and 33% for simple and overhanging beams, respectively. Furthermore, CFRP could absorb greater fracture energy than steel by 22% and 40% for simple overhanging beams, respectively. Hence, CFRP can be an effective alternative to steel.
- As the reinforcement ratio increases, CFRP showed greater load carrying capacity than other FRP types. For simple beams, the load capacity for CFRP bars increased by 80% more than GFRP, 37.5% more than AFRP, and 120% more than BFRP. Similarly, for overhanging beams, CFRP had a 130% greater load than GFRP, 50% than AFRP, and 87.5% than BFRP. This enhancement in the load capacity is attributed to higher stiffness, tensile strength, and their modulus of elasticity compared to other FRP types.
- As the reinforcement ratio increases, the increase in the ultimate load capacity for GFRP and BFRP bars was insignificant due to the low modulus of elasticity compared to other FRP types.
- As the bar diameter increases, the bond strength for FRP bars decreases. Thus, all the FRP types could be considered alternatives to steel when low bars sizes are utilized.

The results obtained from the current study showed that CFRP is an adequate alternative to steel, especially at low reinforcement ratios. Interestingly, the stiffness of beams with CFRP bars was quite near to the reinforced concrete beams. The deflections of BFRP and GFRP beams were typically more significant due to the low elastic modulus and various bond properties. For future research, examining the bond behavior of FRP bars is recommended. Moreover, researchers could explore advanced methods to deal with the brittle behavior of FRP bars and examine the shear behavior of beams using various FRP types. Finally, we recommend including the cost estimation for concrete beams reinforced with FRP bars in future comparative studies.

Supplementary Materials: The following supporting information can be downloaded at: https://www.mdpi.com/article/10.3390/buildings13020374/s1. Table S1. Ultimate load, fracture energy and crack pattern for simple beams with ratio 0.15% reinforcement ratio. Table S2. Ultimate load, fracture energy and crack pattern for simple beams with 0.27% reinforcement ratio. Table S3. Ultimate load, fracture energy and crack pattern for simple beams with 0.42% reinforcement ratio. Table S4. Ultimate load, fracture energy and crack pattern for simple beams with 0.60% reinforcement ratio. Table S5. Ultimate load, fracture energy and crack pattern for overhanging beam with 0.15% reinforcement ratio. Table S6. Ultimate load, fracture energy and crack pattern for overhanging beam with 0.27% reinforcement ratio. Table S7. Ultimate load, fracture energy and crack pattern for overhanging beam with 0.42% reinforcement ratio. Table S8. Ultimate load, fracture energy and crack pattern for overhanging beam with 0.60% reinforcement ratio.

Author Contributions: Conceptualization, N.K.S.A., S.Y.M. and N.H.A.; methodology, N.K.S.A., S.Y.M. and N.H.A.; Software, N.K.S.A. and S.Y.M.; investigation, N.K.S.A., S.Y.M. and N.H.A.; data curation, N.K.S.A. and S.Y.M.; original draft preparation—writing, N.K.S.A., S.Y.M. and N.H.A.; review and editing—writing, N.K.S.A., S.Y.M. and N.H.A.; visualization, N.K.S.A., S.Y.M. and N.H.A.; supervision, N.H.A. and S.Y.M.; project administration, N.H.A. and S.Y.M.; funding acquisition, N.K.S.A. All authors have read and agreed to the published version of the manuscript.

Funding: This research received no external funding.

Institutional Review Board Statement: None applicable.

Informed Consent Statement: None applicable.

Data Availability Statement: Upon reasonable request, the FE data used to support the study's conclusions are available.

Acknowledgments: We acknowledge the technical support and encouragement of Eng. Ahmed Bahgat Tawfik and Eng. Mohammed Rady Ewis Deif.

Conflicts of Interest: The authors declare no conflict of interest.

References

1. Rady, M.; Mahfouz, S.Y. Effects of Concrete Grades and Column Spacings on the Optimal Design of Reinforced Concrete Buildings. *Materials* **2022**, *15*, 4290. [CrossRef] [PubMed]
2. Rady, M.; Mahfouz, S.Y.; Taher, S.E. Optimal Design of Reinforced Concrete Materials in Construction. *Materials* **2022**, *15*, 2625. [CrossRef]
3. Aidy, A.; Rady, M.; Mashhour, I.M.; Mahfouz, S.Y. Structural Design Optimization of Flat Slab Hospital Buildings Using Genetic Algorithms. *Buildings* **2022**, *12*, 2195. [CrossRef]
4. Benmokrane, B.; Theriault, M.; Masmoudi, R.; Rizkalla, S. Effect of Reinforcement Ratio on Concrete Members Reinforced with FRP Bars. In Proceedings of the International SAMPE Symposium and Exhibition, Anaheim, CA, USA, 4–8 May 1997; SAMPE. Diamond Bar, CA, USA, 1997; Volume 42, pp. 87–98.
5. Mustafa, S.A.A.; Hassan, H.A. Behavior of Concrete Beams Reinforced with Hybrid Steel and FRP Composites. *HBRC J.* **2018**, *14*, 300–308. [CrossRef]
6. Thamrin, R.; Kaku, T. Bond Behavior of CFRP Bars in Simply Supported Reinforced Concrete Beam with Hanging Region. *J. Compos. Constr.* **2007**, *11*, 129–137. [CrossRef]
7. Bank, L.C. *Composites for Construction: Structural Design with FRP Materials*; John Wiley & Sons: Hoboken, NJ, USA, 2006; p. 560.
8. Yue, Q.; Yang, Y. Overview of durability research on fiber reinforced composite strengthened structures. *J. Build. Struct.* **2009**, *30*, 8–15.
9. Elgabbas, F.M. Development and Structural Testing of New Basalt Fiber-Reinforced-Polymer (Bfrp) Bars in Rc Beams and Bridge-Deck Slabs. Ph.D. Thesis, Université de Sherbrooke, Sherbrooke, QC, Canada, 2016.
10. Panahi, M.; Izadinia, M. A Parametric Study on the Flexural Strengthening of Reinforced Concrete Beams with Near Surface Mounted FRP Bars. *Civ. Eng. J.* **2018**, *4*, 1917. [CrossRef]
11. Sun, W.; Chen, H.; Luo, X.; Qian, H. The effect of hybrid fibers and expansive agent on the shrinkage and permeability of high-performance concrete. *Cem. Concr. Res.* **2001**, *31*, 595–601. [CrossRef]

12. Chung, D.D.L. Cement reinforcedwithshort carbonfibers:a multifunctional material. *Compos. Part B Eng.* **2000**, *31*, 511–526. [CrossRef]
13. Kwon, H.; Ferron, R.P.; Akkaya, Y.; Shah, S.P. Cracking of fiber-reinforced self-compacting concrete due to restrained shrinkage. *Int. J. Concr. Struct. Mater.* **2007**, *1*, 3–9.
14. Tejaswini, T.; Raju, M.V.R. Analysis of RCC Beams Using ABAQUS. *Int. J. Innov. Eng. Technol. Anal.* **2015**, *5*, 248–255.
15. Mohammed, R.S.; Fangyuan, Z. Numerical Investigation of the Behavior of Reinforced Concrete Beam Reinforced with FRP Bars11. *Civ. Eng. J.* **2019**, *5*, 2296–2308. [CrossRef]
16. Al Hasani, S.; A., N.H.; Abdulraeg, A.A. Numerical Study of Reinforced Concrete Beam by Using ABAQUS Software. *Int. J. Innov. Technol. Interdiscip. Sci.* **2021**, *4*, 733–741.
17. Shirmardi, M.; Mohammadizadeh, M. Numerical Study on the Flexural Behavior of Concrete Beams Reinfroced by GFRP Bars. *J. Rehabil. Civ. Eng.* **2019**, *4*, 88–99. [CrossRef]
18. Shanour, A.S.; Adam, M.; Mahmoud, A.; Said, M. Experimental Investigation of Concrete Beams Reinforced with Gfrp Bars. *Int. J. Civ. Eng. Technol.* **2014**, *5*, 154–164.
19. Krasniqi, C.; Kabashi, N.; Krasniqi, E.; Kaqi, V. Comparison of the Behavior of GFRP Reinforced Concrete Beams with Conventional Steel Bars. *Pollack Period.* **2018**, *13*, 141–150. [CrossRef]
20. Sirimontree, S.; Keawsawasvong, S.; Thongchom, C. Flexural Behavior of Concrete Beam Reinforced with GFRP Bars Compared to Concrete Beam Reinforced with Conventional Steel Reinforcements. *J. Appl. Sci. Eng.* **2021**, *24*, 883–890. [CrossRef]
21. Arivalagan, S. Engineering Performance of Concrete Beams Reinforced with GFRP Bars and Stainless Steel. *Glob. J. Res. Eng.* **2012**, *12*, 35–40.
22. Saraswathy, T.; Dhanalakshmi, K. Investigation of Flexural Behaviour of RCC Beams Using GFRP Bars. *Int. J. Sci. Eng. Res.* **2014**, *5*, 333–338.
23. Murugan, R.; Kumaran, G. Experiment on Rc Beams Reinforced with Glass Fibre Reinforced Polymer Reinforcements. *Int. J. Innov. Technol. Explor. Eng.* **2019**, *8*, 3541. [CrossRef]
24. Hamdy, G.A.; Arafa, D.F. Experimental Investigation Of Concrete Beams Reinforced With Basalt FRP Bars. *J. Eng.* **2021**, *4*, 154–167. [CrossRef]
25. Zhang, L.; Sun, Y.; Xiong, W. Experimental study on the flexural deflections of concrete beam reinforced with Basalt FRP bars. *Mater. Struct. Mater. Constr.* **2015**, *48*, 3279–3293. [CrossRef]
26. Ashour, A.F.; Habeeb, M.N. Continuous concrete beams reinforced with CFRP bars. *Struct. Build.* **2008**, *161*, 349–357. [CrossRef]
27. Mashrei, M.A.; Makki, J.S.; Sultan, A.A. Flexural Strengthening of Reinforced Concrete Beams Using Carbon Fiber Reinforced Polymer (CFRP) Sheets with Grooves. *Lat. Am. J. Solids Struct.* **2019**, *16*, 1–13. [CrossRef]
28. Ahmed, E.; Sobuz, H.R.; Sutan, N.M. Flexural Performance of CFRP Strengthened RC Beams with Different Degrees of Strengthening Schemes. *Int. J. Phys. Sci.* **2011**, *6*, 2229–2238.
29. Soudki, K.; El-Salakawy, E.; Craig, B. Behavior of CFRP Strengthened Reinforced Concrete Beams in Corrosive Environment. *J. Compos. Constr.* **2007**, *11*, 291–298. [CrossRef]
30. Esfahani, M.R.; Kianoush, M.R.; Tajari, A.R. Flexural behaviour of reinforced concrete beams strengthened by CFRP sheets. *Eng. Struct.* **2007**, *29*, 2428–2444. [CrossRef]
31. Issa, M.S.; Elzeiny, S.M. Flexural Behavior of Cantilever Concrete Beams Reinforced with Glass Fiber Reinforced Polymers (GFRP) Bars. *J. Civ. Eng. Constr. Technol.* **2011**, *2*, 33–44.
32. Hsu, L.S.; Hsu, C.-T. Complete stress—Strain behavior of high-strength concrete under compression. *Mag. Concr. Res.* **1994**, *46*, 301–312. [CrossRef]
33. Nayal, R.; Rasheed, H.A. Tension Stiffening Model for Concrete Beams Reinforced with Steel and FRP Bars. *J. Mater. Civ. Eng.* **2006**, *18*, 831–841. [CrossRef]
34. Gilbert, R.; Warner, R. Tension stiffening in reinforced concrete slabs. *J. Struct. Div.* **1978**, *104*, 1885–1900. [CrossRef]
35. Tawfik, A.B.; Mahfouz, S.Y.; Taher, S.E.-D.F. Nonlinear ABAQUS Simulations for Notched Concrete Beams. *Materials* **2021**, *14*, 7349. [CrossRef] [PubMed]
36. Abdelwahed, B.S.; Kaloop, M.R.; El-Demerdash, W.E.; Parisi, F. Nonlinear Numerical Assessment of Exterior Beam-Column Connections with Low-Strength Concrete. *Buildings* **2021**, *11*, 562. [CrossRef]
37. Kobraei, M.; Jumaat, M.Z.; Shafigh, P. An Experimental Study on Shear Reinforcement in RC Beams Using CFRP-Bars. *Sci. Res. Essays* **2011**, *6*, 3447–3460. [CrossRef]
38. Abbood, I.S.; Odaa, S.A.; Hasan, K.F.; Jasim, M.A. Properties Evaluation of Fiber Reinforced Polymers and Their Constituent Materials Used in Structures—A Review. *Mater. Today Proc.* **2021**, *43*, 1003–1008. [CrossRef]
39. Ahmed, A. Modeling of a Reinforced Concrete Beam Subjected to Impact Vibration Using ABAQUS. *Int. J. Civ. Struct. Eng.* **2014**, *4*, 227–236. [CrossRef]
40. Cosenza, E.; Manfredi, G.; Realfonzo, R. Bond characteristics and anchorage length of FRP rebars. In Proceedings of the 2nd International Conference on Advanced Composite Materials in Bridges and Structures (ACMBS-II), Montreal, QC, Canada, 11–14 August 1996.

Disclaimer/Publisher's Note: The statements, opinions and data contained in all publications are solely those of the individual author(s) and contributor(s) and not of MDPI and/or the editor(s). MDPI and/or the editor(s) disclaim responsibility for any injury to people or property resulting from any ideas, methods, instructions or products referred to in the content.

Article

Properties of Concrete Columns Strengthened by CFRP-UHPC under Axial Compression

Bo Wang [1,2,*], Gejia Liu [1] and Jiayu Zhou [1]

[1] School of Civil Engineering, Jilin Jianzhu University, Changchun 130118, China
[2] Jilin Structural and Earthquake Resistance Technology Innovation Centre, Changchun 130118, China
* Correspondence: bo_wang@126.com

Abstract: Ultra-high-performance concrete (UHPC) is a kind of structural material with ultra-high strength, extremely low porosity, and excellent durability, which has extremely broad application prospects. In order to promote the application of UHPC constrained by carbon fiber-reinforced polymer (CFRP) sheets as strengthening material in practical engineering, a total of nine specimens were designed, and two kinds of UHPC strengthening layer thickness (35 mm and 45 mm, respectively) were designed. By changing the constraint form of the UHPC strengthening layer (longitudinal reinforcements and ordinary stirrups, longitudinal reinforcements and spiral stirrups, and CFRP sheets, respectively), the axial compression performance of the strengthened column was explored. The study shows that compared with the without strengthened column, the uplift of carrying capacity of the strengthened test column is 277–561%. The reinforcement form of the strengthening layer has little influence on the lifting capacity. Among the three different strengthening methods, the wrapped CFRP has the best improvement effect on carrying capacity and ductility, followed by longitudinal reinforcements and spiral stirrups in the strengthening layer. With the increase of CFRP layers from two to five, the maximum carrying capacity increases by 21.3%. The carrying capacity of three different types of UHPC-strengthened columns is theoretically calculated, and the accuracy of the theoretical calculation method is verified by comparing the test value with the theoretical value, which provides a theoretical basis for the application of UHPC-strengthened columns in the future.

Keywords: ultra-high-performance concrete (UHPC); carbon fiber-reinforced polymer (CFRP); concrete column; axial compression performance

1. Introduction

Ultra-high-performance concrete (UHPC) is a fiber-reinforced cement-based composite with ultra-high impermeability and mechanical properties, which is based on micro-scale optimization of fine aggregate and ultra-fine aggregate (silica fume and sand), mixed with high-efficiency superplasticizer to reduce the water-cement ratio, and mixed with high-strength steel fiber [1]. As a new type of cement-based structural engineering material, it has ultra-high strength and toughness and can be used to construct new lightweight and high-strength structures. Good durability makes the UHPC structure have a longer life and lower maintenance costs [2].

At present, research on reinforced concrete members strengthened by UHPC has been carried out [3], Tanarslan et al. [4] enhanced the interfacial bonding ability between prefabricated UHPC thin layer and reinforced concrete beams by implanted bars, epoxy resin interfacial agent, and compression. Al-Osta et al. [5] carried out research on UHPC strengthening of reinforced concrete beams, considering the influence of different interface treatment methods and different strengthening positions on the beam strengthening effect. Using UHPC material as the strengthening layer of the ordinary concrete short column can improve the ultimate compressive capacity and ductility of the short column. The test results of Alsomiri and Xie et al. [6,7] show that increasing the thickness of the UHPC

strengthening layer can significantly improve the compressive capacity of the cylinder, which provides a theoretical basis and reference for practical strengthening engineering.

UHPC usually exhibits weak ductility under axial compression, which can be improved by fiber-reinforced polymer confinement. Zeng [8–10] proposed using fiber-reinforced polymer (FRP) grids as the reinforcement material of UHPC and embedding FRP grids in UHPC to improve its strength and ductility. The experimental results show that the confinement of FRP grids improves the strength and ductility of UHPC. Carbon fiber-reinforced polymer (CFRP) sheets have the advantages of high strength and durability and are lighter than other strengthening materials. Many scholars have studied the mechanical properties, thermal properties, and alkaline resistance of CFRP materials [11,12]. Guo et al. [13] performed accelerated aging experiments on carbon fiber-/glass fiber-reinforced hybrid rods. The water absorption and diffusion behavior of the fiber matrix, the change of interfacial strength, and dynamic thermodynamic properties were investigated. The long-term life evaluation shows that the interface shear strength of the hybrid rod shell has a fast degradation rate and reaches a stable level of 62%. The advantages of CFRP are also reflected in the seismic restoration and reinforcement of existing reinforced concrete structures [14]. Many scholars have studied the compressive properties of FRP-constrained UHPC [15–17]; the tests show that the ultimate strength and strain of UHPC columns are significantly improved by increasing the thickness of the FRP constraint layer, showing better ductility. Huang [18] conducted experiments on FRP-constrained UHPC columns under cyclic axial compression and proposed its stress-strain model. Abadel et al. [19] explored the influence of CFRP sheets strengthening on the compressive strength of UHPC at 400 °C. After CFRP sheets material strengthening, the compressive strength has been improved. The model of FRP-constrained concrete column to predict the axial stress-strain and ultimate strength of FRP-constrained UHPC column is not accurate [20–24] due to the limited test results, and a large number of test studies are needed. The study on the strengthening of UHPC with FRP sheets is of great significance to the engineering application of UHPC materials.

Some scholars have studied the axial compression performance of UHPC columns by combining FRP sheets and stirrups. Chang et al. [25,26] used high-strength spiral stirrups constrained UHPC and stirrups and CFRP sheets constrained UHPC to carry out axial compression tests. With the increase of stirrups volume ratio, the ductility of UHPC was improved, and the prediction model of transverse strain and axial strain curves of UHPC constrained by CFRP sheets and stirrups was established. Lu [27] conducted the axial compression test on reinforced concrete columns strengthened with fiber high-performance concrete by using the enlarged section method. The effects of the thickness of the strengthening layer, the stirrups in the strengthening layer, and the stirrup spacing on the axial compression performance of the FRP-strengthened columns were evaluated. Based on the confined concrete theory, a theoretical calculation model was established, and the axial carrying capacity of the column was discussed. The results showed that the carrying capacity and ductility of the column can be improved by adding stirrups in the strengthening layer.

In conclusion, UHPC, as a new type of composite material, has been widely recognized for its superior mechanical properties in the practical application of civil engineering. The performance and durability of reinforced concrete members strengthened UHPC are better than those reinforced concrete members, so it has better engineering application value to strengthen concrete members by UHPC. Due to the limitation of the test equipment and the test site, to avoid the stress concentration phenomenon of the square column, the shortened cylindrical column is used as the research object. Whether the relevant conclusions can be applied to long columns, remain to be studied. In order to further explore the compressive performance of the UHPC strengthening layer, this paper uses CFRP sheets to wrap the UHPC strengthening layer, adds longitudinal reinforcements and ordinary stirrups to the UHPC strengthening layer, and adds longitudinal reinforcements and spiral stirrups to the UHPC strengthening layer. In this paper, a total of nine specimens were designed to study the effects of the thickness of the UHPC strengthening layer, strengthening layer reinforcement form, and confinement form on the axial compression performance of the strengthening column, providing a theoretical basis for the application of UHPC as strengthening material in engineering.

2. Test Program

2.1. Design of Test

Nine reinforced concrete columns were made; the height of the specimen column was 300 mm, the diameter was 150 mm, the concrete strength grade was C35, the longitudinal reinforcement was configured as 6Φ12 mm, the stirrup was configured as Φ8@60 mm, and the thickness of the concrete cover was 15 mm. The parameters of the specimens are shown in Table 1.

Table 1. Test column parameters.

Number	Strengthening Layer Confinement Form	Thickness of Strengthening Layer (mm)	Strengthening Layer Longitudinal Reinforcements Arrangement Form	Strengthening Layer Stirrups Arrangement Form
Z1	-	-	-	-
Z2	longitudinal reinforcements and ordinary stirrups	35	6Φ12	Φ8@40
Z3	longitudinal reinforcements and ordinary stirrups	45	6Φ12	Φ8@40
Z4	longitudinal reinforcements and spiral stirrups	35	6Φ12	Φ8@40
Z5	longitudinal reinforcements and spiral stirrups	45	6Φ12	Φ8@40
Z6	2 layers of CFRP	35	-	-
Z7	5 layers of CFRP	35	-	-
Z8	2 layers of CFRP	45	-	-
Z9	5 layers of CFRP	45	-	-

For each cylinder, 6Φ12 steel bars 270 mm in length were first cut and bound for stirrups. After the steel cage was made, the steel cage was put into the prefabricated mold with an inner diameter of 150 mm and a height of 300 mm. Then concrete was poured, and six C35 concrete cube test blocks (150 mm × 150 mm × 150 mm) were reserved, which were naturally cured to maturity with the specimens. After the column was cured, according to the code for the design of strengthening concrete structure, the column was chipped with a chipping machine, and the surface was cleaned. The column, after chipping, was connected with the longitudinal reinforcement of the strengthening layer through the anchor bar, and then it was put into the prefabricated strengthening layer mold. After removing the mold, it can be seen that the interface between the UHPC strengthening layer and the new and old concrete of the strengthening column was well combined, and the appearance of the UHPC strengthening layer was smooth and dense. The UHPC surface was polished after the strengthening column specimens were cured to the age, the impregnating glue was evenly applied to the CFRP sheets by wet adhesive method, and the CFRP sheets were seamlessly wound around the surface of the strengthening column.

2.2. Details of Material Mechanical Properties

The concrete design strength grade was C35, and the mix design and preparation are shown in Table 2. Six standard cube test blocks (150 mm × 150 mm × 150 mm) were reserved during concrete pouring, and their compressive strength was measured after 28d of laboratory curing together with the specimens. Their mechanical properties are shown in Table 3. Table 4 shows the UHPC mix design and preparation, and Table 5 shows its mechanical properties. Table 6 shows the mechanical properties of steel bars, HPB300 strength grades stirrup, and HRB400 strength grades longitudinal reinforcement. Table 7 shows the mechanical properties of CFRP sheets. Table 8 shows the mechanical properties of CFRP sheets impregnating glue.

Table 2. Mixture proportions of concrete (kg/m^3).

Strength Class	Cement	Water	Sand	Coarse Aggregate	Superplasticizer
C35	380	180	648	1198	2.8

Table 3. Mechanical properties of concrete.

Strength Class	f_{cu} (MPa) [1]	f_c (MPa) [1]	f_t (MPa) [1]	E_c (N/mm^2) [1]
C35	37.36	28.4	2.891	31946

[1] The cubic compressive strength f_{cu} is the measured value. $f_c, f_t,$ and E_c are calculated according to the literature [28].

Table 4. Mixture proportions of UHPC (kg/m^3).

Cement	Fine Sand	Silica Fume, Fly Ash	Additive	Water	Steel Fiber
1025	700	315	8	226	202

Table 5. Mechanical properties of UHPC.

Number	1	2	3	Average
Cube compressive strength f_{cu} (MPa)	125.88	110.8	123.28	124
Axial compressive strength f_c [1] (MPa)	110.8	108.2	108.5	109.1

[1] According to the literature [29], the axial compressive strength of UHPC is $f_c = 0.88 f_{cu}$.

Table 6. Mechanical properties of steel bars.

Reinforced Type	Diameter (mm)	Yield Strength (MPa)	Tensile Strength (MPa)
HRB400	12	468	630
HPB300	8	323	453

Table 7. CFRP sheet parameters.

Name	Type	Density (g/cm^3)	Thickness (mm)	Tensile Strength (MPa)	Elastic Modulus (GPa)
UT70-20G	Unilateral high-strength fabric	1.8	0.111	3400	245

Table 8. Mechanical parameters of adhesive.

Name	Tensile Strength (MPa)	Elastic Modulus (MPa)	Bending Strength (MPa)	Compressive Strength (MPa)	Ultimate Elongation (%)
Sikadur®-330CN	≥40	≥3000	≥60	≥70	≥1.5

2.3. Loading Device and Measuring Point Layout

2.3.1. Test Setup

It was loaded monotonically at a rate of 0.3 mm/min. Before loading, the end of the specimen was polished by an angle grinder and made level with fine sand to ensure that the whole section of the specimen was compressed. The specimens were preloaded, and the preloaded load value was 10% of the calculated ultimate carrying capacity. Checked whether the strain box, testing machine, and other equipment were working normally before starting the formal test. Figure 1 shows the test loading device.

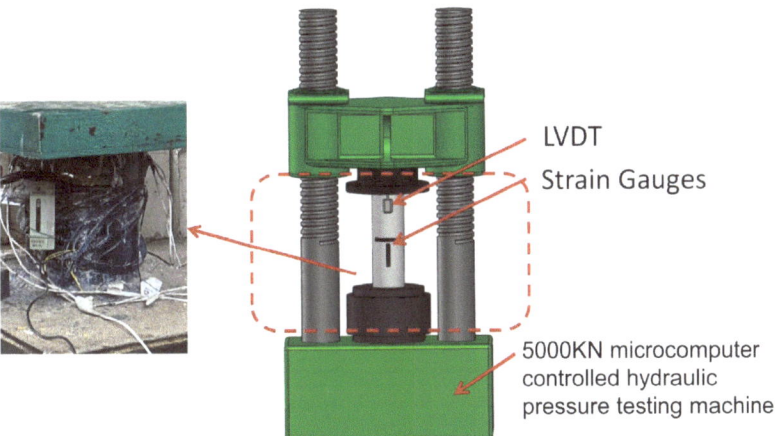

Figure 1. Setup of the test.

2.3.2. Loading Device and Measuring Point Layout

Longitudinal strain gauges and hoop strain gauges were pasted on two symmetrical sides in the middle of the concrete; longitudinal strain gauges and hoop strain gauges were pasted on two symmetrical sides of the middle part of the strengthening layer of UHPC; three longitudinal reinforcements were spaced apart, and strain gauges were pasted in the middle; two strain gauges were symmetrically pasted on the stirrups. Figure 2 shows the position of strain gauges.

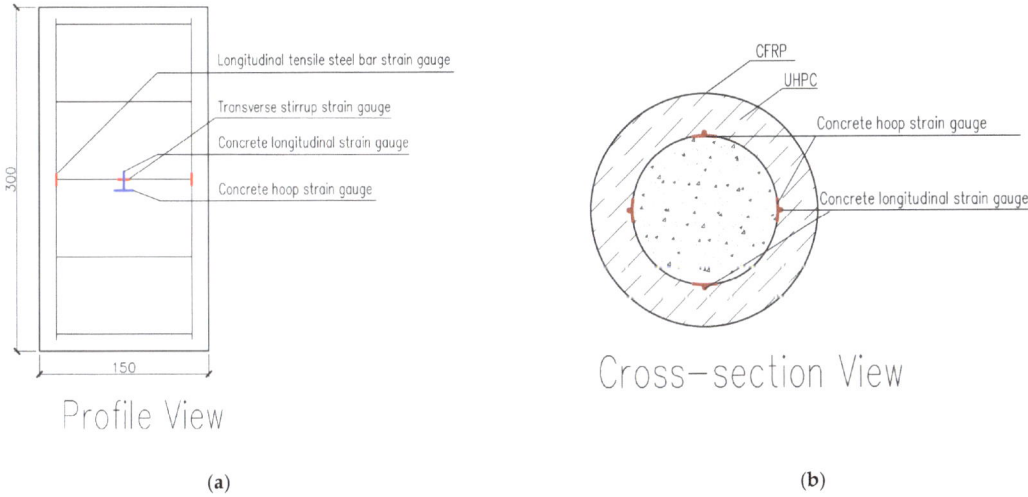

Figure 2. Location of the strain gauge. (**a**) Profile view. (**b**) Cross-section view.

3. Test Result and Analysis

3.1. Test Observations

Z1 is the without-strengthening column, and the strain of the whole section was uniformly distributed under the axial load. At the initial stage of loading, the steel bars and concrete were in the elastic stage. In this stage, the compression deformation of the test column increased with the increase of the load, and the load of the steel bars and concrete was proportional to the displacement. When the load continued to increase, the concrete began to appear in small vertical cracks. When approaching the failure load, the cracks gradually extended, and obvious longitudinal penetrating cracks appeared. The concrete fell off, and the longitudinal reinforcements and stirrups were deformed. The failure form is shown in Figure 3.

Figure 3. Z1, Z2, and Z3 failure forms.

The UHPC-strengthened columns with longitudinal reinforcements and ordinary circular stirrups in the strengthening layers of Z2 and Z3 had the same elastic stage in the initial loading period as Z1, and the load had a linear relationship with the axial deformation. The load continued to increase, and the increased rate of axial deformation exceeded the increased rate of load. When the load was close to the peak, the sound of steel fiber pulling out in UHPC could be heard. When the load reached the peak load, the load did not drop precipitously but showed a slow decline. After the longitudinal crack was connected, it could still bear a large load. Finally, the crack of the UHPC strengthening layer was connected, and the failure form is shown in Figure 3.

Z4–Z5 are UHPC-strengthened columns equipped with longitudinal reinforcements and spiral stirrups in the strengthening layer, and the test phenomenon is similar to that of Z2 and Z3. When the confinement capacity of spiral stirrups completely lost its effect, the load of the test column decreased significantly, and the failure form is shown in Figure 4.

Z6–Z9 are UHPC-strengthened columns with CFRP sheets, which are in the elastic stage in the early stage of the test. With the gradual increase of load, the surface of CFRP sheets did not change, and the sound of concrete cracking was heard, and the sound of steel fiber pulling out was always heard. The crack of the UHPC strengthening layer made a sound, and the slope of the load-displacement curve decreased, which was still linear. The original pictures and failure forms of CFRP-constrained strengthening columns are shown in Figure 5.

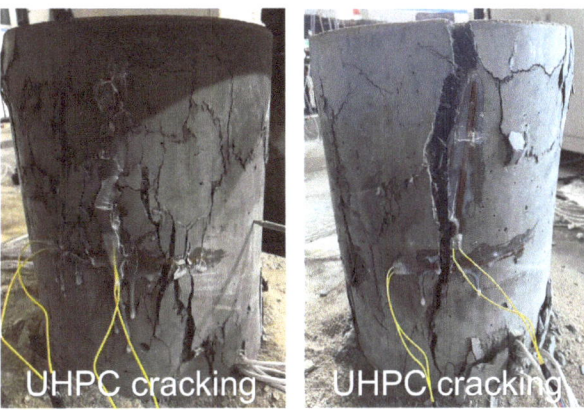

Figure 4. Z4 and Z5 failure forms.

Figure 5. Z6–Z9 original pictures and failure forms. (**a**) Original pictures. (**b**) Failure forms.

3.2. Ultimate Carrying Capacity

The carrying capacity of the test columns with the three strengthening methods was greatly improved compared with that of the without strengthening column. Table 9 and Figure 6 show the ultimate carrying capacity of each test column. According to the different strengthened methods, the three strengthened columns were compared with the without strengthened columns, and the three different strengthening methods were compared.

Table 9. Carrying capacity of test columns.

Number	Test Carrying Capacity N_u (kN)	Calculate Carrying Capacity N_{th} (kN)	$\eta_1 = (N_u - N_1)/N_1$	$\eta = N_{th}/N_u$	Index of Ductility (DI)
Z1	686.12	748.78	-	1.09	1.00
Z2	2583.73	2787.24	2.77	1.08	1.09
Z3	3167.48	3417.00	3.62	1.08	1.32
Z4	2714.05	2947.67	2.96	1.09	1.10
Z5	3259.46	3538.62	3.75	1.09	1.30
Z6	3278.23	3345.73	3.78	1.02	1.00
Z7	3976.31	3720.59	4.80	0.94	1.66
Z8	4089.81	4141.86	4.96	1.01	1.08
Z9	4533.41	4494.80	5.61	0.99	1.33

Note: η_1 represents the increase in the carrying capacity of the strengthened column Z2–Z9 relative to the without-strengthened column Z1. η is the ratio of the theoretical and test carrying capacity of the test column.

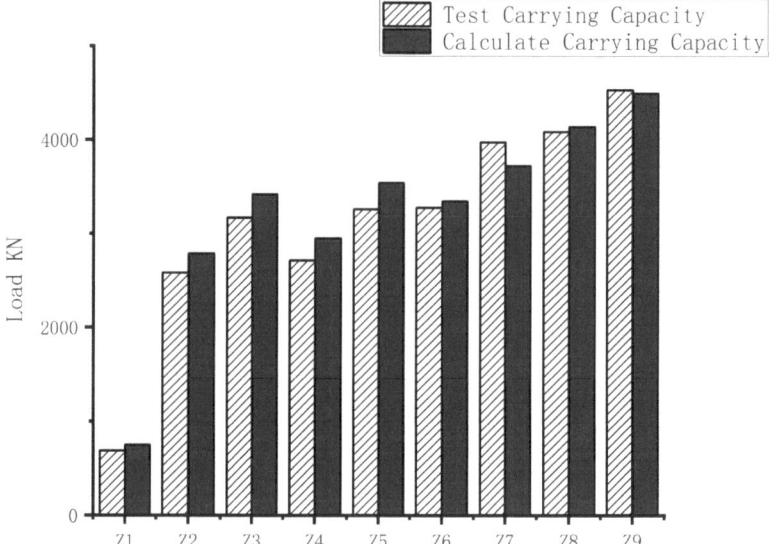

Figure 6. Ultimate carrying capacity of test columns.

Compared with the without-strengthening column, the carrying capacity of the test column was greatly improved by using UHPC strengthening, and the carrying capacity increased more with the increase of the thickness of the strengthening layer. The strengthening method of Z2 and Z3 is the UHPC column with longitudinal reinforcements and ordinary circular stirrups in the strengthening layer. With the increase of the thickness of the strengthening layer, the lifting range of the carrying capacity of the strengthening column also increased. Compared with Z1, the carrying capacity increased by 277% and 362%, respectively. For UHPC columns in Z4 and Z5 with longitudinal reinforcements and spiral stirrups in the strengthening layer, when the thickness of the strengthening layer increased, the carrying capacity of the strengthened column also increased. Compared with Z1, the carrying capacity of Z4 and Z5 increased by 296% and 375%, respectively.

The longitudinal reinforcements and circular stirrups were used in the strengthening layer, and the carrying capacity of Z3 with a thickness of 45 mm increased by 22.6% compared with that of Z2 with a thickness of 35 mm. The carrying capacity of Z5 with a thickness of 45 mm strengthened by longitudinal reinforcements and spiral stirrups was 20.1% higher than that of Z4 with a thickness of 35 mm strengthening layer. When the strengthening layer is thick, the strengthening efficiency can be improved by changing the reinforcement form of the strengthening layer. Under the condition that the thickness of the strengthening layer was the same and the stirrup spacing of the strengthening layer was the same, when the thickness of the strengthening layer was 35 mm, the carrying capacity of Z4 of the strengthening layer with longitudinal reinforcements and spiral stirrups was 5% higher than Z2 of the strengthening layer with longitudinal reinforcements and circular stirrups. When the thickness of the strengthening layer was 45 mm, the carrying capacity of Z5 with longitudinal reinforcements and spiral stirrups in the strengthening layer was 2.9% higher than that of Z3 with longitudinal reinforcements and circular stirrups in the strengthening layer.

By comparing the carrying capacity of three different strengthened column confinement forms, when the thickness of the strengthening layer was 35 mm, the Z6 carrying capacity of the strengthening layer with two layers of CFRP sheets was 26.9% higher than the Z2 of the strengthening layer with longitudinal reinforcements and circular stirrups, and the Z7 carrying capacity of the strengthening layer with five layers of CFRP sheets was 53.9% higher than Z2 of the strengthening layer with longitudinal reinforcements and circular stirrups. When the thickness of the strengthening layer was 45 mm, the Z8 of the strengthening layer with two layers of CFRP sheets was 29.1% higher than the Z3 carrying capacity of the strengthening layer with longitudinal reinforcements and circular stirrups, and the Z9 of the strengthening layer with five layers of CFRP sheets was 43.1% higher than the Z3 carrying capacity of the strengthening layer with longitudinal reinforcements and circular stirrups. When the thickness of the strengthening layer was the same, the maximum increase in the carrying capacity of the wrapped CFRP sheets was 53.9% among the three strengthening methods.

For the CFRP-constrained UHPC strengthening column, Z6 and Z7 were strengthened with CFRP sheets on the basis of 35 mm UHPC strengthening layers. Z8 and Z9 were strengthened with CFRP sheets based on a 45 mm UHPC strengthening layer. Compared with the without strengthened column Z1, the carrying capacity of Z6, Z7, Z8 and Z9 increased by 378%, 480%, 496% and 561%, respectively. The thickness of the strengthening layer of Z6 and Z7, Z8 and Z9 were 35 mm and 45 mm, respectively. With the increase of the thickness of the strengthening layer, the carrying capacity of Z8 with the thickness of the strengthening layer of 45 mm was 24.8% higher than that of Z6 with the thickness of the strengthening layer of 35 mm. Using five layers of CFRP sheets to reinforce the specimen, the carrying capacity of Z9 with a thickness of 45 mm was 14.0% higher than that of Z7 with a thickness of 35 mm. The maximum carrying capacity increased by 24.8% with the increase in strengthening layer thickness. The carrying capacity can be greatly improved by using CFRP constraints. With the increase of CFRP layers from two to five layers, the carrying capacity of Z7 was 21.3% higher than that of Z6, and that of Z9 was 10.8% higher than that of Z8.

All three strengthening methods can greatly improve the carrying capacity of the column. The optimal carrying capacity of the column is strengthened by UHPC constrained by wrapped CFRP. The carrying capacity of the column strengthened by longitudinal reinforcements and spiral stirrups in the UHPC strengthening layer is the second, and the carrying capacity of the column strengthened by longitudinal reinforcements and ordinary circular stirrups in the UHPC strengthening layer is the worst.

The axial force of ordinary reinforced concrete columns is borne by both steel bars and concrete, as shown in Equation (1).

$$N = f_{co}A_{co} + f_y A_s \qquad (1)$$

f_{co} is the axial compression strength of concrete; A_{co} is the section area of concrete; f_y is the compression strength of longitudinal reinforcement; A_s is the cross-section area of longitudinal reinforcement.

According to the literature [30], the axial compressive carrying capacity of UHPC strengthened column with longitudinal reinforcements and ordinary stirrups are shown in Equation (2).

$$N = f_{co}A_{co} + f_y A_s + \alpha_{cs}(f_c A_c + f'_y A'_s) \tag{2}$$

α_{cs} is the correction coefficient considering the degree of strength utilization of newly added concrete and steel bars; f_c is the axial compression strength of UHPC; A_c is the section area of UHPC; f'_y is the compression strength of the longitudinal reinforcement in the strengthening layer; A'_s is the cross-section area of longitudinal reinforcement in the strengthening layer.

According to the literature [31], the calculation formula of carrying capacity of UHPC strengthened column with longitudinal reinforcements and spiral stirrup are shown in Equations (3) and (4).

$$N = f_{co}A_{co} + f_y A_s + \alpha_{cs}(f_c A_c + 2\alpha f_{y,s} A_{ss0} + f'_y A'_s) \tag{3}$$

$$A_{ss0} = \frac{\pi d_{cor} A_{ss1}}{s} \tag{4}$$

α is the reduction coefficient of the constraint of a spiral stirrup on concrete; A_{ss0} is the conversion area of spiral stirrups; d_{cor} is the diameter of the core area, determined according to the inner surface of the spiral stirrup; A_{ss1} is the section area of a single screw stirrup; s is the spacing of spiral stirrups; $f_{y,s}$ is the compression strength of spiral stirrups.

There are few research results on the axial compression test of CFRP-constrained UHPC concrete columns. Due to the strain difference between UHPC and concrete being very small, the confining pressure effect of UHPC on concrete is not considered in the calculation of the carrying capacity, but only the confining effect of CFRP sheets on UHPC and concrete is considered. The calculation of the carrying capacity of the CFRP-constrained UHPC column is given in reference [32], and CFRP constrained concrete column is given in reference [33]. Equations (5) and (6) are used to calculate the confinement stress of the CFRP sheets. The calculation formula of confinement stress f'_l is shown in Figure 7 and Equation (7).

$$f_{cc} = f_{co} + 3.3 f'_l \tag{5}$$

$$f'_{cc} = f'_{co} + 2.53 f'^{0.32}_l \tag{6}$$

$$f'_l = \frac{2 E_{FRP} \varepsilon_{h,rup} t_{frp}}{d'_{cor}} \tag{7}$$

$$\varepsilon_{h,rup} = 0.586 \varepsilon_{frp} \tag{8}$$

$$N = N_C + N_U + N_S \tag{9}$$

$$N = f_{cc} A_{cc} + f'_{cc} A'_{cc} + f_y A_s \tag{10}$$

E_{FRP} is the elastic modulus of CFRP sheets; $\varepsilon_{h,rup}$ is the measured hoop rupture strain of CFRP sheets, and the calculation method is shown in Equation (8); ε_{frp} is the CFRP sheets' ultimate tensile strain; d'_{cor} is the diameter of the confinement members; t_{frp} is the thickness of CFRP sheets; f_{cc} is the confinement stress of CFRP on concrete; f_{co} is the axial compression strength of concrete; f'_{cc} is the confinement stress of CFRP on UHPC; f'_{co} is the axial compression strength of UHPC; A_{cc} is the cross-section area of concrete; A'_{cc} is the cross-section area of UHPC; f_y is the compression strength of longitudinal reinforcement; A_s is the cross-section area of longitudinal reinforcement. The carrying capacity calculation formula is shown in (9) and (10). The calculation results are shown in Table 10.

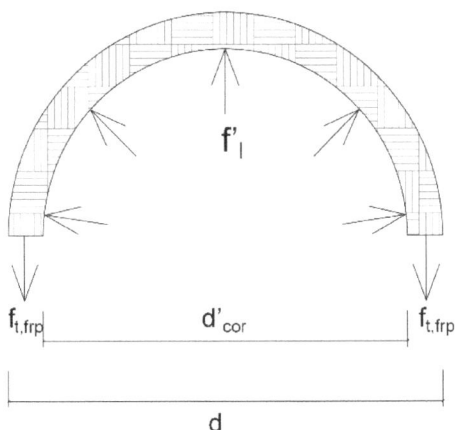

Figure 7. Distribution of confinement stress of UHPC-concrete column by CFRP sheets.

Table 10. Results of calculation.

Number	Test Carrying Capacity N_u (kN)	Calculate Carrying Capacity N_{th} (kN)	$\eta = N_{th}/N_u$
Z6	3278.23	3345.73	1.02
Z7	3976.31	3720.59	0.94
Z8	4089.81	4141.86	1.01
Z9	4533.41	4494.80	0.99

3.3. Load-Displacement Curve

Z2 and Z3 are UHPC-strengthened columns equipped with longitudinal reinforcements and ordinary circular stirrups in the strengthening layer; Z4 and Z5 are UHPC-strengthened columns equipped with longitudinal reinforcements and spiral stirrups in the strengthening layer; Z6–Z9 are UHPC-strengthened columns constrained by CFRP sheets. The carrying capacity of strengthened columns Z2–Z9 was greatly improved compared with that of without strengthened Z1. Compared with Z1, the carrying capacity of the columns strengthened with wrapped with CFRP sheets could be improved by up to 561%. When the wrapped two layers of CFRP sheets Z6 and Z8 (the thickness of the strengthening layer is 35 mm and 45 mm, respectively) reached the peak load, the carrying capacity decreased rapidly. When wrapped in five layers of CFRP sheets, compared with Z6 and Z8, the curve trend of Z7 and Z9 was relatively smooth after the load approached the peak load, and the strengthening effect was better. The displacement corresponding to the peak load of each strengthened column was larger than the displacement corresponding to the ultimate load of the without-strengthened column Z1 because the peak strain of UHPC was larger than that of ordinary concrete. For the strengthened column with steel bars or CFRP sheets, when the peak load was reached except for Z6 and Z8 with fewer CFRP sheets, other strengthened columns had greater deformation capacity.

Figure 8 shows the load-displacement curve of Z1–Z9. The curve of Z2–Z5 shows a straight line at the initial stage of loading, and when it is close to the peak load, the curve shows an inflection point. At this time, large cracks appeared in the UHPC strengthening layer, and then the peak load was reached. Compared with the brittle failure of Z1, Z2–Z5 still has greater deformation ability and better ductility after passing the inflection point. At the initial stage of loading, columns Z6–Z9 constrained by CFRP sheets presented a straight line, which rose to about the peak load of columns Z2–Z5 constrained by longitudinal reinforcement and stirrups. At this stage, there was no large crack in UHPC. When the load exceeded the peak load of Z2–Z5, the strengthening layer of UHPC cracked and generated transverse deformation. At this time, the CFRP sheets began to play a restraint role, and the strengthening layer wrapped by the CFRP sheets was in a state of three-dimensional compression, which could continue to bear the load. After Z6 and Z8 reached the peak load, the curve had an inflection point, and then the specimen had a brittle failure. The brittle failure of Z6 and Z8 columns occurred due to the small number of CFRP sheets. It can be seen from the curve of Z7 and Z9 that with the increase of load, the transverse deformation of the strengthening layer increased, and the restraint effect of CFRP sheets was fully exerted. Due to the large thickness of CFRP sheets, the failure mode was in which CFRP sheets gradually ruptured from the inside out. There is no obvious inflection point in Z7 and Z9 curves, and the load-displacement curve fluctuates until the end of loading. After the test, it was observed that the failure form of Z6–Z9 was that the concrete in the strengthening layer was crushed, and the CFRP sheets were ruptured. Due to the wrapping of the CFRP sheets, it was difficult to monitor the compression state of the internal UHPC and concrete at any time during the test. Therefore, the maximum load generated by the press was recorded as the ultimate carrying capacity of the test column.

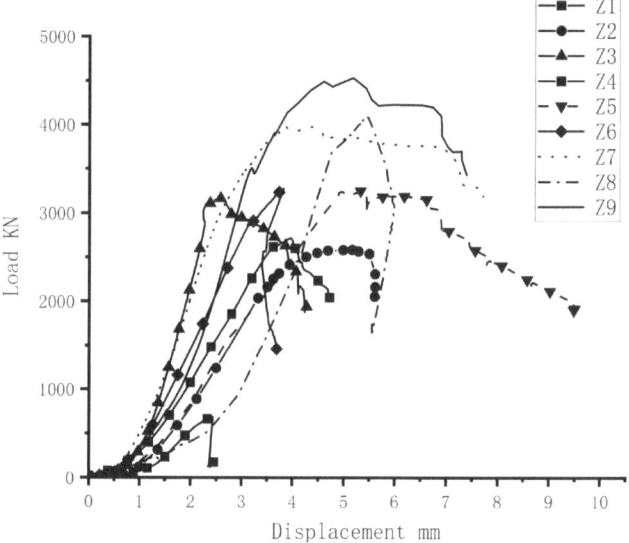

Figure 8. Load-displacement curve of test columns.

3.4. Strain Analysis

Figures 9 and 10 show the load-strain curves of the columns with longitudinal reinforcements and circular stirrups in the strengthening layer and longitudinal reinforcements and spiral stirrups in the strengthening layer. At the initial stage of load application, the load-strain curve of concrete was close to the slope of UHPC. In this case, the difference between the axial strain of UHPC and concrete was small, indicating that there was no relative slip between UHPC and concrete with the increase of load. When the load approached the peak load, the strengthening layer was the strengthening column with longitudinal reinforcements and stirrups, and the concrete reached the axial and transverse limit strain before the UHPC, then the load decreased after the UHPC reached the limit strain. In Figures 9 and 10, the load-strain curves of UHPC and concrete gradually decreased with the increasing load, indicating that the growth rate of axial strain and transverse strain of UHPC and concrete exceeded the increase rate of load when the load approached the peak load.

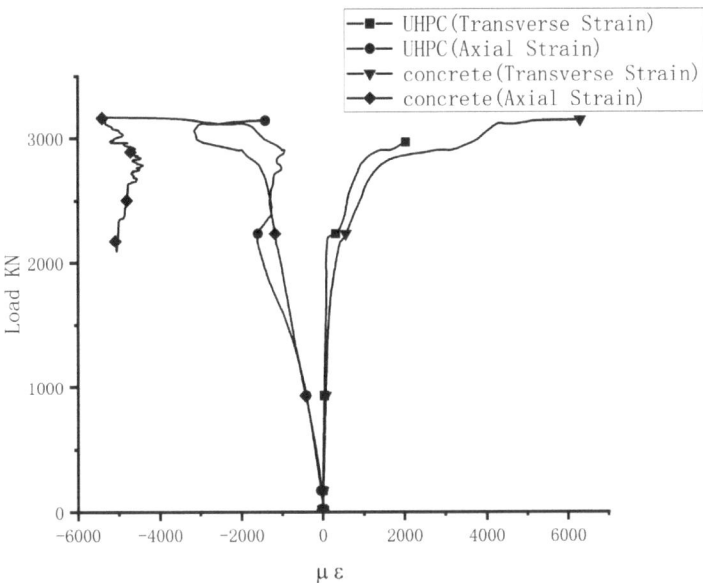

Figure 9. The strengthening layer is the load-strain curve of UHPC and concrete longitudinal reinforcement and ordinary stirrup.

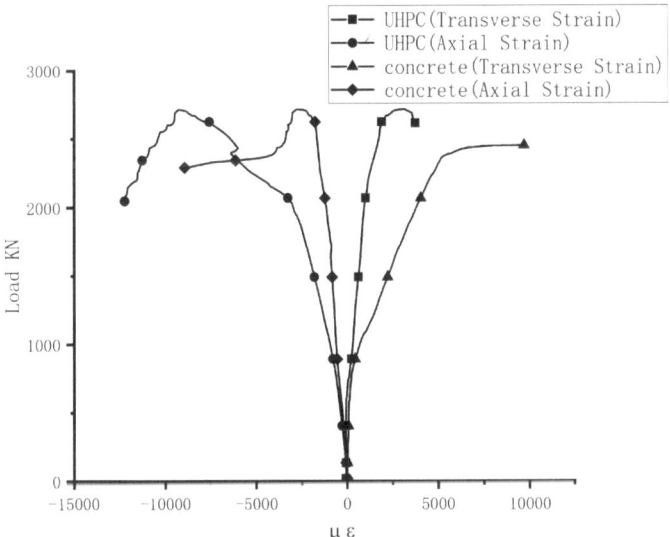

Figure 10. The strengthening layer is the load-strain curve of UHPC and concrete longitudinal reinforcement and spiral stirrup.

Figure 11 shows the load-strain curves of CFRP and concrete of UHPC strengthened column constrained by CFRP. At the initial stage of loading, the slope of the two was close, showing a linear relationship. The strain change was relatively small, indicating that CFRP and concrete had good cooperation. No slip phenomenon occurred, and the overall performance of the strengthened column was good. As the load increased, the concrete reached the ultimate transverse strain before CFRP. The slope of the load-strain curve of CFRP and concrete decreased gradually with the continuous increase of load, indicating that the growth rate of axial strain and transverse strain of CFRP and concrete exceeded the increase rate of load when the load approached the peak load.

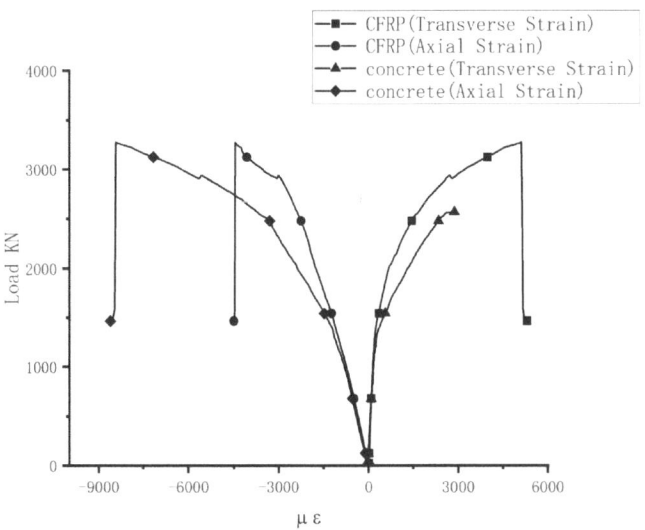

Figure 11. The strengthening columns with CFRP sheets and load-strain curves of CFRP and concrete.

3.5. Ductility Analysis

Ductility is the deformation capacity during which the carrying capacity does not decrease significantly after reaching the ultimate carrying capacity. The ratio DI of displacement $\delta_{0.85}$ when the reinforced column reaches 85% of the residual bearing capacity after the peak load and displacement δ_u when the peak load is adopted as the ductility index. The greater the DI value, the better the ductility. Table 9 shows the DI values of each column. Brittle failure occurs when column Z1 reaches the peak load and the DI value is 1. The DI values of strengthened columns Z2 and Z3 are 1.09 and 1.32, respectively. When the strengthening layer is equipped with longitudinal and stirrups, the ductility can be increased by increasing the thickness of the strengthening layer. The DI values of the strengthened columns Z4 and Z5 are 1.10 and 1.30, respectively. For Z6 and Z8, brittle fracture occurs due to the small number of wrapped CFRP layers, the DI values are 1.00 and 1.08, and the ductility is low. Compared with Z6 and Z8, when the number of CFRP layers is five, the DI values of Z7 and Z9 are 1.66 and 1.33. With the increase of CFRP layers, the ductility of columns will be improved.

According to the load-displacement curves of the three strengthening methods in Figures 8 and 12, it can be seen that the DI value of the column strengthened by longitudinal reinforcement and circular stirrup is higher. Compared with the load-displacement curve of the column strengthened with spiral stirrup and CFRP sheets, the curve of the column strengthened with longitudinal reinforcement and circular stirrup decreases faster and is steeper after reaching the ultimate carrying capacity, indicating that the column strengthened with spiral stirrup and CFRP confinement has better ductility. In the strengthening layer, the longitudinal reinforcements and stirrups have the same influence factors on the ductility of the strengthening column; that is, the thicker the strengthening layer is, the better the ductility is. The ductility of CFRP sheets confined strengthened column is influenced as the more CFRP sheets layers, the better ductility. CFRP sheets confined strengthened column should pay attention to the CFRP sheets; when the number of wrapped CFRP sheets layer is small, brittle fracture is easy to occur.

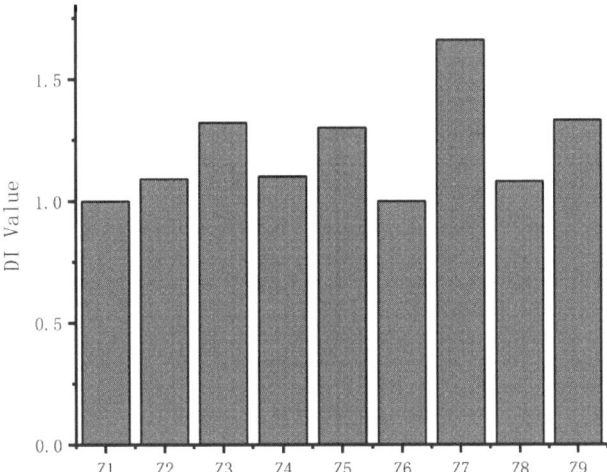

Figure 12. DI values of ductility of test columns.

4. Conclusions

In this paper, the axial compressive properties of three UHPC-strengthened columns were studied which are strengthened columns with longitudinal reinforcements and ordinary circular stirrups in the strengthening layer, strengthened columns with longitudinal reinforcements and spiral stirrups in the strengthening layer, and strengthened columns with CFRP sheets. The failure form and carrying capacity of strengthened columns were studied, and the following conclusions were obtained:

(1) The thicker the UHPC strengthening layer is, the greater the uplift of carrying capacity is. Compared with the without strengthened column Z1, the uplift of carrying capacity of the strengthened test column Z2–Z9 is 277–561%.

(2) For the three different strengthening methods, the wrapped CFRP sheets have the best effect on improving the carrying capacity and ductility, followed by longitudinal reinforcements and spiral stirrups. Compared with the strengthening form in the strengthening layer, the carrying capacity of the wrapped CFRP sheets increases by 53.9% at most, and the reinforcement form of the strengthening layer has little influence on the increase of the carrying capacity. The carrying capacity of the strengthening column with the strengthening form of longitudinal reinforcements and spiral stirrups increases by 2.9% to 5% compared with the strengthening form of longitudinal reinforcements and circular stirrups.

(3) Due to the confinement effect of UHPC and CFRP, the internal core concrete and UHPC strengthening layer are in a state of three-dimensional compression, which greatly improves the strength of concrete and UHPC, and the ultimate carrying capacity of UHPC strengthened concrete column with CFRP confinement is greatly improved.

(4) Calculate the carrying capacity of three different strengthening methods, and the calculated value is similar to the test value, which provides a reference for the calculation of the carrying capacity of columns strengthened with UHPC.

Author Contributions: Conceptualization, B.W.; Resources, B.W.; Writing—review and editing, B.W.; Supervision, B.W.; Funding acquisition, B.W.; Validation, G.L.; Writing—original draft, G.L.; Writing—review and editing, G.L.; Visualization, G.L.; Investigation, J.Z.; Visualization, J.Z. All authors have read and agreed to the published version of the manuscript.

Funding: This research was funded by Jilin Province Innovation and Entrepreneurship Talent Funding Project; the National Key Research and Development Program, grant number 2017YFC0806100; National Natural Science Foundation of China, grant number 51178206.

Data Availability Statement: Not applicable.

Conflicts of Interest: The authors declare no conflict of interest.

References

1. Bajaber, M.A.; Hakeem, I.Y. UHPC evolution, development, and utilization in construction: A review. *J. Mater. Res. Technol.* **2021**, *10*, 1058–1074. [CrossRef]
2. Ullah, R.; Qiang, Y.; Ahmad, J.; Vatin, N.I.; El-Shorbagy, M.A. Ultra-High-Performance Concrete (UHPC): A State-of-the-Art Review. *Materials* **2022**, *15*, 4131. [CrossRef] [PubMed]
3. Zhu, Y.P.; Zhang, Y.; Hussein, H.H.; Chen, G.D. Flexural strengthening of reinforced concrete beams or slabs using ultra-high performance concrete (UHPC): A state of the art review. *Eng. Struct.* **2020**, *205*, 110035. [CrossRef]
4. Tanarslan, H.M. Flexural strengthening of RC beams with prefabricated ultra high performance fibre reinforced concrete laminates. *Eng. Struct.* **2017**, *151*, 337–348. [CrossRef]
5. Al-Osta, M.A.; Isa, M.N.; Baluch, M.H.; Rahman, M.K. Flexural behavior of reinforced concrete beams strengthened with ultra-high performance fiber reinforced concrete. *Constr. Build. Mater.* **2017**, *134*, 279–296. [CrossRef]
6. Alsomiri, M.; Jiang, X.F.; Liu, Z. Elastic Confinement Effect of Concrete Circular Columns with Ultrahigh-Performance Concrete Jackets: An Analytical and Experimental Study. *Materials* **2021**, *14*, 3278. [CrossRef]
7. Xie, J.; Fu, Q.H.; Yan, J.B. Compressive behaviour of stub concrete column strengthened with ultra-high performance concrete jacket. *Constr. Build. Mater.* **2019**, *204*, 643–658. [CrossRef]
8. Zeng, J.J.; Long, T.W. Compressive Behavior of FRP Grid-Reinforced UHPC Tubular Columns. *Polymers* **2022**, *14*, 125. [CrossRef]
9. Zeng, J.J.; Pan, B.Z.; Fan, T.H.; Zhuge, Y.; Liu, F.; Li, L.J. Shear behavior of FRP-UHPC tubular beams. *Compos. Struct.* **2023**, *307*, 116576. [CrossRef]
10. Zeng, J.J.; Zeng, W.B.; Ye, Y.Y.; Liao, J.J.; Zhuge, Y.; Fan, T.H. Flexural behavior of FRP grid reinforced ultra-high-performance concrete composite plates with different types of fibers. *Eng. Struct.* **2022**, *272*, 115020. [CrossRef]
11. Xian, G.J.; Guo, R.; Li, C.G.; Wang, Y.J. Mechanical performance evolution and life prediction of prestressed CFRP plate exposed to hygrothermal and freeze-thaw environments. *Compos. Struct.* **2022**, *293*, 115719. [CrossRef]
12. Pan, Y.F.; Yan, D.M. Study on the durability of GFRP bars and carbon/glass hybrid fiber reinforced polymer (HFRP) bars aged in alkaline solution. *Compos. Struct.* **2021**, *261*, 113285. [CrossRef]
13. Guo, R.; Li, C.G.; Xian, G.J. Water absorption and long-term thermal and mechanical properties of carbon/glass hybrid rod for bridge cable. *Eng. Struct.* **2023**, *274*, 115176. [CrossRef]
14. Kobatake, Y. A seismic retrofitting method for existing reinforced concrete structures using CFRP. *Adv. Compos. Mater.* **1998**, *7*, 1–22. [CrossRef]
15. Wang, W.Q.; Wu, C.Q.; Liu, Z.X.; Si, H.L. Compressive behavior of ultra-high performance fiber-reinforced concrete (UHPFRC) confined with FRP. *Compos. Struct.* **2018**, *204*, 419–437. [CrossRef]
16. Liao, J.J.; Zeng, J.J.; Gong, Q.M.; Quach, W.M.; Gao, W.Y.; Zhang, L.H. Design-oriented stress-strain model for FRP-confined ultra-high performance concrete (UHPC). *Constr. Build. Mater.* **2022**, *318*, 126200. [CrossRef]
17. Liao, J.J.; Yang, R.T.; Zeng, J.J.; Quach, W.M.; Ye, Y.Y.; Zhang, L.H. Compressive behavior of FRP-confined ultra-high performance concrete (UHPC) in circular columns. *Eng. Struct.* **2021**, *249*, 113246. [CrossRef]
18. Huang, L.; Xie, J.H.; Li, L.M.; Xu, B.Q.; Huang, P.Y.; Lu, Z.Y. Compressive behaviour and modelling of CFRP-confined ultra-high performance concrete under cyclic loads. *Constr. Build. Mater.* **2021**, *310*, 124949. [CrossRef]
19. Abadel, A.A.; Alharbi, Y.R. Confinement effectiveness of CFRP strengthened ultra-high performance concrete cylinders exposed to elevated temperatures. *Mater. Sci. -Pol.* **2021**, *39*, 478–490. [CrossRef]
20. Zohrevand, P.; Mirmiran, A. Stress-Strain Model of Ultrahigh Performance Concrete Confined by Fiber-Reinforced Polymers. *J. Mater. Civ. Eng.* **2013**, *25*, 1822–1829. [CrossRef]
21. Zohrevand, P.; Mirmiran, A. Behavior of Ultrahigh-Performance Concrete Confined by Fiber-Reinforced Polymers. *J. Mater. Civ. Eng.* **2011**, *23*, 1727–1734. [CrossRef]
22. Guler, S.; Copur, A.; Aydogan, M. Nonlinear finite element modeling of FRP-wrapped UHPC columns. *Comput. Concr.* **2013**, *12*, 243–259. [CrossRef]
23. Guler, S. Axial behavior of FRP-wrapped circular ultra-high performance concrete specimens. *Struct. Eng. Mech.* **2014**, *50*, 709–722. [CrossRef]
24. Lam, L.; Huang, L.; Xie, J.H.; Chen, J.F. Compressive behavior of ultra-high performance concrete confined with FRP. *Compos. Struct.* **2021**, *274*, 114321. [CrossRef]
25. Chang, W.; Hao, M.J.; Zheng, W.Z. Compressive behavior of UHPC confined by both spiral stirrups and carbon fiber-reinforced polymer (CFRP). *Constr. Build. Mater.* **2020**, *230*, 117007. [CrossRef]
26. Chang, W.; Zheng, W.Z. Lateral response of HPC confined by both spiral stirrups and CFRP under axial compression. *Mater. Struct.* **2021**, *54*, 81. [CrossRef]
27. Lu, C.L.; Ouyang, K.; Wang, Q.; Zhu, W.X.; Chen, H.Q.; Lei, Z.X.; Qin, Z.J. Axial behavior of RC column strengthened with laterally reinforced FRHPC jacket. *Struct. Concr.* **2022**, *23*, 1718–1734. [CrossRef]

28. *GB/T50081-2002*; Standard for Test Method of Mechanical Properties on Ordinary Concrete. China Architecture & Building Press: Beijing, China, 2003. (In Chinese)
29. Qu, W.J.; Wu, S.J.; Qin, Y.H. Mechanical Property Tests of Reactive Powder Concrete. *J. Archit. Civ. Eng.* **2008**, *25*, 13–18. (In Chinese)
30. *GB 50367-2013*; Code for Design of Strengthening Concrete Structure. China Architecture & Building Press: Beijing, China, 2014. (In Chinese)
31. *GB 50010-2010*; Code for Design of Concrete Structures. China Architecture & Building Press: Beijing, China, 2011. (In Chinese)
32. Deng, Z.C. Effects of thicknesses and types of fiber reinforced polymer tubes on the uniaxial compressive behaviors of confined UHPC specimen. *J. Harbin Eng. Univ.* **2016**, *37*, 218–222. (In Chinese)
33. Lam, L.; Teng, J.G. Design-oriented stress-strain model for FRP-confined concrete. *Constr. Build. Mater.* **2003**, *17*, 471–489. [CrossRef]

Disclaimer/Publisher's Note: The statements, opinions and data contained in all publications are solely those of the individual author(s) and contributor(s) and not of MDPI and/or the editor(s). MDPI and/or the editor(s) disclaim responsibility for any injury to people or property resulting from any ideas, methods, instructions or products referred to in the content.

Article

Assessment of Flexural Performance of Reinforced Concrete Beams Strengthened with Internal and External AR-Glass Textile Systems

Rana A. Alhorani [1], Hesham S. Rabayah [1], Raed M. Abendeh [1,*] and Donia G. Salman [2]

[1] Department of Civil and Infrastructure Engineering, Al-Zaytoonah University of Jordan, Amman 11733, Jordan; r.alhourani@zuj.edu.jo (R.A.A.); h.rabayah@zuj.edu.jo (H.S.R.)
[2] Department of Civil Engineering, University of Mississippi, Oxford, MS 38677, USA; dgsalman@go.olemiss.edu
* Correspondence: r.abendeh@zuj.edu.jo

Abstract: This paper is an experimental study of the effectiveness of using internal and external alkali-resistant glass fabric textile (AR-GT) layers for flexural strengthening of reinforced concrete (RC) beams. The experimental work compares internal single and triple layers of AR-GT as supplemental flexural reinforcement with textile-reinforced mortar (TRM) in RC beams subjected to four-point bending loading. In addition, a control beam specimen is cast with no AR-GT fabric. Monitoring the load–deflection curves, crack patterns, and strengthening layer performance showed that using AR-GT for internal and external layers increased the load-carrying capacity of RC beams. The failure patterns of beams with one external AR-GT layer and three internal AR-GT layers showed a similar trend, with higher loading capacity and lower deflections than the other beams. Three internal textile AR-GT layers recorded higher flexural strength (52%) than one internal layer (6.3%), compared to the control beam specimen. Moreover, using one layer of external AR-GT fabric exhibited higher flexural strength than using one or three internal layers (56.8%).

Keywords: AR-glass textile; textile-reinforced mortar; load-carrying capacity; flexural strengthening; reinforced concrete beams; repair

---

Citation: Alhorani, R.A.; Rabayah, H.S.; Abendeh, R.M.; Salman, D.G. Assessment of Flexural Performance of Reinforced Concrete Beams Strengthened with Internal and External AR-Glass Textile Systems. *Buildings* **2023**, *13*, 1135. https://doi.org/10.3390/buildings13051135

Academic Editor: Fabrizio Greco

Received: 26 March 2023
Revised: 9 April 2023
Accepted: 22 April 2023
Published: 24 April 2023

Copyright: © 2023 by the authors. Licensee MDPI, Basel, Switzerland. This article is an open access article distributed under the terms and conditions of the Creative Commons Attribution (CC BY) license (https://creativecommons.org/licenses/by/4.0/).

1. Introduction

Textile-reinforced concrete (TRC) is an innovative technique which may replace numerous traditional approaches of repairing or strengthening existing concrete structures (steel jackets, bonding of sheets of fiber-reinforced polymer (FRP), shotcrete, etc.). FRP is a popular method for rehabilitation or strengthening concrete elements due to its large ratio of strength to weight, ease of application, low thermal conductivity, and durability in a severe environment [1]. In addition, the use of FRP in damaged concrete elements is efficient, since it enhances the load-carrying capacity and ductility. It is worth mentioning that FRP composites have been used to reinforce RC beams against seismic [2–4] and impact loads [5–7]. Nevertheless, it is reported that FRP methods have some disadvantages, such as application cost, low performance at high temperatures, weak integration between the concrete surface and the binder, and adhesion on wet surfaces [8–10].

In order to overcome these disadvantages, attention towards the use of TRC has been growing as a reinforcing material for concrete elements in buildings, as another option for FRP techniques. TRC is usually made of fibers woven or stitched in two orthogonal orientations, producing an open mesh. TRC can enhance the mechanical strength, energy absorption, and ductility, as well as reduce application cost, weight, and emissions of carbon dioxide of concrete members [11,12]. TRC is made of high-strength materials such as carbon, AR-glass, or basalt fibers embedded in inorganic materials, e.g., cement-based mortars, when it is used as an external layer on a concrete surface, and it is known as textile-reinforced mortar (TRM) or fabric-reinforced cementitious matrix (FRCM).

Many researchers have compared TRC and FRP techniques for external strengthening of flexural or shear capacities of concrete members. In general, the limitations of some composite materials, such as their incompatibility with sustainable environmental requirements, their brittleness, and their low fire resistance, have slowed their development and use for strengthening/repairing purposes. Furthermore, the literature has highlighted that the performance of hybrid TRC solutions is similar to that of CFRP under service limit states, but using TRC alone to strengthen RC beams revealed lower capacity gain performance under strength limit states. In particular, as noted by Larbi et al. [13], beams strengthened with TRC exhibited crack kinematics similar to that of undamaged RC beams, while no effect of TRC composite strengthening on the qualitative development of crack opening was observed. Verbruggen et al. [14] studied the effect of using external CFRP and TRC systems to test small-scale reinforced concrete beams that were strengthened for flexure. The results showed that both external systems cause the concrete beams to maintain high initial stiffness despite crack initiation. This is until reaching the cracking loads, which were found to be in excess of the calculated loads. The experiment showed that the number of cracks is independent of the type of external strengthening systems, but more than twice that of the reference beams. The researchers concluded that using CFRP or TRC for the external reinforcing layers of RC beams has a beneficial effect on the crack width, which was smaller, thus protecting the reinforcing bars by reducing moisture penetration. The crack widths were comparable to TRC- and CFRP-reinforced beams, up to 75% of the failure loads. Finally, the authors concluded that the pre-cracking of RC beams does not affect failure mode, ultimate load, crack number, and crack width compared to the performance of uncracked beams. The difference was the loss of the initial stiffness of the beams due to the opening of the existing cracks.

The flexural strengthening of beams using TRC has been conducted in different studies to investigate various parameters such as the material of the textile fiber, including carbon fiber textile [15–17], polyparaphenylene benzobisoxazole (PBO) fiber textile [16–18], and basalt fiber textile [19], the number of textile fiber layers used [16–21], the strengthening configuration [16], and the compressive strength of concrete [20]. It was concluded that using various textile fiber for reinforced concrete beams improved their flexural capacity, and increasing the number of textile layers increased flexural capacity and changed the failure mode.

Some studies have made numerical simulations of RC beams in order to assess the load-carrying capacity and crack resistance. Maio et al. [22] used an integrated numerical fracture model to model the damage phenomena of FRP-strengthened RC beams. Rimkus et al. [23] simulated concrete cracking of RC beams by using a smeared crack approach. The numerical analysis included the influence of the bond, fracture energy, and mesh of finite elements.

Ohno and Hannant and Peled et al. [24,25] initiated research to classify TRC composite structures according to their tensile strength properties. Then, Triantafillou and Papanicolaou [26] and Brückner et al. [27] initiated studies focusing on using TRC to strengthen and repair concrete elements. Thus, several experimental and numerical studies have been performed to evaluate the technical feasibility of TRC to determine the mechanical performance of composite structures compared to conventional solutions incorporating CFRP [28]. Elsanadedy et al. [19] conducted experimental and numerical investigations concerning textile-reinforced mortar (TRMs) effectiveness in the enhancement of the flexural capacity of RC beams. Basalt-based textiles were used to study variables including mortar types, number of TRM layers, and TRM types versus CFRP composites. The researchers concluded that the TRM strengthening system was less effective as it increased the tested beams' flexural strengths by 7.2% only but provided 61% higher ductility than FRP systems. The experimental tests in their study showed that using polymer-modified cementitious mortar to install TRM layers on concrete provides better bonds in the composite structure than using cementitious mortar. Additionally, reinforcing the concrete beams with layers of basalt-reinforced mortar resulted in a significant increase in the flexural strength ranging from 39% to 91%.

AR-glass textile fabric is a high-strength reinforced structural fabric made from alkali-resistant glass fibers with a special reactive coating. The general characteristics of AR-GT include its bidirectional configuration, high tensile strength, high ductility, and durability. It is removable without damaging the structure, easy to use, resists the alkaline environment of mortars due to its ZrO_2 content, and is fully compatible with mortars based on hydraulic cement or lime. Furthermore, it has high adhesion properties with a special reactive coating, and is available in different weight and mesh size options. Structural applications of AR-GT fabric include repairing and strengthening of concrete structures.

Experiments by Giese et al. [29] were carried out to study the flexural strengthening of RC beams using AR-glass textiles (two, three, and four layers) with variable TRM ages (3, 7, and 28 days) for different pre-cracking levels (no pre-cracking, 50%, and 100% of yielding loads obtained from the corresponding control specimens). The study concluded that all TRM beams showed increasing ultimate loads within the service limit state. The beams' cracking and yielding loads were affected by different TRM ages (increased by 49% for 28-day TRMs) and pre-cracking levels (from 35% to 72% for the uncracked beams compared to the control beams). Nevertheless, the same had no significant effect on the ultimate loads. However, the TRM strengthening systems significantly improved the beams' ultimate loads when increasing the number of layers, by 31%, 54%, and 72% for using two-, three-, and four-layer glass fabrics, respectively. It was observed that the beams with a pre-crack level of 50% exhibited a decrease of 10.9%, while there was a decrease of 41.4% for pre-cracked beams with a level of 100%. However, the TRM external strengthening of beams enhanced their behavior in the second stage of the load–deflection curves to record yielding loads equivalent to those of the uncracked beams, but reduced their ductility, as the average ductility ratio was 2.45–3.25 for the strengthened beams versus 4.02 for the control beams.

A few studies have investigated the embedded fabric textile in reinforced concrete members to strengthen flexural capacity. Limited researchers have studied the use of internal layers of AR-GT in concrete prisms and slabs. They reported that the inclusion of bidirectional glass grids could improve the flexural capacity of reinforced foam concrete prisms under a three-point bending test [30]. Furthermore, using lateral reinforcements of AR-GT grids under the four-point loading test improved the shear resistance of the polypropylene fiber-reinforced foam concrete beams [31]. Applying glass fiber grids and polypropylene grids for concrete slabs improved the punching capacity, and better behavior was observed at the interface between concrete and glass fiber grids compared to the glued fiber-reinforced polymer plates on the surface of the slab [32].

Accordingly, it is clear that study of the feasibility of using internal fabric textile layers for flexural strengthening of RC beams has not been performed. This article aims to conduct and evaluate the mechanical characteristics of solutions based on AR-GT fabric. In comparison to the conventional external AR-GT application, this paper investigates the qualitative and quantitative effectiveness of enhancing the flexural strength and ductility of reinforced concrete beams by applying internal layers of AR-GT fabric. Thus, the RC beam specimens prepared with one internal layer and three internal layers of AR-GT fabric were tested. Furthermore, one strengthened beam specimen was prepared with one external layer of AR-GT fabric using the TRM technique. Four RC beams were inspected by four-point flexural loading to monitor load-bearing capacity, load–deflection curves, crack propagation, ductility index, and failure pattern.

2. Experiment Work

2.1. Materials

2.1.1. Concrete and Steel

The beams were prepared using ready-mix concrete, which has been tested according to ASTM standards. Three cubes with side dimensions of 150 mm were tested for compressive strength of concrete and obtained an average value of 50 MPa. The tensile strength of concrete splitting was determined based on the test of two cylinders with a diameter of 150 mm and a height of 300 mm, with an average value of 4.5 MPa. The measured yield strengths of the main and shear reinforcement were 517 MPa and 280 MPa, respectively.

2.1.2. AR-Glass Textile and Mortar Matrix

The general characteristics of the AR-glass fabric listed on the manufacturer's data sheet show 81% fiberglass content with a 19% alkali-resistant treatment. The web width (warp) was shown to be 4.15 mm ± 5%, and the web width (weft) was 3.8 mm ± 5%. The mechanical properties of the glass fabric demonstrate a tensile strength (warp and weft) of more than 35 N/mm, with an elongation of 5%. Details of AR-glass fabric textiles are shown in Figure 1.

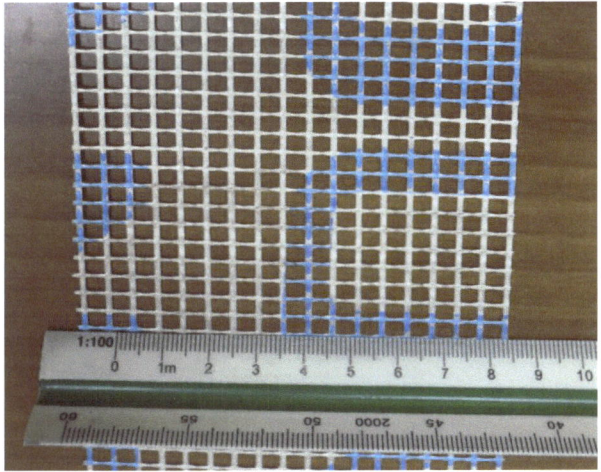

Figure 1. AR-glass fabric textile.

A hydraulic cement-based mortar was used to install the GT layers. It consists of a cementitious powder with a density of 1.6 kg/L which has a high polymer content, specific silicon/quartz mineral charges, and additives. The mortar mixture has an initial bond strength of 2.1 MPa and can be used up to 15 mm in thickness.

2.2. Preparation of Beams

The mortar mixture was prepared by mixing 1 kg of mortar and 0.483 L of water using an electric mixer and then left for 5–10 min before re-mixing to bed in the layers of the GT fabrics. The stages of fixing the external glass fabric involved firstly spreading a 3 mm thick layer of mortar using a trowel at the bottom of the beams to gradually lay the GT fabric over the mortar, then covering the fabric with a 2–3 mm layer of mortar. Figure 2 illustrates applying the external AR-GT fabric on a reinforced concrete beam.

To install the internal textiles, the glass fabric was placed directly under the stirrups of the specimens. For the three-layer strengthening beams, the first and second layers were placed on top of each other below the stirrups. The third layer of fabric was installed directly on top of a 30 mm layer of concrete above the main reinforcing bars, with a distance of 18 mm between the first two layers and the third layer. Figure 3 depicts the installation technique of one layer of internal GT fabric in reinforced concrete beams.

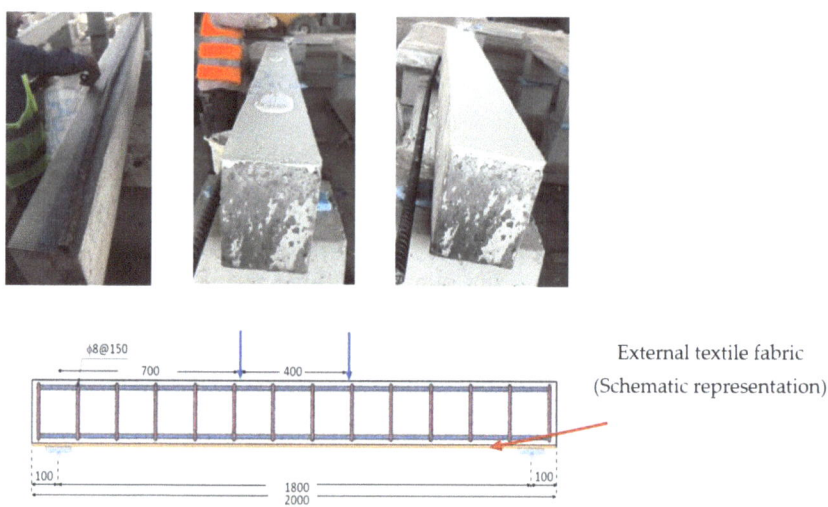

Figure 2. Application of external AR-GT fabric.

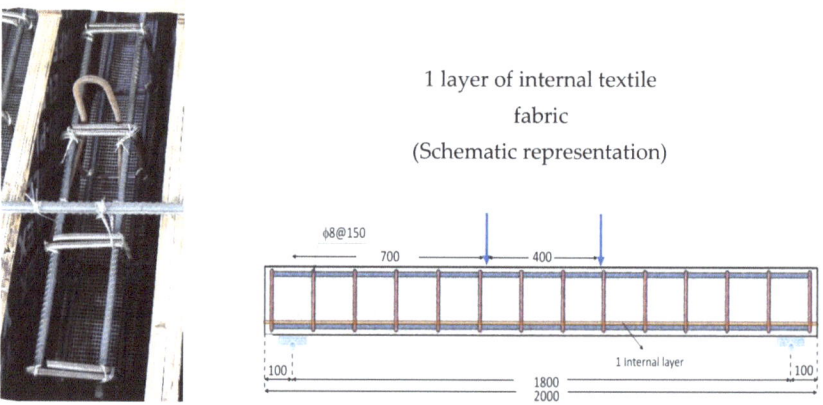

Figure 3. Installation of internal AR-GT fabric.

2.3. The Experiment Setup and Specimen Details

The study was completed based on the test of four RC beams 150 mm wide, 200 mm deep, and 2000 mm long under four-point loading. Details of the reinforcement of the beams are shown in Figure 4. The experimental program included the preparation of a control beam specimen (CTRL), and three strengthened beams as follows: two beams with one layer and three inner GT fabric layers marked with (INT1L) and (INT3L), respectively, and a single beam with one outer GT textile layer (EXT1L).

The beam specimens were simply supported on solid concrete blocks with a center-to-center supported distance of 1800 mm. Deflection measurements were taken every 5 kN incremental loading using a linear variable differential transducer (LVDT) positioned at the center of the supported specimen length.

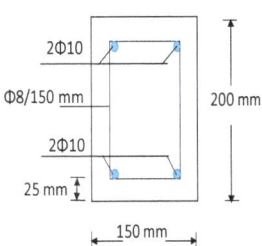

Figure 4. The test setup and beam reinforcement details.

2.4. Setup of Beams

The beam specimens were placed for testing as shown in Figure 5. The supports were placed 100 mm from the edge of the beams; thus, the centers of supports were distanced at 1800 mm. The supports were placed on rigid concrete blocks at the two edges. The load was applied gradually through a heavy-duty load cell, and deflection values were recorded every 5 kN increment. The deflection measurements were taken with a linear variable differential transformer (LVDT) placed beneath the center of the beams. Figure 6 shows schematic representations of the experimental setup of the four tested beams.

Figure 5. Experimental setup of RC beam specimen.

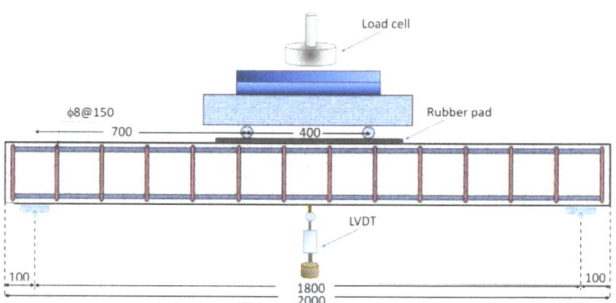

Figure 6. Schematic representations of the experimental setup of RC beams.

3. Experiment Results and Discussion

Figure 7 displays the load–deflection curves for the four beams (CTRL, INT1L, INT3L, and EXT1L) along with the ultimate flexural loads (kN) and the associated deflections (mm). Table 1 lists the experimental results for the cracking load (Pcr) and deflection (Δcr), the yield load (Py) and deflection (Δy), the ultimate load (Pu) and deflection (Δu), the failure load (Pf) and deflection (Δf), the ductility indices, and the strengthening ratios. The failure modes are presented in Figure 8, where cracks and loading are recorded for visual inspection.

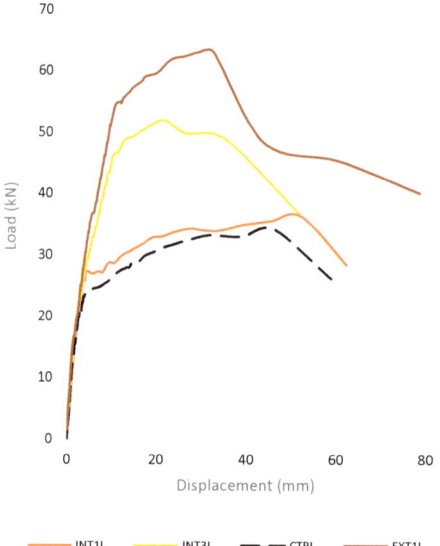

Figure 7. Load–deflection curves for four RC beam specimens.

Table 1. The experimental results.

Specimen	Pcr (kN)	Δcr (mm)	Py (kN)	Δy (mm)	Pu (kN)	Δu (mm)	Pf (kN)	Δf (mm)	Ductility Index (Δu/Δy)	Strengthening Ratio for (Pu, Δu) (%)
CTRL	17.71	2.24	23.06	3.71	34.13	46.13	26.13	58.84	12.43	-
INT1L	16.84	2.05	27.23	4.32	36.28	52.14	28.39	62.28	12.07	(6.3, -)
INT3L	17.99	2.02	43.45	9.34	51.87	21.78	36.77	51.77	2.33	(52.0, 52.8)
EXT1L	17.89	1.82	54.87	11.55	63.01	32.36	40.06	78.53	2.80	(84.62, 29.9)

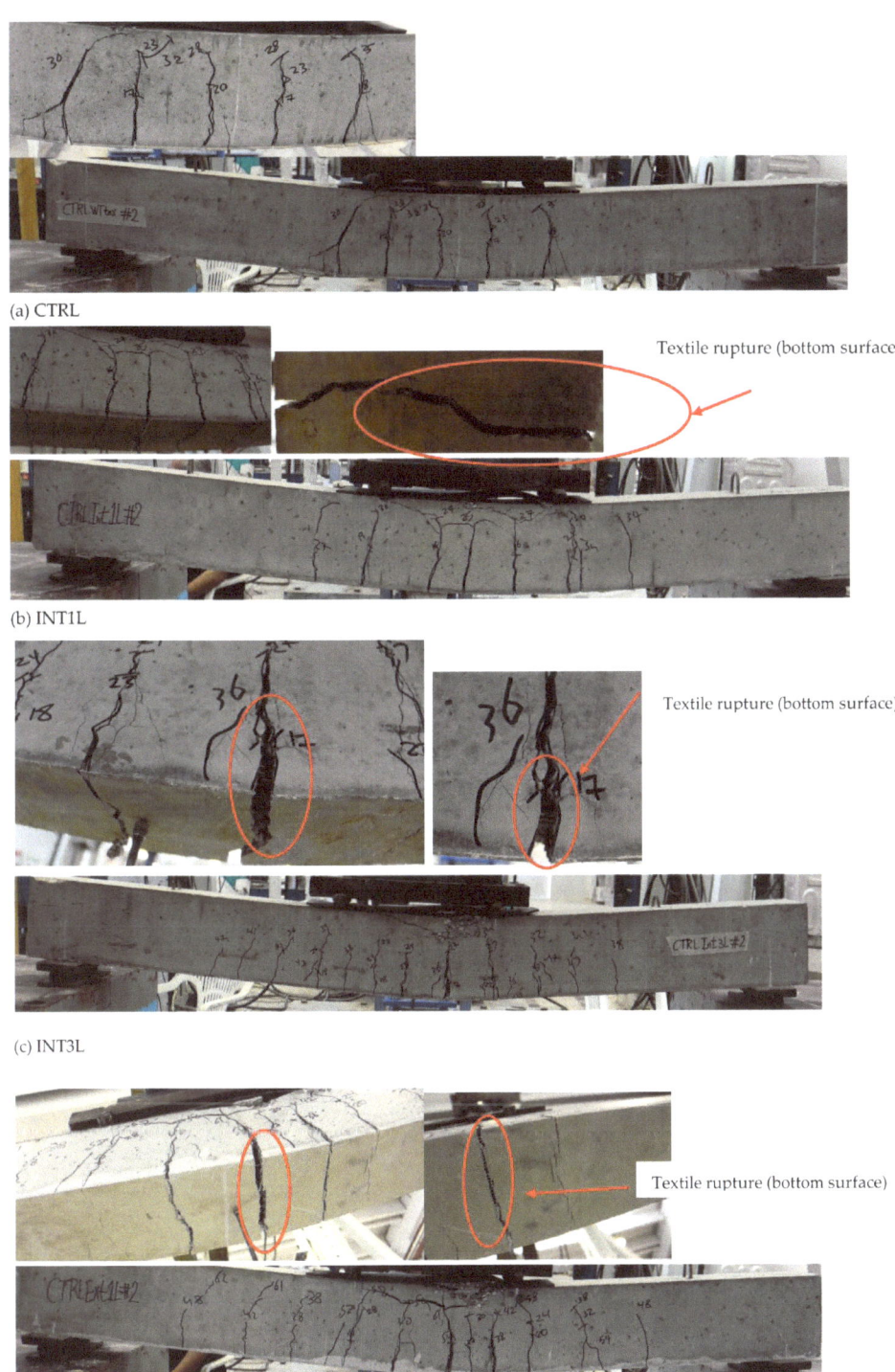

Figure 8. Failure patterns of specimens.

3.1. Failure Patterns

Overall, there is a relationship between the developed cracking pattern and failure and the configuration of AR-GT fabric applied to strengthen the flexural capacity of the beams (Figure 8 and Table 1). Schematic representations of the recorded failure patterns are also displayed in Figure 9. All the strengthened beams (INT1L, INT3L, and EXT1L) exhibited a flexural crack pattern similar to the un-strengthened specimen (CTRL). The loss of strengthening action occurred due to AR-GT fabric rupture prior to the beams' failure by concrete crushing in the compression zone.

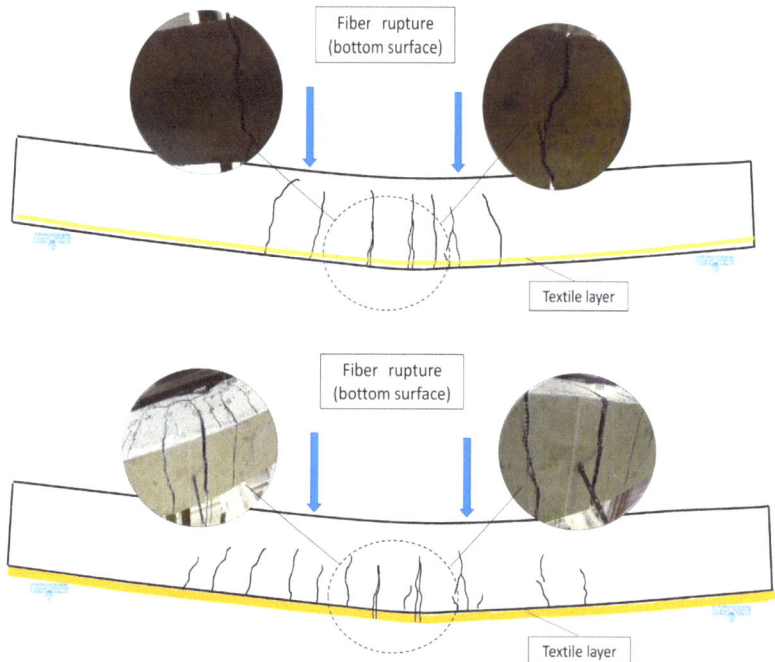

Figure 9. Schematic failure modes of specimens.

The control beam (CTRL) failed in flexure after the formation of flexural cracks in the constant moment span. The failure was due to the post-yielding response and rupture of the tensile reinforcement bars (Figure 8a). This type of failure mode is typical for under-reinforced beams. This type of failure was consistent with the results obtained by Giese et al. and Sen and Reddy [29,33]. Sen and Reddy [33] tested two control beams and observed major vertical cracks developed at the mid-span in the lower face of the RC beams and extended towards the top face.

All AR-GT-strengthened beams also failed in flexure at loads substantially higher than the control beam (Table 1). Thus, the contribution of AR-GT strengthening fabrics in increasing the flexural capacity was 6.3%, 52.0%, and 84.62%, for INT1L, INT3L, and EXT1L, respectively. Similar behavior was observed in the literature [19,21,34]. Raoof et al. [21] reported that all FRP-strengthened beams failed in flexure and rupture of the fibers and had an ultimate load higher than the control beam. The RC beam strengthened with TRM recorded an ultimate load of 43.2 kN, whereas the control beam recorded 34.6 kN. The main failure mode for INT1L, INT3L, and EXT1L specimens was a textile rupture, in which textiles are damaged because cracks on the extreme concrete surface open while increasing load. Flexural cracks occurred until the yielding load was reached. This increases the beam deflection and causes the concrete to crush in the compression zone at the ultimate loads.

Figure 10 presents the total number of visible cracks in the beam specimens and change percentage in the number of cracks relative to the CTRL beam. In all specimens, approximately 50% of the total visible cracks occurred in the constant moment span. A similar behavior trend was observed in previous studies [35–37]. Park et al. [35] found that the number of cracks in the pure moment zone of the TRM beam appeared more than in the control beam. This result indicates that AR-GT textile fabric is beneficial for the uniform distribution of cracks and effectively enhances flexural capacity.

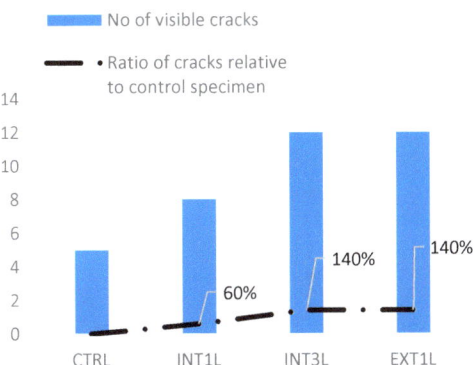

Figure 10. Visible crack numbers and percentage change in crack number relative to CTRL.

The following sections discussed the flexural capacity, ductility, and comparisons of the load–deflection behavior developed by beam specimens strengthened with different configurations (internal or external) and numbers of AR-GT fabrics.

3.2. Flexural Strengthening and Load–Deflection Relationship

In the literature, load–deflection curves obtained by beam flexural tests have been simplified to illustrate the effect of strengthening systems used. Three linear branches up to the ultimate load describe the flexural behavior of the tested beams in three phases [26], namely, the uncracking phase up to the first cracking of the concrete, the cracking phase up to steel yielding, and the plastic hinge phase in the case of un-strengthened elements or the full activation phase of the fabric until the ultimate load in the case of strengthened specimens.

The slope of the straight branch of the load–deflection curves in Figure 7 describes the flexural stiffness of the uncracked beams in the flexural tests. The AR-GT-reinforced beams showed stiffness behavior almost the same as the control beam specimen. In this loading stage, with uncracked beam sections, the deflection is slight because of the full section stiffness capacity of the beams. Figure 7 indicates that crack loads for all specimens occurred at approximately the same load level (17–18 kN), indicating that the AR-GT fabrics in the tensile zone were not activated prior to concrete cracking [38].

The second branch of the load–deflection curves in Figure 7 reveals the behavior of cracked concrete beams with decreasing stiffness and, thus, increasing deflection. The load–deflection curves differ due to applying different strengthening configurations of the AR-GT. In this stage, multiple crack modes of concrete resulted in AR-GT fabric layer activation. Therefore, relatively, a stiffer flexural behavior compared to the control specimen was observed in INT3L and EXT1L beams, along with increased loads at the yielding stage. Therefore, from the steel yielding point and beyond, the contribution of AR-GT fabrics to the beam flexural resistance has become significant. Any additional load after that point is expected to be carried almost solely by the AR-GT layers until failure occurs and the ultimate deflection is reached.

As shown in Figure 7 and presented in Table 1, the control beam (CTRL) supported an ultimate load of 34.13 kN, causing a deflection of 46.13 mm. The specimen failed at 26.13 kN and deflection capacity of 58.84 mm.

The AR-GT-reinforced beams INT1L and INT3L exhibited approximately the same shape of load–deflection curves as the control beam. The beam strengthened with a single internal layer of GT fabric had an ultimate load of 36.28 kN, with a slight increase in its flexural load (6.3%), indicating a negligible effect of the AR-GT fabric layer. However, using a single internal GT fabric slightly enhanced the beam deflection capacity by increasing its ultimate deflection (Figure 7 and Table 1) and increasing the number and spread of flexural cracks (Figures 8 and 10).

When three internal layers of AR-GT fabric (INT3L) were used, the load–deflection curve in Figure 7 shows that the ultimate load increased to 51.87 kN, resulting in a 52% increase in the flexural load capacity compared to the control beam. Additionally, the beam exhibited a significant decrease in its mid-span deflection caused by the ultimate load. This was associated with an increase in the width of the cracks. This behavior can be attributed to the increased yield loads, approximately from 23.06 kN for CTRL to 27.23 kN and 43.45 kN for INT1L and INT3L, respectively. A textile rupture at the failure stage was observed in INT1L and INT3L. Therefore, it can be expected that the number of internal GT layers to three layers can positively enhance the flexural load capacity of a strengthened beam with a decrease in its ultimate deflection, 52.0% and 52.8%, respectively.

The highest flexural capacity was found for the RC beam with TRM of a single external layer of AR-GT fabric (EXT1L), which was 63.01 kN with 84.62% flexural strengthening for this beam compared to the control beam. Furthermore, the yield load for EXT1L was 54.87 kN, resulting in a 138% enhancement compared to the control beam. These findings were similar to the studies by D'Ambrisi and Focacci and Raoof et al. [16,21], with a few exceptions. Raoof et al. [21] used TRM with seven layers of glass-fiber reinforcement to strengthen RC beams. They found that the RC beams failed in flexure due to rupture of fibers at the constant moment region at loads higher than the reference beams with increasing in flexural capacity of 39.3%. The failure of specimens is attributed to the loss of the strengthening action, which can be either progressive or abrupt in the mode of concrete crushing cases or shear failure. After a significant loss of strength, the residual flexural capacity of the strengthened specimens approaches the plastic moment capacity of the control specimen. The combination of concrete crushing in the compression zone and shear cracks can be seen from the failure pattern of EXT1L in Figure 8d as well as from the load–deflection curve in Figure 7. Furthermore, rupture of the external AR-GT fabric layer was observed with no separation occurring at the interfaces between the fabric, concrete, and mortar. In EXT1L, with the layer positioned at the extreme face of the beam section, textile fibers in the region of the maximum moment are expected to reach high tension stresses that exceed their tensile capacity. This mechanism is brittle, resulting in a sudden drop in the beam load capacity [19–21].

It should be noted that a higher deflection reading was obtained in EXT1L than in INT3L, comparatively. This flexural response can be attributed to the gradual decrease in the flexural stiffness of the beam with increasing crack spacing, as indicated by the failure pattern and the load–deflection curve of EXT1L in Figures 8d and 7, respectively.

3.3. Ductility Index

The ductility index is the ratio of ultimate deflection to the deflection at the yielding of the tensile reinforcement bar. Figure 11 shows that all of the strengthened specimens developed lower levels of ductility index than that in the control beam. This is consistent with the results from Ebead et al. [17]. The authors found that the strengthening led to a reduction in the ductility index compared to the control specimen. This can be explained by the fact that the deflection of the strengthened beams at yielding loads was generally higher than that of the control beam, while the control beam's ultimate deflection was higher than that of the beams strengthened with AR-GT fabrics, except for INT1L. This means that the use of AR-GT fabrics as a flexural strengthening reduced the increase in deflection that occurred when the applied load increased from the yield stage to the ultimate stage.

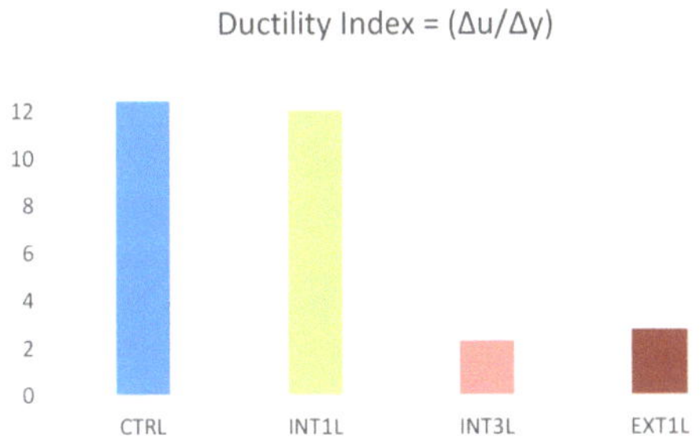

Figure 11. Ductility of the tested RC beam specimens.

Moreover, the results presented in Figure 11 indicate a correlation between the decrease in the ductility levels of the tested beams and the increase in the number of visible cracks relative to CTRL (Figure 10). The beam specimens that developed more cracks resulted in lower ductility than the CTRL specimen of 77.5–81.26%. Furthermore, strengthening the beam with a single external layer of AR-glass textile fabric showed a slightly higher level of ductility than strengthening the beam with three layers of internally fixed AR-glass fabric. This is due to the relatively high deflection (78.53 mm) at the ultimate load in the EXT1L beam specimen. The high deflection and the ultimate load recorded in the EXT1L beam can be attributed to its longer lever arm of the AR-GT layer compared with the lever arm of internal AR-GT layers.

4. Conclusions

This article experimentally studied the behavior of alkali-resistant glass textile fabric in the flexural strengthening of RC beams. Two main parameters were examined in four full-scale RC beams under the four-point flexural test: (a) internal and external GT fabric and (b) the number of GT layers. Based on the load–deflection curves, mode of failure, and strengthening layer behavior, the following conclusions can be drawn:

- Generally, using AR-glass textile fabric in reinforced concrete beams increased the load-bearing capacity.
- The embedded AR-GT as an internal supplementary reinforcement layer in RC beams enhanced not only the flexural strength but also substantially increased the cracking and post-yielding stiffness (up to 52%) compared to the un-strengthened beam.
- The flexural capacity is sensitive to the number of internal AR-GT fabric layers used. Using one internal layer of AR-GT fabric recorded a flexure capacity enhancement of only 6.3%, whereas using three layers of AR-GT resulted in an enhancement of 52% in load-bearing capacity.
- The use of textile concrete mortar systems increased the beam's flexural capacity. The strengthened RC beam with one external layer displayed an increase of 56.8% in flexural capacity with respect to the control specimen.
- The load deflection response of the two beams made with internal AR-GT fabrics was similar to that of the control beam. This behavior may be due to the fact that AR-GT layers are embedded at almost the same level as the main steel reinforcement bars.
- When the textile fabric was used as external strengthening, the beam specimen exhibited a different load–deflection behavior than the control beam specimen because the AR-GT fabric worked as additional tensile resisting reinforcement with a larger lever arm.
- The failure patterns of RC beams strengthened with one external AR-GT fabric layer and three internal fabric layers showed a similar trend with higher load-bearing capacity and lower deflections compared to the other beams.
- Using many layers of internal AR-GT fabric can be used in strengthening RC beams and may efficiently replace using a TRM technique.

- All the strengthened RC beams exhibited lower levels of ductility index than that in the control beam. This means that the use of AR-GT fabrics as a flexural strengthening reduced the increase in deflection that occurred when the applied load increased from the yield stage to the ultimate stage.

The abovementioned conclusions were based on a limited number of RC beams. For future research, it is recommended to use RC beam specimens made with AR-GT by considering different parameters such as different concrete strengths, different AR-GT fabric configurations, and subjecting the specimens to harsh environments or a wide range of high temperatures.

Author Contributions: Methodology, R.A.A.; Resources, D.G.S.; Data curation, H.S.R.; Writing—original draft, R.M.A. and R.A.A. All authors have read and agreed to the published version of the manuscript.

Funding: This research was funded by Al-Zaytoonah University of Jordan, Amman, Jordan. Grant No. 08/23/2019-2020.

Acknowledgments: The authors thank the Deanship of Scientific Research at Al-Zaytoonah University of Jordan, Amman-Jordan, for funding this research under Grant No. 08/23/2019–2020.

Conflicts of Interest: The authors declare no conflict of interest.

References

1. Hamilton, H.R.; Benmokrane, B.; Dolan, C.W.; Sprinkel, M. Polymer materials to enhance performance of concrete in civil infrastructure. *Polym. Rev.* **2009**, *49*, 1–24. [CrossRef]
2. Li, M.; Shen, D.; Yand, Q.; Cao, X.; Liu, C.; Kang, J. Rehabilitation of seismic-damaged reinforced concrete beam-column joints with different corrosion rates using basalt fiber-reinforced polymer sheets. *Compos. Struct.* **2022**, *289*, 115397. [CrossRef]
3. Shen, D.; Li, M.; Kang, J.; Liu, C.; Li, C. Experimental studies on the seismic behavior of reinforced concrete beam-column joints strengthened with basalt fiber-reinforced polymer sheets. *Constr. Build. Mater.* **2021**, *287*, 122901. [CrossRef]
4. Attari, N.; Youcef, Y.; Amziane, S. Seismic performance of reinforced concrete beam–column joint strengthening by frp sheets. *Structures* **2019**, *20*, 353–364. [CrossRef]
5. Huang, Z.; Chen, W.; Tran, T.; Pham, T.; Hao, H.; Chen, Z.; Elchalakani, M. Experimental and numerical study on concrete beams reinforced with Basalt FRP bars under static and impact loads. *Compos. Struct.* **2021**, *263*, 113648. [CrossRef]
6. Dlugosch, M.; Fritsch, J.; Lukaszewicz, D.; Hiermaier, S. Experimental Investigation and Evaluation of Numerical Modeling Approaches for Hybrid-FRP-Steel Sections under Impact Loading for the Application in Automotive Crash-Structures. *Compos. Struct.* **2017**, *174*, 338–347. [CrossRef]
7. Jahami, A.; Temsah, Y.; Khatib, J.; Baalbaki, O.; Kenai, S. The behavior of CFRP strengthened RC beams subjected to blast loading. *Mag. Civ. Eng.* **2021**, *103*, 10309. [CrossRef]
8. Elshazli, M.; Nick Saras, N.; Ibrahim, A. Structural response of high strength concrete beams using fiber reinforced polymers under reversed cyclic loading. *J. Sustain. Struct.* **2022**, *2*, 000018. [CrossRef]
9. Wu, Z.; Wang, X.; Zhao, X.; Noori, M. State-of-the-art review of FRP composites for major construction with high performance and longevity. *Int. J. Sustain. Mater. Struct. Syst.* **2014**, *1*, 201–231. [CrossRef]
10. Almasaeid, H.H.; Suleiman, A.; Alawneh, R. Assessment of high-temperature damaged concrete using non-destructive tests and artificial neural network modelling. *Case Stud. Constr. Mater.* **2022**, *16*, e01080. [CrossRef]
11. Triantafillou, T.C. *Textile Fibre Composites in Civil Engineering*; Woodhead Publishing: Sawston, UK, 2016.
12. Peled, A.; Mobasher, B.; Bentur, A. *Textile Reinforced Concrete*; CRC Press: Boca Raton, FL, USA, 2017.
13. Larbi, A.S.; Contamine, R.; Hamelin, P. TRC and hybrid solutions for repairing and/or strengthening reinforced concrete beams. *Eng. Struct.* **2012**, *45*, 12–20. [CrossRef]
14. Verbruggen, S.; Tysmans, T.; Wastiels, J. TRC or CFRP strengthening for reinforced concrete beams: An experimental study of the cracking behaviour. *Eng. Struct.* **2014**, *77*, 49–56. [CrossRef]
15. Triantafillou, T.C.; Papanicolaou, C.G. Textile Reinforced Mortars (TRM) versus Fiber Reinforced Polymers (FRP) as Strengthening Materials of Concrete Structures. *Spec. Publ.* **2005**, *230*, 99–118.
16. D'Ambrisi, A.; Focacci, F. Flexural strengthening of RC beams with cement-based composites. *J. Compos. Constr.* **2011**, *15*, 707–720. [CrossRef]
17. Ebead, U.; Shrestha, K.C.; Afzal, M.S.; El Refai, A.; Nanni, A. Effectiveness of fabric-reinforced cementitious matrix in strengthening reinforced concrete beams. *J. Compos. Constr.* **2017**, *21*, 04016084. [CrossRef]
18. Ombres, L. Debonding analysis of reinforced concrete beams strengthened with fibre reinforced cementitious mortar. *Eng. Fract. Mech.* **2012**, *81*, 94–109. [CrossRef]

19. Elsanadedy, H.M.; Almusallam, T.H.; Alsayed, S.H.; Al-Salloum, Y.A. Flexural strengthening of RC beams using textile reinforced mortar–Experimental and numerical study. *Compos. Struct.* **2013**, *97*, 40–55. [CrossRef]
20. Babaeidarabad, S.; Loreto, G.; Nanni, A. Flexural strengthening of RC beams with an externally bonded fabric-reinforced cementitious matrix. *J. Compos. Constr.* **2014**, *18*, 04014009. [CrossRef]
21. Raoof, S.M.; Koutas, L.N.; Bournas, D.A. Textile-reinforced mortar (TRM) versus fibre-reinforced polymers (FRP) in flexural strengthening of RC beams. *Constr. Build. Mater.* **2017**, *151*, 279–291. [CrossRef]
22. Maio, U.; Gaetano, D.; Greco, F.; Lonetti, P.; Pranno, A. The damage effect on the dynamic characteristics of FRP-strengthened reinforced concrete structures. *Compos. Struct.* **2023**, *309*, 116731. [CrossRef]
23. Rimkus, A.; Cervenka, V.; Gribniak, V.; Cervenka, J. Uncertainty of the smeared crack model applied to RC beams. *Eng. Fract. Mech.* **2020**, *233*, 107088. [CrossRef]
24. Ohno, S.; Hannant, D. Modeling the stress-strain response of continuous fber reinforced cement composites. *Mater. J.* **1994**, *91*, 306–312.
25. Peled, A.; Bentur, A.; Yankelevsky, D. Woven Fabric Reinforcement of Cement Matrix. *Adv. Cem. Based Mater. J.* **1994**, *1*, 216–223. [CrossRef]
26. Triantafillou, T.C.; Papanicolaou, C.G. Shear strengthening of reinforced concrete members with textile reinforced mortar (TRM) jackets. *Mater. Struct.* **2006**, *39*, 93–103. [CrossRef]
27. Brückner, A.; Ortlepp, R.; Curbach, M. Textile reinforced concrete for strengthening in bending and shear. *Mater. Struct.* **2006**, *39*, 741–748. [CrossRef]
28. Amir, S.L.; Raphael, C.; Emmanuel, F.; Patrice, H. Flexural strengthening of reinforced concrete beams with textile reinforced concrete (TRC). In *Advances in FRP Composites in Civil Engineering, Proceedings of the 5th International Conference on FRP Composites in Civil Engineering (CICE 2010), Beijing, China, 27–29 September 2010*; Springer: Berlin/Heidelberg, Germany, 2011; pp. 665–667.
29. Giese, A.C.H.; Giese, D.N.; Dutra, V.F.P.; Da Silva Filho, L.C.P. Flexural behavior of reinforced concrete beams strengthened with textile reinforced mortar. *J. Build. Eng.* **2021**, *33*, 101873. [CrossRef]
30. Falliano, D.; De Domenico, D.; Ricciardi, G.; Gugliandolo, E. Improving the flexural capacity of extrudable foamed concrete with glass-fiber bi-directional grid reinforcement: An experimental study. *Compos. Struct.* **2019**, *209*, 45–59. [CrossRef]
31. Al-Kasasbeh, T.; Allouzi, R. Behavior of polypropylene fiber reinforced foam concrete beams laterally reinforced with/without glass fiber grid. *Int. J. Struct. Integr.* **2020**, *12*, 439–453. [CrossRef]
32. Bouzeboudja, F.; Ahmed, C.A. Modeling of the interface between the concrete and the fibers grid in concrete slab. *J. Build. Mater. Struct.* **2018**, *5*, 137–146. [CrossRef]
33. Sen, T.; Reddy, H. Strengthening of RC beams in flexure using jute fibre textile reinforced composite system and its comparative study with CFRP and GFRP strengthening systems. *Int. J. Sustain. Built Environ.* **2013**, *2*, 41–55. [CrossRef]
34. Park, J.; Park, S.K.; Hong, S. Experimental study of flexural behavior of reinforced concrete beam strengthened with prestressed textile-reinforced mortar. *Materials* **2020**, *13*, 1137. [CrossRef] [PubMed]
35. Park, J.; Hong, S.; Park, S.K. Experimental study on flexural behavior of TRM-strengthened RC beam: Various types of textile-reinforced mortar with non-impregnated textile. *Appl. Sci.* **2019**, *9*, 1981. [CrossRef]
36. Yin, S.; Xu, S.; Lv, H. Flexural behavior of reinforced concrete beams with TRC tension zone cover. *J. Mater. Civ. Eng.* **2014**, *26*, 320–330. [CrossRef]
37. Liu, L.; Du, Y.; Zhou, F.; Pan, W.; Zhang, X.; Zhu, D. Flexural Behaviour of Carbon Textile-Reinforced Concrete with Prestress and Steel Fibres. *Polymers* **2018**, *10*, 98.
38. Koutas, L.N.; Tetta, Z.; Bournas, D.A.; Triantafillou, T.C. Strengthening of concrete structures with textile reinforced mortars: State-of-the-art review. *J. Compos. Constr.* **2019**, *23*, 03118001. [CrossRef]

Disclaimer/Publisher's Note: The statements, opinions and data contained in all publications are solely those of the individual author(s) and contributor(s) and not of MDPI and/or the editor(s). MDPI and/or the editor(s) disclaim responsibility for any injury to people or property resulting from any ideas, methods, instructions or products referred to in the content.

Article

Nondestructive Testing (NDT) for Damage Detection in Concrete Elements with Externally Bonded Fiber-Reinforced Polymer

Jesús D. Ortiz [1,*], Seyed Saman Khedmatgozar Dolati [2,*], Pranit Malla [2], Armin Mehrabi [2] and Antonio Nanni [1]

1 Department of Civil and Architectural Engineering, University of Miami, Coral Gables, FL 33146, USA; nanni@miami.edu
2 Department of Civil and Environmental Engineering, Florida International University, Miami, FL 33174, USA; pmall011@fiu.edu (P.M.); amehrabi@fiu.edu (A.M.)
* Correspondence: jesus.ortiz@miami.edu (J.D.O.); skhed004@fiu.edu (S.S.K.D.)

Abstract: Fiber-reinforced polymer (FRP) composites offer a corrosion-resistant, lightweight, and durable alternative to traditional steel material in concrete structures. However, the lack of established inspection methods for assessing reinforced concrete elements with externally bonded FRP (EB-FRP) composites hinders industry-wide confidence in their adoption. This study addresses this gap by investigating non-destructive testing (NDT) techniques for detecting damage and defects in EB-FRP concrete elements. As such, this study first identified and categorized potential damage in EB-FRP concrete elements considering where and why they occur. The most promising NDT methods for detecting this damage were then analyzed. And lastly, experiments were carried out to assess the feasibility of the selected NDT methods for detecting these defects. The result of this study introduces infrared thermography (IR) as a proper method for identifying defects underneath the FRP system (wet lay-up). The IR was capable of highlighting defects as small as 625 mm^2 (1 in.^{2}) whether between layers (debonding) or between the substrate and FRP (delamination). It also indicates the inability of GPR to detect damage below the FRP laminates, while indicating the capability of PAU to detect concrete delamination and qualitatively identify bond damage in the FRP system. The outcome of this research can be used to provide guidance for choosing effective on-site NDT techniques, saving considerable time and cost for inspection. Importantly, this study also paves the way for further innovation in damage detection techniques addressing the current limitations.

Keywords: CFRP laminates; externally bonded FRP; NDT methods; inspection; damage detection

Citation: Ortiz, J.D.; Dolati, S.S.K.; Malla, P.; Mehrabi, A.; Nanni, A. Nondestructive Testing (NDT) for Damage Detection in Concrete Elements with Externally Bonded Fiber-Reinforced Polymer. *Buildings* 2024, 14, 246. https://doi.org/10.3390/buildings14010246

Academic Editors: Bo Wang, Rui Guo, Muye Yang, Weidong He and Chuntao Zhang

Received: 15 December 2023
Revised: 3 January 2024
Accepted: 11 January 2024
Published: 16 January 2024

Copyright: © 2024 by the authors. Licensee MDPI, Basel, Switzerland. This article is an open access article distributed under the terms and conditions of the Creative Commons Attribution (CC BY) license (https://creativecommons.org/licenses/by/4.0/).

1. Introduction

1.1. FRP Composites

In steel-reinforced concrete (RC) structures, corrosion poses a significant problem, causing the loss of cross-sectional area, deterioration of the rebar-to-concrete bond, and deterioration of the concrete cover [1,2]. Various methods exist to prevent or mitigate corrosion, along with techniques for strengthening, repairing, and retrofitting deteriorated structures. Fiber-reinforced polymer (FRP) composites offer an alternative to steel for strengthening purposes due to their mechanical properties and chemical resistance [3–5]. They are favored for their strength-to-weight ratio, ease of installation, and adaptability to curved surfaces. FRP composites can be produced using various types of fibers, such as glass, carbon, basalt, and aramid [6–9]. Thermosetting resins, such as polyester, epoxy, and vinyl ester, are commonly employed with fibers in the production of FRP composites.

Extensive research efforts have been dedicated to examining the long-term durability of FRP materials. The findings indicate that FRP materials exhibit minimal degradation over extended periods [10]. Additionally, when compared to traditional materials, FRP composites display superior resistance to salt, water, and various chemicals. Notably, substances like oil and other heavy hydrocarbons have a comparatively reduced impact on FRP composites in comparison to their effects on conventional materials [11–14]. The aging process typically results in the need of repairs in conventional steel-RC structures. Furthermore, structural elements often require strengthening and retrofitting due to design and construction errors, damage from exceptional events or accidental impacts, natural disasters, and functional modifications [15,16].

FRP laminates (i.e., wraps/fabrics, strips, and plates) are the most common externally bonded FRP (EB-FRP) systems for strengthening existing structures. The application techniques can be categorized into 'wet lay-up', 'prepreg', and 'precured' systems. In a 'wet lay-up' system, the resin impregnation occurs on site, while in a 'prepreg' system, it takes place at the manufacturer's facility, with the resin matrix being partially cured beforehand. Conversely, 'precured' FRP systems, which are manufactured off site, are available in multiple forms [17]. Commonly, the same polymer resin is also employed to act as an adhesive in plates or as a primer, putty coat, and saturant in 'wet lay-up' systems [18].

1.2. Application of EB-FRP Systems

The utilization of FRP composites in concrete bridges has experienced substantial growth in recent decades. EB-FRP systems are generally applied to the tension side of the concrete girders, beams, and slabs to enhance their flexural strength. They can also be used to provide additional shear strength when applied on the sides of beams and girders. In seismic zones, FRP wraps can be used for columns to increase pseudo-ductility due to the induced confinement of the concrete [19]. Although, research studies have indicated substantial increases in flexural ultimate strength, often ranging from 40% to 95%, and stiffness, typically showing improvements of 17% to 95% [20–22], design guidelines impose strengthening limits to guard against collapse of the structure. The effectiveness of these applications depends on factors like proper anchorage systems, reinforcement configuration, and FRP type. Due to its ease of installation, the wet-layup system is preferred over other systems when used as an externally bonded strengthening material. Figure 1 shows a typical application procedure for EB-FRP systems. Before strengthening, the extent of deficiency and suitability of FRP strengthening should be evaluated. Surface preparation, which removes contamination and weak surface layers, is one of the most important steps in adhesive bonding of composite laminates to the concrete elements [23–25]. Improper surface preparation can lead to premature failure of bonded FRP sheets due to rupture/debonding [26]. It is crucial to ensure that fibers are thoroughly wetted, and the amount of resin is maintained at the minimum level as per the manufacturer's recommendations.

While a wide variety of fiber and resin combinations exist, carbon FRP (CFRP) with epoxy resin stands out as the most employed type for external applications (strengthening) in RC elements within the US market. It has been extensively studied in the available literature and it has consistently demonstrated superior performance in aggressive environments (typically, there is less than a 10% reduction in tensile strength when subjected to harsh environmental conditions), and its higher stiffness, compare to other types of FRPs [27], makes it more suitable for strengthening applications.

Figure 1. Application of an Externally Bonded FRP system [28]; (**a**) surface preparation; (**b**) application of resin and FRP sheets; (**c**) coating and finishes.

Several organizations have developed guidelines for the design of reinforced concrete structures externally bonded with FRP composites [29–31]. For external application, different systems can be used for strengthening, retrofitting, or repairing of RC elements that are normally reinforced with conventional steel. However, there are currently no established inspection protocols for evaluation and maintaining structures exposed to demanding environments or dealing with potential issues in the FRP system. These issues in FRP applications can be traced back to various causes, encompassing mechanical, environmental, and design considerations, as well as fabrication and workmanship [27].

1.3. Inspection of EB-FRP Concrete Elements

Nondestructive testing (NDT), in general, is defined as "an examination, test, or evaluation performed on any type of test object without changing or altering that object in any way, in order to determine the absence or presence of conditions or discontinuities that may have an effect on the usefulness or serviceability of that object." [32]. NDT methods have been increasingly used for quality control, quality assurance and quality assessment of both new and old structures [33–35]. Many NDT techniques, such as visual inspection (**VT**) [36,37], tap testing (**TT**) [38–42], impact echo testing (**IE**) [43–47], microwave testing (**MW**) [48–53], ground penetrating radar (**GPR**) [54–58], ultrasonic testing (**UT**) or phased array ultrasonic testing (**PAU**) [59–64], infrared thermography testing (**IR**) [56–64], acoustic emission testing (**AE**) [65–68], laser testing (**LT**) [69–74], radiographic testing (**RT**) [75], etc., have been studied for detecting damage in the externally applied FRP composites.

The external application of FRP composites to strengthen/repair RC structures involves three materials: internal reinforcement (typically steel rebars, with the potential for future adoption of FRP rebars), and concrete and FRP composites (wet lay-up system), along with the different interfaces. Defects associated with the external application of FRP composites can occur within either of the three materials or at the interface between them [15].

The present study was divided in the following tasks: The initial step involved identifying and classifying the location of potential damage and defects in externally bonded FRP (EB-FRP) concrete elements. This included determining where the damage occurs on the element (i.e., rebar, concrete, or interface, as seen in Figure 2). Understanding "where" the potential defects can occur was crucial to further investigation, as it facilitated precise targeting of the specific location. Along with the "where", it was essential to address the timing and reasons behind the occurrence of damage. This involved investigating the factors that contribute to defects/damage in FRP external applications and when they manifest (i.e., during fabrication or service life). Understanding the "why" and "when" helped uncover the root causes of the potential defects/damage. Building upon the understanding of the location, timing, and causes, the subsequent step was to conduct a literature review focusing on the damage and defects of EB-FRP concrete elements. The objective was to identify and classify observed and potential damage comprehensively. This step aimed to answer the question of "what" potential defects/damage exist. Finally, an experimental phase was carried out to assess the selected NDTs. Two small-scale slabs were fabricated with different rebars and internal/external defects to evaluate the feasibility of the chosen NDT method. The findings of this research offer a reference for inspectors in choosing the most suitable on-site NDT techniques, with the prospect of saving both time and cost significantly.

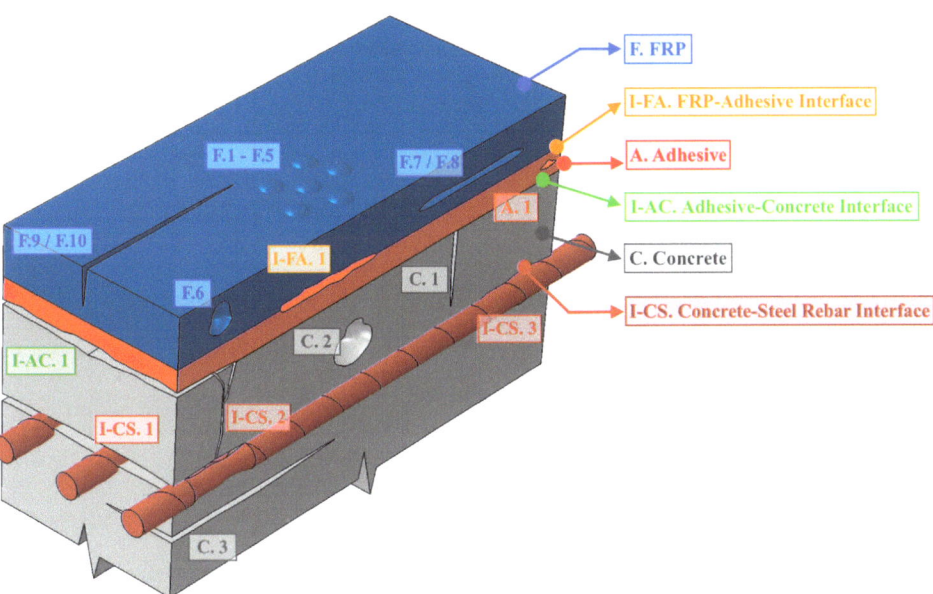

Figure 2. Defects in EB-FRP concrete elements [15].

2. Location of Potential Defects or Damage

The term "defect" can be defined as "discontinuity whose size, shape, orientation or location (1) makes it detrimental to the useful service of its host object or (2) exceeds an accept/reject criterion of an applicable specification" [76], while damage can be defined as "changes introduced into a system that adversely affect its current or future performance" [77]. Defects denote material-level anomalies, while damage encompasses the combination of these material-level imperfections, ultimately evolving into system-level deterioration. Breaking down the EB-FRP concrete element into its distinct components enables a focused approach to examine the potential defects and damage that may arise during its service life. Initially, potential damage and defects in EB-FRP concrete elements were categorized based on their likely locations, i.e., where they occur, typically falling into three distinct groups: (1) defects in FRP composites (i.e., FRP composites (F)); (2) bond defects (i.e., FRP–adhesive interface (I-FA), adhesive (A), and adhesive–concrete interface (I-AC); and (3) defects in concrete (i.e., concrete (C) and concrete–steel rebar interface (I-CS)) [15]. In a prior study, the authors classified damage based on its location and initiation time, while also identifying its sources. Figure 2 illustrates the location of the most likely defects in EB-FRP concrete elements [15].

3. Damage and Defects in EB-FRP Concrete Elements

Once the locations of potential defects have been determined and the question "where do they occur?" has been answered, it is important to establish "what" damage and defects exist that affect the performance of the EB-FRP concrete element. Tables 1–3 present the defects shown in Figure 2 with brief descriptions. These defects were identified based on the available literature [6,33,78–87] and can be further explored in Malla et al. (2023) [15] for more in-depth information.

Table 1. Defects in FRP composites (F—FRP composites).

Defects	Description
F.1 Surface Defects—Blisters	Blisters are observed as bubble-like formations on the surfaces of the EB-FRP system because of the combined action of freeze–thaw cycles and entrapped moisture. However, since their effects are primarily limited to the surface, this imperfection have minimal impact on the structural performance of the structure.
F.2 Surface Defects—Wrinkling	Wrinkling appears as creases or folds on the surface of the FRP composites, often occurring at corners and curves of the structure. It is caused by improper installation practices. The safety of the structure is compromised only if they result in insufficient surface contact of the FRP composites with the substrate.
F.3 Surface Defects—Scratches	Scratches represent marks or wounds on the surface of the FRPs and can occur at any point during the installation and service life of the structure. They become detrimental when they evolve into full-depth cracks.
F.4 Surface Defects—Discoloration	Discoloration manifests as stains on the FRP composites and is primarily induced by exposure to UV rays, heat, chemicals, fire, excessive strain, subsurface defects, voids, and moisture penetration. These stains serve as indicators of composite degradation, frequently preceding the occurrence of cracks and embrittlement.
F.5 Surface Defects—Fiber Exposure	Improper handling and installation of FRP composites results in exposed fibers of FRP composites. These exposed fibers serve as entry points for moisture and contamination into the composite, leading to the deterioration of its properties.
F.6 Voids in FRP	Voids are cavities that exist at the fiber–matrix interface, formed as a result of entrapped air within the layers of the composites. They can also occur due to the overlapping of fabrics during fabrication or installation. They cand lead to a reduction in their laminar shear strength.
F.7 Debonding	Debonding within FRP composites refers to the separation at the interface between the two components of the composite: the fiber and the matrix. This separation is primarily triggered by the presence of surface moisture on the fibers. The consequences of debonding encompass a loss of composite action.
F.8 Delamination in FRP Layers	Delamination in FRP involves the separation at the interface between the layers. It is frequently induced by factors such as moisture, foreign object contamination, and trapped air between the FRP layers. The repercussions are significant and can result in a substantial reduction in the material's shear transfer capacity.
F.9 Cracks	Cracks in FRP composites primarily occur parallel to fiber layers due to factors like trapped air, uneven resin distribution, and exposure to impact and service loads. Failure risk increases as cracks deepen and widen under sustained or dynamic loading.
F.10 Impact Damage in FRP	Impact damage can happen from both slow-moving and fast-moving objects. Slow-moving objects may harm the internal structure, while fast-moving ones cause severe surface damage. Regardless, impact damage harms the system's structural integrity.

Table 2. Bond defects (**I-FA:** FRP–adhesive interface; **A:** adhesive; **I-AC:** adhesive–concrete interface).

Defects	Description
I-FA.1 FRP–Adhesive Debonding	FRP–adhesive debonding between laminates can occur due to factors such as the use of an inappropriate adhesive, improper mixing, poor adhesive application, or insufficient curing of the adhesive. These factors can lead to a weakened bond between the FRP layers, reducing the effectiveness of the composite material.
A.1 Voids in Adhesive	Voids are areas where FRP composites lack contact with the concrete substrate. They result from trapped air, contaminants in the resin, or substrate irregularities, and can sometimes resemble "bubbles." Voids create stress concentrations, weakening the bond strength of the FRP application.
I-AC.1 Adhesive–Concrete Debonding	Debonding is the separation of externally applied FRP from the concrete substrate, often due to factors like high loads, improper installation, inadequate resin curing, or surface moisture. Excessive debonding can lead to brittle concrete fracture, as the composite loses its ability to transfer stresses to the substrate.

Table 3. Defects in concrete (**C:** concrete; **I-CS:** concrete–steel rebar interface).

Defects	Description
C.1 Cracks in Concrete	Obscured cracks in the concrete substrate, hidden beneath the externally applied FRP, result from various factors such as shrinkage, thermal stresses, chemical exposure, and more. They can lead to structural failure by allowing corrosive chemicals to attack steel reinforcement and weaken the bond between FRP and concrete.
C.2 Voids in Concrete	Concrete voids, unrelated to external FRP application, stem from inadequate design and construction practices during casting. Causes include improper vibration, concrete quality issues, rebar congestion, consolidation problems, and irregular aggregates. These voids lead to gradual structural deterioration.
C.3 Delamination/Spalling in Concrete	Delamination is caused by the relatively weaker nature of concrete compared to the adhesive and FRP materials. It occurs when high stresses in the FRP material pull the concrete apart, typically near cracks or the ends of the FRP system where stress buildup is significant. Delamination failures are sudden and brittle, posing a serious structural risk.
I-CS.1 Cover Separation	Cover separation differs from delamination and occurs deeper within the concrete, extending to the cover distance of internal reinforcement. This separation happens as cracks near the internal reinforcement propagate horizontally due to high stresses from external FRP. Like delamination, it is a sudden, brittle failure.
I-CS.2 Corrosion in Steel Reinforcement	External FRP strengthening is typically applied to steel-reinforced concrete elements. Although it can reduce the corrosion rate of steel reinforcement, it does not completely stop it. As a result, corrosion continues over time, and it is essential to monitor corrosion activity in concrete elements even after applying strengthening measures.
I-CS.3 Concrete Reinforcement Debonding	Due to environmental and load factors, the bond may gradually weaken over time, resulting in bond failure of the steel-reinforced concrete element. Debonding might compromise the structure integrity and tensile resistance, making it susceptible to more damage.

4. Source of Damage

After locations and potential defects ("where" and "what") have been determined, it is important to establish the possible sources of these defects/damage ("why") as well as the timing during the service life when they occur ("when"). Based on the existing literature reports on similar studies [88], the causes of defects in FRP application can be categorized into four main sources. These sources include fabrication and workmanship, design factors, mechanical factors, and environmental factors, as presented in Table 4. The initial two sources are associated with the construction process of the EB-FRP concrete element, encompassing the manufacturing of the FRP composite, the design of the RC element, and its subsequent construction. The latter two occur during the in-service stage.

Table 4. Source of damage and defects in FRP applications [15].

Fabrication and Workmanship	Design Factors	Environmental Factors	Mechanical Factors
• Manufacturing • Transportation • Storage • Handling • Installation	• Unreasonable Design/Lack of Specification and Code • Calculation/Design Errors • Inadequate Installation/Construction Details (Constructability) • Improper Composite Choice	• Water Exposure • Saline Exposure • Alkaline Exposure • UV Exposure • Elevated Temperature Exposure • Freeze–thaw Cycles Exposure	• Fatigue • Creep Rupture • Shrinkage • Impact • Service Loads

The classification of damage/defects in EB-FRP concrete elements and their sources are depicted in Figure 3. The damage is classified based on "when" they could occur. Malla et al. (2023) [15] and Ortiz et al. (2023) [27] provide more in-depth information about the source of the damage. Each defect is rationally related to its possible cause.

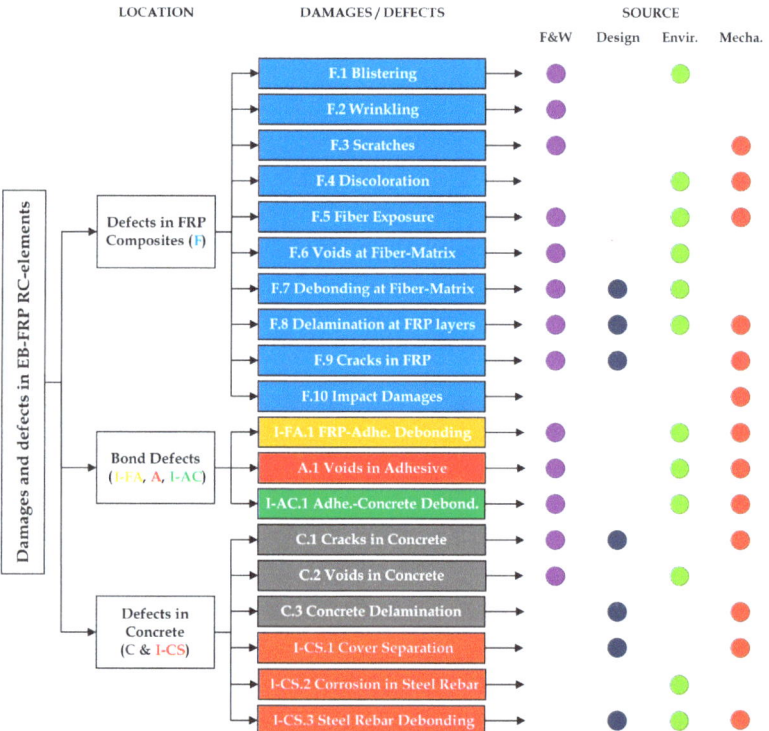

Figure 3. Source of damage in EB-FRP concrete elements. Note: colors are related to location given in Figure 1 and source given in Table 4.

5. NDT Methods Applicable to EB-FRP Concrete Elements

A previous study by the authors reviewed the applicability of available non-destructive testing (NDT) methods for detecting damage in structural elements reinforced or strengthened with FRP [89]. An extensive literature survey was conducted, encompassing over 100 past studies on the application of NDT methods in detecting damage in FRP for external applications. The damage detectability was divided into seven groups: **i.** surface Anomalies (F.1–F.5), **ii.** defects within FRP composite layers (F.6–F.10), **iii.** bond defects (I-FA, A, I-AC), **iv.** cracks in concrete (C.1), **v.** voids in concrete (C.2), **vi.** delamination in concrete (C.3), **vii.** steel reinforcement defects (I-CS). The most promising methods were selected and are summarized in Table 5, along with the percentage of applicability in the available literature for detecting each of the seven groups of damage. Khedmatgozar Dolati et al. (2023) [89] can be consulted for a deeper explanation on each NDT method. The following is a brief description of the most used NDT methods found in the available literature.

Table 5. Applicability of NDT methods for EB-FRP concrete elements in the available literature [89].

NDT Method	i. Surface Anomalies *	ii. FRP Composite	iii. Bond Defects	iv. Cracks in Concrete	v. Voids in Concrete	vi. Concrete Delamination	vii. Rebar Defects
Tap testing (**TT**)	-	12%	6%	0%	0%	0%	0%
Impact echo testing (**IE**)	-	5%	5%	15%	20%	20%	10%
Ground-penetrating radar (**GPR**) and microwave testing (**MW**)	-	7%	17%	12%	30%	42%	70%
Ultrasonic testing (**UT**) and phased array ultrasonic testing (**PAU**)	-	27%	16%	37%	15%	9%	10%
Infrared thermography testing (**IR**)	-	26%	38%	8%	9%	5%	0%
Acoustic emission testing (**AE**)	-	4%	5%	0%	0%	0%	0%
Laser testing method (**LT**)	-	7%	11%	0%	0%	0%	0%
Radiography testing (**RT**)	-	12%	2%	5%	9%	1%	0%
Impulse response testing (**IRT**)	-	0%	0%	20%	17%	23%	5%
Magnetic flux leakage (**MFL**)	-	0%	0%	3%	0%	0%	5%

* Visual inspection (**VT**) can be used for qualitative and quantitative detection of almost all surface anomalies.

- **Visual Inspection (VT):** A common, versatile, and straightforward NDT method, is used to identify surface defects in EB-FRP concrete elements. Although some researchers do not consider VT as an NDT method, it completely fits the definition of NDT method as described earlier in this paper. In any case, it is a fast and cost-effective method, and provides real-time results, serving as a baseline for other NDT techniques. Based on its findings, decisions can be made about further inspection. However, it can only detect surface defects and may be subjective, depending on individual perception.
- **Tap Testing (TT):** This method detects defects by analyzing changes in stiffness and sound frequency upon impact. It is a quick, cost-effective, and user-friendly approach for inspecting large areas in real-time, but its results are subjective and can vary due to differences in applied force, angle, and equipment. Misinterpretations may occur due to ambient noise and geometric changes.
- **Impact Echo Testing (IE):** This method relies on stress waves from an impact to identify subsurface defects in materials, particularly in concrete. It is effective for evaluating issues like cracks and delamination. By using lower frequencies, it can penetrate deeper and requires access to only one surface for testing. However, its applicability is limited to materials up to 40 inches thick. Skilled operators are needed, and it may have difficulty detecting smaller cracks and discontinuities.
- **Ground-Penetrating Radar (GPR):** This method uses radio waves to pass through a material and detects reflections from any interfaces between materials or subsurface defects like voids, cracks, debonding, and delamination. It can go beyond concrete–air interfaces, inspecting features below, and identifying defects at greater depths than some other NDT methods. It is not effective for detecting air-filled defects.

- **Ultrasonic Testing (UT) and Phased Array Ultrasonic Testing (PAU):** This method uses the reflection of ultrasonic waves at material interfaces with differing acoustic impedances to locate defects. It excels in identifying defects in concrete and composites due to the strong reflection caused by these flaws. It offers fast and field-friendly testing with good resolution, capable of penetrating materials and detecting various defects. It necessitates highly trained personnel for conducting and interpreting tests and is primarily suitable for materials of limited thickness. **PAU** uses multiple transducer elements in a phased array probe to enable precise control.
- **Infrared Thermography Testing (IR):** This method relies on differences in thermal properties between anomalies and sound areas within the material. By measuring surface temperature, it can detect subsurface defects to some extent. It is particularly suitable for inspecting larger surface areas quickly and cost-effectively, with real-time data interpretation. However, it is not reliable for detecting water-filled defects, has limitations in identifying deep-seated defects in concrete, and necessitates specific environmental conditions for optimal results.

The results from their literature review indicated that **IR**, **GPR**, and **UT** can be considered the most applicable methods for detecting bond defects [89]. For damage detection within FRP composites (e.g., debonding, voids, delamination at layers), **UT**, **IR**, and **TT** have been recognized as the most suitable ones. For concrete damage detection, **UT**, **IRT**, **GPR**, and **IE** have emerged as the most promising approaches. For all FRP surface anomalies, visual testing (**VT**) is proposed. Overall, **GPR** has been selected as the most effective NDT method for detecting different types of damage in FRP strengthened concrete elements followed by **UT**, as shown in Table 5.

GPR with high frequency antennas of about 2 GHz was able to easily detect debonding and delamination. A 1.5 GHz ground-coupled GPR antenna was effective for water-filled voids (as small as 50 × 50 × 1.5 mm^3). Air-filled voids could not be detected because of CFRP's higher electrical conductivity that leads to higher attenuation and smaller echoes [90]. However, it exhibits limitations in accuracy and frequency dependencies when assessing various structural defects. **UT** was able to detect debonding of 6.3 mm in diameter and qualitatively detect debonding and voids. Studies confirmed its efficacy in detecting and locating typical FRP defects. It was able to detect flaws as small as 0.8 mm with a penetration depth of 25 mm. **IR** detected air-filled debonding of sizes 75 mm × 75 mm, 50 mm × 50 mm, and 35 mm × 35 mm along with water-filled debonding. It also detected near-surface voids (<10 cm from surface). Furthermore, it provided qualitative detection of delamination.

6. Experimental Verification—Inspection of EB-FRP Concrete Elements

6.1. Materials and Constrcution

In order to assess the most promising NDT methods for damage detection in EB-FRP concrete elements (i.e., **VT**, **TT**, **GRP**, **UT**, **IR**), two small-scale slab specimens were fabricated (as shown in Figures 4–6). Slab M' was 760 mm (30 in.) wide by 760 mm (30 in.) long and 178 mm (7 in.) deep, and Slab Q' was 904 mm (36 in.) wide by 904 mm (36 in.) long and 178 mm (7 in.) deep. The concrete mix used to cast the slab specimens was 'Class II 4500 Bridgedeck' concrete, as per the Florida Department of Transportation (FDOT). This class specified a guaranteed compressive strength of 31 MPa (4500 psi). Type II Cement was used with a water to a cementitious material ratio (w/m) of 0.44, and #57 stone and silica sand were used as coarse and fine aggregate, respectively. To obtain the actual strength value, concrete cylinders were fabricated and tested according to ASTM C39 [91]. An average compressive strength of 31.7 MPa (4600 psi) was obtained with a standard deviation of 0.69 MPa (100 MPa).

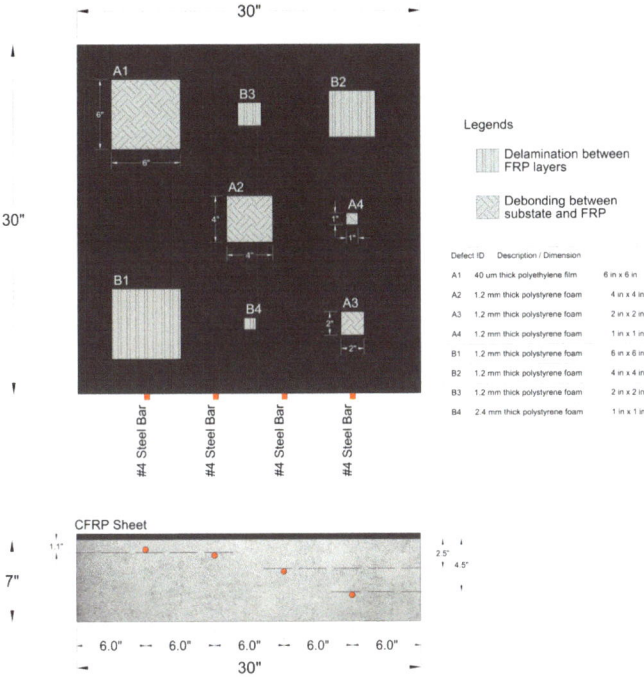

Figure 4. Slab M′, specimen with steel rebars and damage in the EB-FRP system. 1 in. = 25.4 mm.

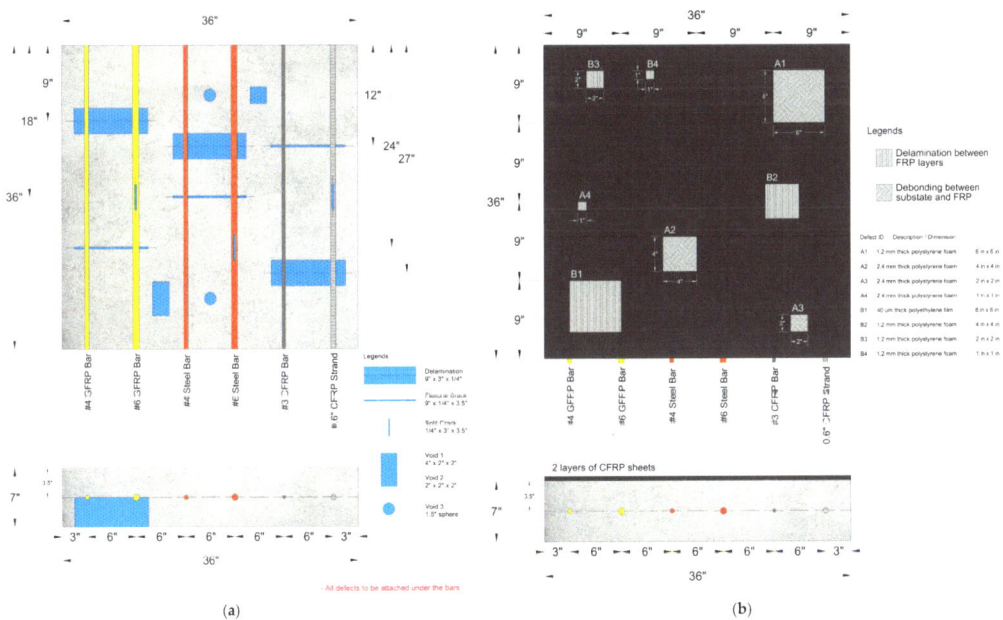

Figure 5. Slab Q′, specimen defects within concrete and in the EB-FRP system. 1 in. = 25.4 mm. (**a**) Before EB-FRP system installation. (**b**) After EB-FRP system installation.

Figure 6. Defects generation in EB-FRP concrete elements. (**a**) Unidirectional CFRP sheet, (**b**) mixing of the resin, (**c**) adhesive–concrete debonding, (**d**) surface impregnation, (**e**) first layer of CFRP, (**f**) strengthened slab.

To eliminate the need for plastic chairs/spacers or internal supports for the rebars, openings were made in the sides of the formwork and the rebars were inserted through them for support. In Slab M′, the internal steel reinforcement was located in different depths in order to target different possible concrete covers (see Figure 4). Additionally, as in a future FRP-RC elements will become more available and they could also require strengthening and retrofitting, Slab Q′ was constructed using different internal reinforcement types (i.e., glass-, carbon-FRP, and steel rebars). In this case, they were all located at the same depth since this slab had defects and damage within the concrete, as shown in Figure 5. The reason to incorporate different reinforcement types was also to evaluate the detectability of these materials when an EB-FRP system is used.

Four types of defects in concrete were simulated to evaluate the feasibility of different NDT methods when an EB-FRP system is applied on the surface of the element. These defects were selected based on the literature review of prior phase (i.e., C1 to C3 in Table 3). Delamination, flexural and split cracks, as well as voids in the concrete, were simulated in Slab Q′ using thin architectural polystyrene foam (thickness of 6.35 mm or 1/4″) held in place with the use of epoxy as shown in Figure 5a.

6.2. Defects Generation for EB-FRP Concrete Elements

External application commonly refers to the installation of an FRP system, typically a wet-layup system where an FRP fabric/sheet is impregnated with resin in situ to facilitate the strengthening process. To simulate the common/potential defects of external FRP applications, two layers of an CFRP unidirectional system was applied on one face of the slab. Two different types of defects were generated to evaluate the feasibility of application of the selected NDT methods. The first, "adhesive–concrete debonding" (or FRP composite–concrete debonding in general) was generated by placing a thin film of 1.2 mm expanded polystyrene foam (EPS) on the surface of the concrete before impregnating it with resin. Figure 6a shows the first layer of defects also with a 40 μm thin film of polyethylene; however, this material was more difficult to bond to the concrete surface and prevent movement during installation. The second defect, FRP–adhesive debonding or delamination between layers of FRP sheets in a composite, was simulated by applying the same thin film (polystyrene foam) between the CFRP layers, thus creating a discontinuity between the layers. Figure 5 shows the strengthening application process.

7. Results and Discussion

7.1. Visual Inspection (VT)

Visual inspections were conducted on the EB-FRP slab specimens to identify various defects and damage such as fiber kinks, waviness, swelling, bubbles, voids, debonding, delamination, peeling, cracking, and fiber breakage (Figure 7), with further tap testing to confirm defects in areas suspected of having air pockets. Inspectors should also watch for signs of internal steel reinforcement damage, like corrosion, indicating potential concrete deterioration beneath the FRP layer. Visual inspection is limited to determining the location and quantity of defects/damage that appear on the surface, while additional NDT like UT, GPR, and IR may be needed to accurately size defects or locate those not visible on the surface, such as voids. The visual inspection on Slab M' and Q' successfully detected defects simulated with polystyrene foam but struggled to identify those created using thinner and denser polyethylene cutouts, suggesting that delamination without bulging of the laminate could be challenging to perceive.

Figure 7. Visual inspection of slabs: (**a**) Slab M', (**b**) Slab Q'.

7.2. Tap Testing (TT)

Tap testing (TT) was conducted over the suspected areas determined from visual inspection and over remaining areas with at least one strike per 0.1 m^2 (1 sq. ft), as shown in Figure 8. The procedure was executed on the external FRP application to discern variations in sound between bonded and unbonded laminate. It is similar to the one performed over concrete elements by bridge inspectors trained to hear the difference between concrete with and without delamination. The tap testing conducted successfully detected all simulated defects and damage in Slab M' and Q'. Small areas of 625 mm^2 (1 in.^{2}) were detected with TT. Typically, delamination less than 1300 mm^2 (2 in.^{2}) is permissible as long as the delaminated area is less than 5% of the total laminate area and there are no more than 10 such delaminations per 1 m^2 (10 ft^2) [30].

Figure 8. Tap Testing on Slab Q′.

7.3. Infrared Thermography (IR)

Infrared thermography was conducted on the slab specimens to identify minor defects and damage beneath the laminate, which may not have been detected through visual inspection and tap testing. The test relied on solar heating to establish a thermal gradient between the defective and intact areas. In cases where the slabs had maintained a uniform temperature across the top surface due to prolonged sun exposure prior to testing, a canopy was employed to induce the necessary thermal gradient. All defects and damage appeared as thermal anomalies or hot spots in the infrared images of the slabs (Figure 9). These anomalies recorded temperatures of about 49 °C (120 °F) in contrast to approximate 40 °C (105 °F) registered in the sound element. The IR was capable of highlighting defects as small as 625 mm^2 (1 in.^{2}), but the most favorable results were observed in the 2500 mm^2 (4 in.^{2}), whether between layers (delamination) or between the substrate and FRP (debonding). The B1 defect (delamination between layers) in Slab Q′ was created using a 40 μm polyethylene film, making it harder to identify the defect. However, this demonstrates that IR could detect delamination without bulging with the adequate expertise of the inspector.

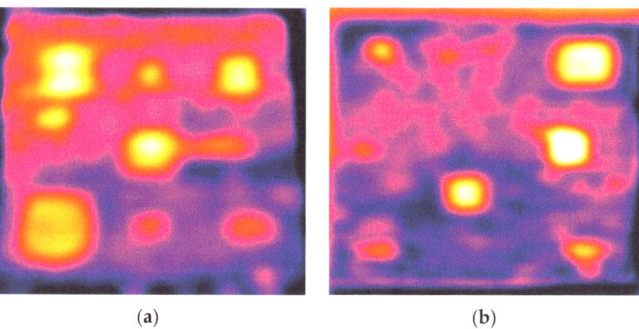

(a) (b)

Figure 9. Infrared image of slabs: (**a**) Slab M′, (**b**) Slab Q′.

7.4. Ground-Penetrating Radar (GPR)

The externally applied CFRP sheets on Slabs M′ and Q′ acted as a reflective surface for GPR devices due to their conductivity. GPR line scans (Figure 10) indicated significant interference, with multiple hyperbolas and reflections, making it challenging to discern any targets beneath the CFRP layer. The distinctive hyperbolas representing the four steel rebars embedded in Slab M′ before the application of the EB-FRP system, were no longer discernible in the line scan conducted over the slab with the EB-CFRP layer, as displayed in Figure 10 (perpendicular to the direction of the rebars). No information about external defects could be detected. No tests were conducted for the external application of GFRP sheets in this study, primarily because the vast majority of external FRP applications utilize CFRP due to its higher strength and stiffness. Therefore, it remains a possibility that GPR could detect anomalies in the concrete substrate if GFRP sheets were employed.

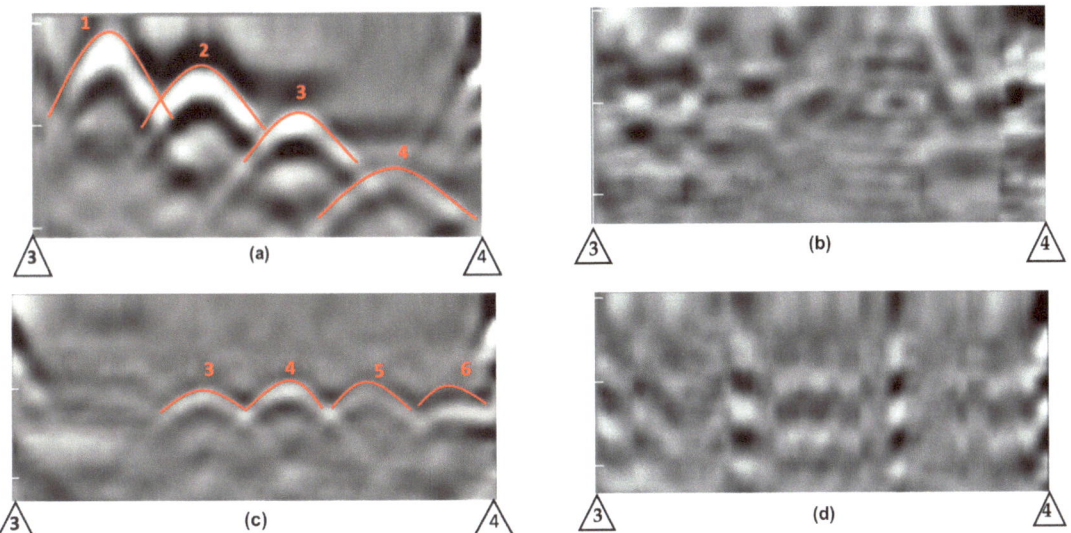

Figure 10. GPR line scans of Slab M′ and Q′. (**a**) Slab M′ before strengthening, (**b**) Slab M′ after strengthening, (**c**) Slab Q′ before strengthening, (**d**) Slab Q′ after strengthening.

7.5. Phased Array Ultrasonic Testing (PAU)

The advantage of employing PAU instead of a GPR device for inspecting externally applied FRP systems is its ability to penetrate conductive FRP layers, such as CFRP layers. The individual B-scan (line scan) of Slab Q′ with externally applied CFRP sheets, obtained during a PAU stripe scan, revealed the detectability of internal features like a 230 mm × 76 mm (9 in. × 3 in.) concrete delamination (Figure 11i) and steel rebars (Figure 11iii), which was not achievable using GPR. Although, in Figure 11ii, it is evident that a debonding introduced between the first layer of CFRP and the concrete substrate is clearly detectable, the size of this defect cannot be clearly established (in this case, it was a 100 mm × 100 mm (4 in. × 4 in.) debonding). However, this defect also acts as a strong reflector for nearly all ultrasonic waves, which explains why Figure 11ii appears uniformly red beneath the top surface. This implies that it is not possible to detect internal features immediately beneath the defects or damage on the top surface using PAU. Furthermore, other simulated defects such as voids and cracks could not be detected using PAU.

The effectiveness of the evaluated NDT methods (i.e., GPR, PAU, VT, TT, and IR) for inspection of EB-FRP concrete elements is presented in Table 6. They were classified into three different categories based on the detectability of the introduce defect: D (Detectable) if the defect was detectable either quantitative or qualitative; LD (Limited Detectability) is based on size of defect/damage; and ND (Not Detectable) if the defect cannot be either quantitatively or qualitatively detected by the technique evaluated.

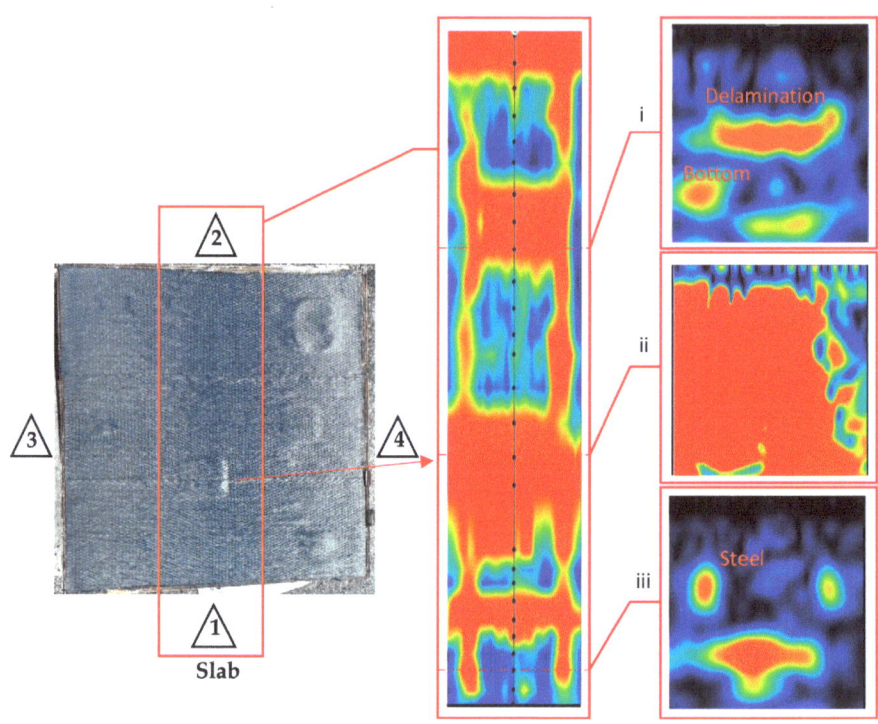

Figure 11. PAU stripe scan of Slab Q'. (**i**) 300 mm, (**ii**) 570 mm and (**iii**) 800 mm from face 2.

Table 6. Effectiveness of selective NDT methods for inspection of EB-FRP concrete elements.

Slab	Parameters [1]		Selected NDTs				
			GPR	PAU	VT	TT	IR
M'	Internal Targets		ND	D [2]	-	-	-
	External defects/damage	Debonding or Delamination	ND	LD [3]	LD [1]	D	D
Q'	Internal Targets		ND	D [2]	-	-	-
	External defects/damage	Debonding or Delamination	ND	LD [3]	LD [1]	D	D

Note: D = Detectable; LD = Limited Detectability; ND = Not Detectable. [1] Limited detection (LD) is based on the size of defect/damage. [2] Concrete delamination is detectable as long as the external applied fabric is sound. [3] Qualitatively detectable.

8. Conclusions

This study examined the types, characteristics, and identification of damage and defects that were either observed or expected in EB-FRP concrete elements. The defects and damage were categorized based on their location, time of initiation, and sources. The inspection of FRP-EB concrete elements can be categorized into three main groups. The first category focuses on visible surface damage and defects within the FRP composite. The second category involves inspecting damage within the FRP composite and at the bond layer between the FRP and concrete. The third category concentrates on identifying damage and defects in the concrete substrate itself. The most promising non-destructive testing (NDT) methods were reviewed and subsequently evaluated in small-scale EB-FRP concrete slabs. By offering a structured framework for inspecting structures utilizing wet lay-up carbon FRP systems, the findings of this study can serve as the foundation for the development of a guide and training materials for the inspection of structures employing wet lay-up carbon FRP systems.

- The externally applied FRP system should be visually examined thoroughly to identify surface anomalies, including blister-like formations, exposed fibers, surface scratches, and cracks. Signs of moisture and water stains near joints or lower areas underneath the structure. Surface anomalies observed in the externally applied FRP may indicate defects within the FRP composite or bonding issues between the FRP and concrete.
- Inspecting FRP composite defects and bond issues may necessitate NDT methods beyond visual inspection (VT). Tap testing (TT) is suitable for detecting bond defects to prevent the separation of externally applied FRP system from the concrete substrate. Additionally, IR can be employed for quantitative defect assessment within the FRP composite or between the FRP and concrete, capable of detecting areas as small as 625 mm^2. PAU can be employed for qualitative assessment of the EB-FRP.
- Inspecting hidden concrete under external FRP is challenging but achievable by noting evidence of internal defects (e.g., detecting FRP tearing due to concrete spalling), observing anomalies deviating from sound FRP (e.g., CFRP bulging indicating underlying cracks), and checking for rust stains (e.g., a sign of embedded steel corrosion). Employment of NDT devices capable of penetrating FRP (e.g., PAU) is desirable for an in-depth investigation. The coupling of these defects potentially adds complexity to accurate defect identification. However, the effectiveness of the device and the technician's expertise play a crucial role in detecting and distinguishing such complex defects. Nevertheless, the presence of damage regardless of the type and complexity should trigger further examination and potentially corrective action.
- In a contrast to the results of a previous literature review, it was determined that GPR could not detect defects or damage introduced into the externally applied CFRP and the internal targets beneath the CFRP layer due to its conductive nature. PAU exhibited relatively better performance in inspecting the external application of FRP, being able to qualitatively detect introduced debonding/delamination in the external FRP and delamination within the concrete. Other NDT techniques, including visual inspection (VT), tap testing (TT), and infrared thermography (IR), were also found to be quite effective in detecting primarily surface anomalies and some bond defects, such as voids.

Author Contributions: Conceptualization, J.D.O. and S.S.K.D.; methodology, J.D.O.; formal analysis, J.D.O., S.S.K.D. and P.M.; investigation, J.D.O., S.S.K.D. and P.M.; resources, A.M. and A.N.; writing—original draft preparation, J.D.O.; writing—review and editing, A.M., A.N., S.S.K.D. and P.M.; visualization, J.D.O., S.S.K.D. and P.M.; funding acquisition, A.M. and A.N. All authors have read and agreed to the published version of the manuscript.

Funding: This research received no external funding.

Data Availability Statement: Data are contained within the article.

Acknowledgments: The authors greatly acknowledge the internal support of the Department of Civil and Environmental Engineering at Florida International University and the Department of Civil and Architectural Engineering at the University of Miami. The contents of this paper reflect the views of the authors, who are responsible for the facts and the accuracy of the information presented herein.

Conflicts of Interest: The authors declare no conflicts of interest.

References

1. Rodrigues, R.; Gaboreau, S.; Gance, J.; Ignatiadis, I.; Betelu, S. Reinforced Concrete Structures: A Review of Corrosion Mechanisms and Advances in Electrical Methods for Corrosion Monitoring. *Constr. Build. Mater.* **2021**, *269*, 121240. [CrossRef]
2. Angst, U.M.; Isgor, O.B.; Hansson, C.M.; Sagüés, A.; Geiker, M.R. Beyond the Chloride Threshold Concept for Predicting Corrosion of Steel in Concrete. *Appl. Phys. Rev.* **2022**, *9*, 011321. [CrossRef]

3. Mugahed Amran, Y.H.; Alyousef, R.; Rashid, R.S.M.; Alabduljabbar, H.; Hung, C.C. Properties and Applications of FRP in Strengthening RC Structures: A Review. *Structures* **2018**, *16*, 208–238. [CrossRef]
4. Liu, T.Q.; Liu, X.; Feng, P. A Comprehensive Review on Mechanical Properties of Pultruded FRP Composites Subjected to Long-Term Environmental Effects. *Compos. B Eng.* **2020**, *191*, 107958. [CrossRef]
5. Benmokrane, B.; Hassan, M.; Robert, M.; Vijay, P.V.; Manalo, A. Effect of Different Constituent Fiber, Resin, and Sizing Combinations on Alkaline Resistance of Basalt, Carbon, and Glass FRP Bars. *J. Compos. Constr.* **2020**, *24*, 04020010. [CrossRef]
6. Zaman, A.; Gutub, S.A.; Wafa, M.A. A Review on FRP Composites Applications and Durability Concerns in the Construction Sector. *J. Reinf. Plast. Compos.* **2013**, *32*, 1966–1988. [CrossRef]
7. Moy, S. Advanced Fiber-Reinforced Polymer (FRP) Composites for Civil Engineering Applications. In *Developments in Fiber-reinforced Polymer (FRP) Composites for Civil Engineering*; Elsevier: Amsterdam, The Netherlands, 2013; pp. 177–204.
8. Hollaway, L.C. A Review of the Present and Future Utilisation of FRP Composites in the Civil Infrastructure with Reference to Their Important In-Service Properties. *Constr. Build. Mater.* **2010**, *24*, 2419–2445. [CrossRef]
9. Teng, J.G.; Chen, J.-F.; Smith, S.T.; Lam, L. *FRP: Strengthened RC Structures*; John Wiley & Sons: Hoboken, NJ, USA, 2002; ISBN 0471487066.
10. Benzecry, V.; Brown, J.; Al-Khafaji, A.; Haluza, R.; Koch, R.; Nagarajan, M.; Bakis, C.; Myers, J.; Nanni, A. *Durability of GFRP Bars Extracted from Bridges with 15 to 20 Years of Service Life*; ACI Foundation: Farmington Hills, MI, USA, 2019.
11. Benmokrane, B.; Wang, P.; Ton-That, T.M.; Rahman, H.; Robert, J.-F. Durability of Glass Fiber-Reinforced Polymer Reinforcing Bars in Concrete Environment. *J. Compos. Constr.* **2002**, *6*, 143–153. [CrossRef]
12. Micelli, F.; Nanni, A. Durability of FRP Rods for Concrete Structures. *Constr. Build. Mater.* **2004**, *18*, 491–503. [CrossRef]
13. Ceroni, F.; Cosenza, E.; Gaetano, M.; Pecce, M. Durability Issues of FRP Rebars in Reinforced Concrete Members. *Cem. Concr. Compos.* **2006**, *28*, 857–868. [CrossRef]
14. Chen, Y.; Davalos, J.F.; Ray, I.; Kim, H.Y. Accelerated Aging Tests for Evaluations of Durability Performance of FRP Reinforcing Bars for Concrete Structures. *Compos. Struct.* **2007**, *78*, 101–111. [CrossRef]
15. Malla, P.; Khedmatgozar Dolati, S.S.; Ortiz, J.D.; Mehrabi, A.; Nanni, A. Damage and Defects in Fiber-Reinforced Polymer Reinforced and Strengthened Concrete Elements. *J. Compos. Constr.* **2023**, *27*. [CrossRef]
16. Das, S.C.; Nizam, M. Applications of Fiber Reinforced Polymer Composites (FRP) in Civil Engineering. *Int. J. Adv. Struct. Geotech. Eng.* **2014**, *3*, 299–309.
17. Wu, Z.; Wu, Y.; Fahmy, M.F.M. *Structures Strengthened with Bonded Composites*; Woodhead Publishing Series in Civil and Structural Engineering; Woodhead Publishing: Sawston, UK, 2020; Volume 1, ISBN 9780128210888.
18. Frigione, M.; Lettieri, M. Durability Issues and Challenges for Material Advancements in FRP Employed in the Construction Industry. *Polymers* **2018**, *10*, 247. [CrossRef]
19. Motavalli, M.; Czaderski, C. FRP Composites for Retrofitting of Existing Civil Structures in Europe: State-of-the-Art Review. In Proceedings of the International Conference of Composites & Polycon, Tampa, FL, USA, 17–19 October 2007; American Composites Manufacturers Association: Tampa, FL, USA, 2007; pp. 17–19.
20. Ritchie, P.A.; Thomas, D.A.; Lu, L.-W.; Connelly, G.M. External Reinforcement of Concrete Beams Using Fiber Reinforced Plastic. Master's Thesis, Lehigh University, Bethlehem, PA, USA, 1988.
21. Arduini, M.; Nanni, A. Parametric Study of Beams with Externally Bonded FRP Reinforcement. *ACI Struct. J.* **1997**, *94*, 493–501.
22. Toutanji, H.; Zhao, L.; Zhang, Y. Flexural Behavior of Reinforced Concrete Beams Externally Strengthened with CFRP Sheets Bonded with an Inorganic Matrix. *Eng. Struct.* **2006**, *28*, 557–566. [CrossRef]
23. Hutchinson, A.R. *Surface Preparation of Component Materials*; Woodhead Publishing Limited: Sawston, UK, 2008; ISBN 9781845694487.
24. Hutchinson, A. Adhesives for Externally Bonded FRP Reinforcement. *ICE Man. Constr. Mater.* **2009**, *2*, 667–674.
25. Hollaway, L.C. *Applications of Advanced Fibre-Reinforced Polymer (FRP) Composites in Bridge Engineering: Rehabilitation of Metallic Bridge Structures, All-FRP Composite Bridges, and Bridges Built with Hybrid Systems*; Woodhead Publishing Limited: Sawston, UK, 2013; ISBN 9780857094186.
26. Alkhrdaji, T.; Nanni, A.; Chen, G.; Barker, M.G. Destructive and Non-Destructive Testing of Bridge J857, Phelps County, Missouri. Volume I-Strengthening and Testing to Failure of Bridge Decks; 2001; Vol. No. RDT01-. Available online: https://spexternal.modot.mo.gov/sites/cm/CORDT/RDT01002A.pdf (accessed on 14 December 2023).
27. Ortiz, J.D.; Khedmatgozar Dolati, S.S.; Malla, P.; Nanni, A.; Mehrabi, A. FRP-Reinforced/Strengthened Concrete: State-of-the-Art Review on Durability and Mechanical Effects. *Materials* **2023**, *16*, 1990. [CrossRef]
28. Horse Construction-Structural Strengthening System Structural Strengthening with Carbon Fiber CFRP Composite System. Available online: www.horseen.com (accessed on 21 October 2023).
29. AASHTO. *Guide Specifications for Design of Bonded FRP Systems for Repair and Strengthening of Concrete Bridge Elements*, 2nd ed.; American Association of State Highway and Transportation Officials: Washington, DC, USA, 2023; ISBN 978-1-56051-807-5.
30. American Concrete Institute (ACI) Committee 440. *ACI 440.2R-17. Guide for the Design and Construction of Externally Bonded FRP Systems for Strengthening Concrete Structures*; American Concrete Institute: Farmington Hills, MI, USA, 2017.
31. JPCI. *Recommendation for Design and Construction of Concrete Structures Using Fiber Reinforced Polymer*; Japan Prestressed Concrete Institute: Tokyo, Japan, 2021.
32. Riccitelli, F.; Mehrabi, A.; Abedin, M.; Farhangdoust, S.; Khedmatgozar Dolati, S.S. *Performance of Existing Abc Projects: Inspection Case Studies*; Accelerated Bridge Construction University Transportation Center (ABC-UTC): Miami, FL, USA, 2020.

33. Khedmatgozar Dolati, S.S.; Malla, P.; Mehrabi, A.; Ortiz Polanco, J.; Nanni, A. Non-Destructive Testing Applications for in-Service FRP Reinforced/Strengthened Concrete Bridge Elements. In *Proceedings of the Nondestructive Characterization and Monitoring of Advanced Materials, Aerospace, Civil Infrastructure, and Transportation XVI, Long Beach, CA, USA, 6 March–11 April 2022*; Wu, H.F., Gyekenyesi, A.L., Shull, P.J., Yu, T., Eds.; SPIE: Long Beach, CA, USA, 2022; Volume 12047, p. 1204708.
34. Malla, P.; Khedmatgozar Dolati, S.S.; Ortiz, J.D.; Mehrabi, A.B.; Nanni, A.; Dinh, K. Feasibility of Conventional Non-Destructive Testing Methods in Detecting Embedded FRP Reinforcements. *Appl. Sci.* **2023**, *13*, 4399. [CrossRef]
35. Khedmatgozar Dolati, S.S.; Caluk, N.; Mehrabi, A.; Khedmatgozar Dolati, S.S. Non-Destructive Testing Applications for Steel Bridges. *Appl. Sci.* **2021**, *11*, 9757. [CrossRef]
36. Yazdani, N.; Garcia, E.C.; Riad, M. *Field Assessment of Concrete Structures Rehabilitated with FRP*; Elsevier Ltd.: Amsterdam, The Netherlands, 2018; ISBN 9780081021811.
37. Ettouney, S.; Alampalli, M. *Infrastructure Health in Civil Engineering*; CRC Press: Boca Raton, FL, USA, 2011; ISBN 9781439866542.
38. Khanal, S. *Review of Modern Nondestructive Testing Techniques for Civil Infrastructure*; West Virginia University: Morgantown, WV, USA, 2020.
39. Wheeler, A.S. *Nondestructive Evaluation of Concrete Bridge Columns Rehabilitiated with Fiber Reinforced Polymers Using Digital Tap Hammer and Infrared Thermography Nondestructive Evaluation of Concrete Bridge Columns*; West Virginia University: Morgantown, WV, USA, 2018.
40. Halabe, U.B.; Joshi, R.M.; Gangarao, H.V.S. Nondestructive Testing of FRP Composite Structural Components and FRP Rehabilitated Bridge Using Digital Tap Testing. *J. Multidiscip. Eng. Sci. Technol.* **2020**, *7*, 11477–11482.
41. Taillade, F.; Quiertant, M.; Benzarti, K.; Aubagnac, C.; Moser, E. Non-Destructive Evaluation (NDE) of Composites: Using Shearography to Detect Bond Defects. In *Non-Destructive Evaluation (NDE) of Polymer Matrix Composites Techniques and Applications*; Woodhead Publishing: Sawston, UK, 2013; pp. 542–556. [CrossRef]
42. Ekenel, M.; Myers, J.J. Nondestructive Evaluation of RC Structures Strengthened with FRP Laminates Containing Near-Surface Defects in the Form of Delaminations. *Sci. Eng. Compos. Mater.* **2007**, *14*, 299–315. [CrossRef]
43. Hsieh, C.T.; Lin, Y. Detecting Debonding Flaws at the Epoxy-Concrete Interfaces in near-Surface Mounted CFRP Strengthening Beams Using the Impact-Echo Method. *NDT E Int.* **2016**, *83*, 1–13. [CrossRef]
44. Crawford, K.C. Non-Destructive Testing of FRP-Structural Systems Applied to Concrete Bridges. In *Nondestructive Testing of Materials and Structures*; Springer: Dordrecht, The Netherlands, 2013; pp. 835–840.
45. Crawford, K.C. NDT Evaluation of Long-Term Bond Durability of CFRP-Structural Systems Applied to RC Highway Bridges. *Int. J. Adv. Struct. Eng.* **2016**, *8*, 161–168. [CrossRef]
46. ACI. *ACI 228.2R-13: Report on Nondestructive Test Methods for Evaluation of Concrete in Structures*; American Concrete Institute: Farmington Hills, MI, USA, 2013.
47. Dong, Y.; Ansari, F. Non-Destructive Testing and Evaluation (NDT/NDE) of Civil Structures Rehabilitated Using Fiber Reinforced Polymer (FRP) Composites. In *Service Life Estimation and Extension of Civil Engineering Structures*; Elsevier: Amsterdam, The Netherlands, 2011; pp. 193–222.
48. Gower, M.; Lodeiro, M.; Aktas, A.; Shaw, R.; Maierhofer, C.; Krankenhagen, R.; Augustin, S.; Rollig, M.; Knazovicka, L.; Blahut, A.; et al. Design and Manufacture of Reference and Natural Defect Artefacts for the Evaluation of NDE Techniques for Fibre Reinforced Plastic (FRP) Composites in Energy Applications. In Proceedings of the 19th World Conference on Non-Destructive Testing (WCNDT 2016), Munich, Germany, 13–17 June 2016; Volume 21.
49. Aboukhousa, M.; Qaddoumi, N. Near-Field Microwave Imaging of Subsurface Inclusions in Laminated Composite Structures. In Proceedings of the 16th World Conference on Nondestructive Testing, Montreal, QC, Canada, 30 August–3 September 2004.
50. Kharkovsky, S.; Ryley, A.C.; Stephen, V.; Zoughi, R. Dual-Polarized near-Field Microwave Reflectometer for Noninvasive Inspection of Carbon Fiber Reinforced Polymer-Strengthened Structures. *IEEE Trans. Instrum. Meas.* **2008**, *57*, 168–175. [CrossRef]
51. Navagato, M.D.; Narayanan, R.M. Microwave Imaging of Multilayered Structures Using Ultrawideband Noise Signals. *NDT E Int.* **2019**, *104*, 19–33. [CrossRef]
52. Ekenel, M.; Stephen, V.; Myers, J.J.; Zoughi, R. *Microwave Nde of Rc Beams Strengthened With Cfrp Laminates Containing Surface Defects and Tested Under Cyclic Loading*; Electrical and Computer Engineering, University of Missouri-Rolla: Rolla, MO, USA, 2004.
53. Akuthota, B.; Hughes, D.; Zoughi, R.; Myers, J.; Nanni, A. Near Field Microwave Detection of Disbond in Carbon Fiber Reinforced Polymer Composites Used for Strengthening Cement-Based Structures and Disbond Repair Verification. *J. Mater. Civ. Eng.* **2004**, *16*, 540–546. [CrossRef]
54. Drobiec, Ł.; Jasiński, R.; Mazur, W. The Use of Non-Destructive Methods to Detect Non-Metallic Reinforcement in Concrete and Masonry. *Preprints* **2019**. [CrossRef]
55. Jackson, D.; Islam, M.; Alampalli, S. Feasibility of Evaluating the Performance of Fiber Reinforced Plastic (FRP) Wrapped Reinforced Concrete Columns Using Ground Penetrating RADAR (GPR) and Infrared (IR) Thermography Techniques. In Proceedings of the Structural Materials Technology IV—An NDT Conference, Atlantic City, NJ, USA, 28 February–3 March 2000.
56. Yazdani, N.; Beneberu, E.; Riad, M. Nondestructive Evaluation of FRP-Concrete Interface Bond Due to Surface Defects. *Adv. Civ. Eng.* **2019**, *2019*, 2563079. [CrossRef]
57. La Malfa Ribolla, E.; Rezaee Hajidehi, M.; Rizzo, P.; Fileccia Scimemi, G.; Spada, A.; Giambanco, G. Ultrasonic Inspection for the Detection of Debonding in CFRP-Reinforced Concrete. *Struct. Infrastruct. Eng.* **2018**, *14*, 807–816. [CrossRef]

58. Hing, C.L.C.; Halabe, U.B. Nondestructive Testing of GFRP Bridge Decks Using Ground Penetrating Radar and Infrared Thermography. *J. Bridge Eng.* **2010**, *15*, 391–398. [CrossRef]
59. Taheri, H.; Hassen, A.A. Nondestructive Ultrasonic Inspection of Composite Materials: A Comparative Advantage of Phased Array Ultrasonic. *Appl. Sci.* **2019**, *9*, 1628. [CrossRef]
60. Taheri, H.; Delfanian, F.; Du, J. Acoustic emission and ultrasound phased array technique for composite material evaluation. In Proceedings of the ASME International Mechanical Engineering Congress and Exposition, Proceedings (IMECE), San Diego, CA, USA, 15–21 November 2013; American Society of Mechanical Engineers (ASME): New York, NY, USA, 2013; Volume 1.
61. Meola, C.; Boccardi, S.; Carlomagno, G.M.; Boffa, N.D.; Monaco, E.; Ricci, F. Nondestructive Evaluation of Carbon Fibre Reinforced Composites with Infrared Thermography and Ultrasonics. *Compos. Struct.* **2015**, *134*, 845–853. [CrossRef]
62. Boychuk, A.S.; Generalov, A.S.; Stepanov, A.V. CFRP Structural Health Monitoring by Ultrasonic Phased Array Technique. In Proceedings of the 7th European Workshop on Structural Health Monitoring, EWSHM 2014—2nd European Conference of the Prognostics and Health Management (PHM) Society, Nantes, France, 8–11 July 2014; pp. 2206–2211.
63. Ryan, T.W.; EricMann, J.; Chill, Z.M.; Ott, B.T. Bridge Inspector's Reference Manual. *FHWA* **2012**, *BIRM 1*, 1020.
64. Galietti, U.; Luprano, V.; Nenna, S.; Spagnolo, L.; Tundo, A. Non-Destructive Defect Characterization of Concrete Structures Reinforced by Means of FRP. *Infrared Phys. Technol.* **2007**, *49*, 218–223. [CrossRef]
65. Taillade, F.; Quiertant, M.; Benzarti, K.; Aubagnac, C.; Moser, E. Shearography Applied to the Non Destructive Evaluation of Bonded Interfaces between Concrete and CFRP Overlays: From the Laboratory to the Field. *Eur. J. Environ. Civ. Eng.* **2010**, *15*, 545–556. [CrossRef]
66. Yang, L. Recent Developments in Digital Shearography for Nondestructive Testing. *Mater. Eval.* **2006**, *64*, 704–709.
67. Qiu, Q.; Lau, D. A Novel Approach for Near-Surface Defect Detection in FRP-Bonded Concrete Systems Using Laser Reflection and Acoustic-Laser Techniques. *Constr. Build. Mater.* **2017**, *141*, 553–564. [CrossRef]
68. Qiu, Q.; Lau, D. Experimental Evaluation on the Effectiveness of Acoustic-Laser Technique towards the FRP-Bonded Concrete System. *Struct. Health Monit. Insp. Adv. Mater. Aerosp. Civ. Infrastruct.* **2015**, *9437*, 943705. [CrossRef]
69. Büyüköztürk, O.; Haupt, R.; Tuakta, C.; Chen, J. Remote Detection of Debonding in FRP-Strengthened Concrete Structures Using Acoustic-Laser Technique. In *Nondestructive Testing of Materials and Structures*; Springer: Dordrecht, The Netherlands, 2013; Volume 6, pp. 19–24.
70. Chen, J.G.; Haupt, R.W.; Büyüköztürk, O. Remote Characterization of Defects in FRP Strengthened Concrete Using the Acoustic-Laser Vibrometry Method. In Proceedings of the ASNT Fall Conference, Las Vegas, NV, USA, 4–7 November 2013.
71. Zhu, Y.K.; Tian, G.Y.; Lu, R.S.; Zhang, H. A Review of Optical NDT Technologies. *Sensors* **2011**, *11*, 7773–7798. [CrossRef]
72. Chen, J.G.; Buyukozturk, O.; Haupt, R.W. Operational and Defect Parameters Concerning the Acoustic-Laser Vibrometry Method for FRP-Reinforced Concrete. *NDT Int.* **2015**, *71*, 43–53. [CrossRef]
73. Malhotra, V.M.; Carino, N.J. *Handbook on Nondestructive Testing of Concrete*; CRC Press: Boca Raton, FL, USA, 2004.
74. Garney, G. Defects Found through Non-Destructive Testing Methods of Fiber Reinforced Polymeric Composites. Master's Thesis, California State University, Fullerton, CA, USA, 2006.
75. Wan, B. *Using Fiber-Reinforced Polymer (FRP) Composites in Bridge Construction and Monitoring Their Performance: An Overview*; Woodhead Publishing Limited: Sawston, UK, 2014; ISBN 9780857097019.
76. American Society for Nondestructive Testing. *Nondestructive Testing Handbook*, 3rd ed.; Moore, D.G., Moore, P.O., Eds.; American Society for Nondestructive Testing: Columbus, OH, USA, 2008; Volume 8, ISBN 9781571171849.
77. Farrar, C.R.; Worden, K. An Introduction to Structural Health Monitoring. *Philos. Trans. R. Soc. A Math. Phys. Eng. Sci.* **2007**, *365*, 303–315. [CrossRef]
78. Böer, P.; Holliday, L.; Kang, T.H.K. Interaction of Environmental Factors on Fiber-Reinforced Polymer Composites and Their Inspection and Maintenance: A Review. *Constr. Build. Mater.* **2014**, *50*, 209–218. [CrossRef]
79. Karbhari, V.M.; Chin, J.W.; Hunston, D.; Benmokrane, B.; Juska, T.; Morgan, R.; Lesko, J.J.; Sorathia, U.; Reynaud, D. Durability Gap Analysis for Fiber-Reinforced Polymer Composites in Civil Infrastructure. *J. Compos. Constr.* **2003**, *7*, 238–247. [CrossRef]
80. Miracle, D.B.; Donaldson, S.L. *ASM Handbook (Volume 21): Composites*; ASM International: Novelty, OH, USA, 2001.
81. Telang, N.; Dumalo, C.; Mehrabi, A.B.; Ciolko, A.T.; Jim, G. *Field Inspection of In-Service FRP Bridge Decks*; National Academies Press: Washington, DC, USA, 2006.
82. Kang, T.H.K.; Howell, J.; Kim, S.; Lee, D.J. A State-of-the-Art Review on Debonding Failures of FRP Laminates Externally Adhered to Concrete. *Int. J. Concr. Struct. Mater.* **2012**, *6*, 123–134. [CrossRef]
83. Yumnam, M.; Gupta, H.; Ghosh, D.; Jaganathan, J. Inspection of Concrete Structures Externally Reinforced with FRP Composites Using Active Infrared Thermography: A Review. *Constr. Build. Mater.* **2021**, *310*, 125265. [CrossRef]
84. Barry, K.F.; Silva, M.; Biagini, G.; Santamarina, J.C.; Wall, J.J.; Marcel, Y.; Lc, R.; Lindberg, J.T.; Cha, M.; Dai, S.; et al. Void Detection System. US patent 2013/0192375 A1, 1 August 2013.
85. Karbhar, V.M.; Kaiser, H.; Navada, R.; Ghosh, K.; Lee, L. *Methods for Detecting Defects in Composite Rehabilitated Concrete Structures*; Federal Highway Administration: Washington, DC, USA, 2005.
86. Da Fonseca, B.S.; Castela, A.S.; Silva, M.A.G.; Duarte, R.G.; Ferreira, M.G.S.; Montemor, M.F. Influence of GFRP Confinement of Reinforced Concrete Columns on the Corrosion of Reinforcing Steel in a Salt Water Environment. *J. Mater. Civ. Eng.* **2015**, *27*, 04014107. [CrossRef]

87. Masoud, S.; Soudki, K. Evaluation of Corrosion Activity in FRP Repaired RC Beams. *Cem. Concr. Compos.* **2006**, *28*, 969–977. [CrossRef]
88. Mehrabi, A.; Farhangdoust, S. *NDT Methods Applicable to Health Monitoring of ABC Closure Joints*; Accelerated Bridge Construction University Transportation Center: Miami, FL, USA, 2019.
89. Khedmatgozar Dolati, S.S.; Malla, P.; Ortiz, J.D.; Mehrabi, A.; Nanni, A. Identifying NDT Methods for Damage Detection in Concrete Elements Reinforced or Strengthened with FRP. *Eng. Struct.* **2023**, *287*, 116155. [CrossRef]
90. Sen, R. Developments in the Durability of FRP-Concrete Bond. *Constr. Build. Mater.* **2015**, *78*, 112–125. [CrossRef]
91. *ASTM Committee C09 C39/C39M-21*; Standard Test Method for Compressive Strength of Cylindrical Concrete Specimens. ASTM International: West Conshohocken, PA, USA, 2021.

Disclaimer/Publisher's Note: The statements, opinions and data contained in all publications are solely those of the individual author(s) and contributor(s) and not of MDPI and/or the editor(s). MDPI and/or the editor(s) disclaim responsibility for any injury to people or property resulting from any ideas, methods, instructions or products referred to in the content.

Article

Modified Constitutive Models and Mechanical Properties of GFRP after High-Temperature Cooling

Junjie Wu [1,2] and Chuntao Zhang [1,2,3,*]

1. Shock and Vibration of Engineering Materials and Structures Key Laboratory of Sichuan Province, Mianyang 621010, China; wujunjie19981228@163.com
2. School of Civil Engineering and Architecture, Southwest University of Science and Technology, Mianyang 621010, China
3. Department of Mechanical Engineering, University of Houston, Houston, TX 77204, USA
* Correspondence: chuntaozhang@swust.edu.cn

Abstract: Many materials are highly sensitive to temperature, and the study of the fire resistance of materials is one of the important research directions, which includes the study of the fire resistance of fiber-reinforced polymer (FRP) composites, but the cooling mode on the change of FRP mechanical properties after high temperature has not been investigated. This study analyzes the mechanical properties of GFRP under various cooling methods after exposure to high temperatures. The tensile strength of GFRP was evaluated through water cooling, firefighting foam cooling, and air cooling within the temperature range of 20–300 °C. Damage modes were investigated at different target temperatures. The results indicate that the tensile strength of air-cooled GFRP is the highest, whereas water cooling yields the lowest retention rate. It indicates that the FRP temperature decreases slowly under air cooling and the better recovery of the damage within the resin matrix, while under water cooling, the damage at the fiber/resin interface is exacerbated because of the high exposed temperature and the water, resulting in a reduction in the strength of GFRP. Between 20 and 150 °C, GFRP essentially recovers its mechanical properties after cooling, with a residual tensile strength factor exceeding 0.9. In the range of 150–250 °C, GFRP exhibits a graded decline in strength. At 300 °C, GFRP loses certain mechanical properties after cooling, with a residual tensile strength factor below 0.1. Furthermore, the analysis of experimental results led to the modification of the Johnson–Cook constitutive model, proposing a model for GFRP under three cooling methods. Additionally, a predictive model for the elastic modulus of GFRP after high-temperature cooling was derived, showing agreement with experimental results.

Keywords: glass fiber-reinforced polymer (GFRP); cooling methods; mechanical properties; modified constitutive model

Citation: Wu, J.; Zhang, C. Modified Constitutive Models and Mechanical Properties of GFRP after High-Temperature Cooling. *Buildings* **2024**, *14*, 439. https://doi.org/10.3390/buildings14020439

Academic Editor: Cedric Payan

Received: 27 December 2023
Revised: 20 January 2024
Accepted: 26 January 2024
Published: 5 February 2024

Copyright: © 2024 by the authors. Licensee MDPI, Basel, Switzerland. This article is an open access article distributed under the terms and conditions of the Creative Commons Attribution (CC BY) license (https://creativecommons.org/licenses/by/4.0/).

1. Introduction

Fiber-reinforced polymer (FRP) composites, known for their lightweight and high strength, offer excellent corrosion resistance and specific strength. They find diverse applications in aerospace, military, civil construction, and infrastructure projects, serving as cost-effective alternatives to traditional steel components [1–4]. Understanding the mechanical behavior of FRP composites at elevated temperatures is crucial for both civil and military applications [5,6]. In addition to being cheaper than AFRP and CFRP, GFRP has better bonding performance and mature technology. As a result, it is used in large-scale applications in various fields. So this paper conducts experimental research using GFRP. These composites, comprised of glass fibers, carbon fibers, and a resin matrix formed through processes like winding, molding, or pultrusion, are particularly sensitive to temperature variations, with the resin component being a key influencing factor.

The resin matrix in FRP commonly consists of thermosetting and thermoplastic resins, both of which are prone to softening at elevated temperatures. As the temperature rises, the matrix undergoes transitions from a glassy to a leathery, rubbery, and eventually decomposed state, leading to a notable decline in FRP's mechanical properties [7,8]. In addition, the tensile strength of FRP decreases significantly with increasing exposed temperature and time [9]. Experimental and theoretical support is essential to ascertaining the post-fire mechanical behavior of FRP, a critical consideration for structures seeking to regain functionality after fire incidents. Consequently, investigating the mechanical properties of FRP at high temperatures becomes imperative.

Exploration into the mechanical properties of FRP at elevated temperatures stands as a pivotal focus within FRP material research. This exploration encompasses three primary directions: the mechanical behavior of FRP at high temperatures [6,10–14], the bond strength of FRP reinforcement and concrete under high temperatures [15–20], and the effects of high temperatures on FRP reinforcing and restraining components [21–25]. Numerous researchers have delved into this subject, utilizing experimental, numerical, and analytical approaches. For instance, Zike Wang [9] presents an investigation on the durability of basalt- and glass-fiber-reinforced polymer bars exposed to an SWSSC environment under different sustained stress levels. It was experimentally obtained that the tensile strength of BFRP and GFRP decreased significantly under the combined effect of sustained stress and exposure temperature. Gang Wu [26] investigated the tensile properties of BFRP bars under four environments: alkaline solution, salt solution, acid solution, and de-ionized water at 25, 40, and 55 °C. The results showed that the exposure temperature and environmental corrosion can damage the fiber/resin interface, resulting in a significant decrease in the tensile strength of BFRP bars. H. Wang et al. [27] conducted dynamic and quasi-static compressive tests on a ceramicized polymer composite, revealing a substantial decrease in compressive strength with increasing temperature. Khaneghahi MH et al. [28] explored the impact of intumescent paint on the mechanical properties of FRP at various temperatures, observing a significant inhibitory effect on the decrease in tensile strength. C. Wang [29] investigated the mechanical properties of FRP as internal reinforcement in concrete structures at high temperatures, noting variations in performance. Despite these valuable insights, existing research predominantly focuses on the mechanical properties of FRP materials at elevated temperatures, leaving a notable gap in the exploration of their behavior after high-temperature cooling—an essential consideration for preserving FRP performance after exposure to elevated temperatures.

By investigating the effects of different cooling methods on the mechanical properties of GFRP after high temperature, the aim of this study is to provide the best fire extinguishing method for GFRP materials after exposure to fire, and to establish the constitutive model of GFRP under different cooling methods after high temperature. This is crucial for the application of GFRP in practical engineering and provides a basis for the reliability assessment of thermal protection structures in extreme environments. In addition to this, providing the failure range of GFRP's mechanical properties after cooling will facilitate the safety assessment of the building.

2. Materials and Methods
2.1. Specimen Design

The Glass Fiber Reinforced Polymer (GFRP), a flat strip composed of unsaturated polyester resin and glass fibers, was chosen for this test. The design approach for the layering structure of GFRP is unidirectional layering. The thermosetting temperature of this resin is approximately 90 °C. According to GB/T 1447-2005 [30], the dimensions and configuration of the GFRP are depicted in Figure 1. With a thickness of 3.7 mm and a fiber volume fraction of 62%, the glass fibers are predominantly aligned along the axial direction of the plate. Table 1 presents the essential mechanical properties of the GFRP.

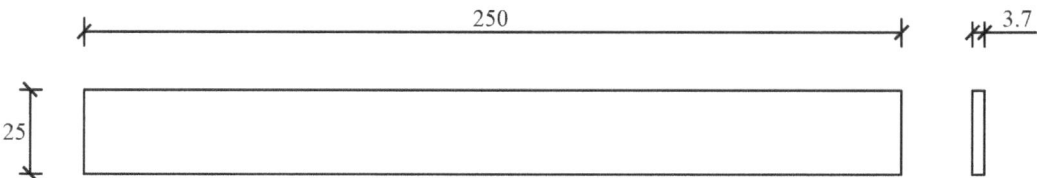

Figure 1. Dimensions of the test specimen (mm).

Table 1. Mechanical Properties of GFRP.

Specimen Labels	Tensile Strength f_u/MPa	Modulus of Elasticity E/GPa	Elongation δ/%
1	427.89	37.05	12.05
2	563.31	39.16	14.33
3	467.92	38.15	12.51
Average	486.37	38.12	12.96
Standard Deviation	56.80	0.86	0.98

2.2. Test Procedures

To investigate the impact of different cooling methods on GFRP after exposure to high temperatures, the test involved five target temperatures and 48 specimens. Three specimens were tested at each target temperature using various cooling methods, and an additional three specimens were kept at ambient temperature as a control group. The high-temperature setting utilized a box-shaped resistance furnace (SX2-4-10 type) which is illustrated in Figure 2a, with target temperatures set at 100 °C, 150 °C, 200 °C, 250 °C, and 300 °C. After reaching the target temperature, the specimens were left for a one-hour constant temperature phase, according to GB/T 38515-2020 [31]. Subsequently, cooling was performed using water, firefighting foam, and the natural cooling method. The entire process, including heating, insulation, and cooling under different target temperatures, is illustrated in Figure 3. After cooling to ambient temperature, a strain gauge was affixed to the center of each specimen to measure the GFRP strain. A static tensile test, according to GB/T 1447-2005 [30], involved controlling tensile loading by displacement with a loading rate of 1 mm/min until specimen fracture. The EMT504D electronic universal testing machine (Figure 2b) is manufactured in Shenzhen Wance Testing Equipment Co., Ltd., Shenzhen, China, and the load cell used is S-TYPE LOAD CELLS, which facilitated the static tensile test. The comparison and analysis of test results under different temperature conditions, including the stress–strain relationship curve of GFRP and related mechanical property parameters, were conducted to assess the influence of elevated temperatures and cooling methods.

(a) Electric furnace

(b) ETM series electronic universal testing machine

Figure 2. Test equipment.

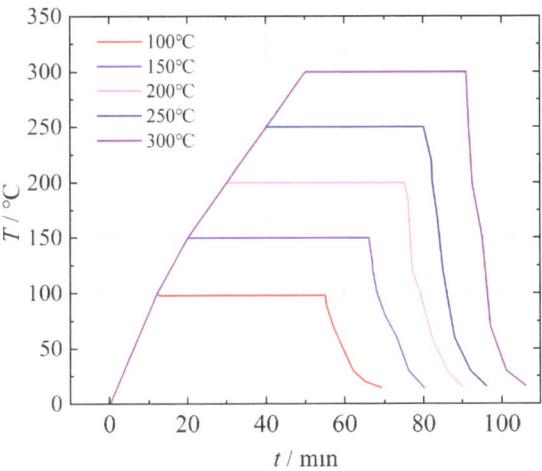

Figure 3. Temperature–time curves.

3. Test Results
3.1. Experimental Phenomena
3.1.1. Surface Characteristics

The surface color variations of the test specimens after exposure to high temperatures are illustrated in Figure 4. It can be seen that the cooling methods slightly affect the characteristics of the specimen. Under natural conditions, the GFRP surface exhibited a yellow hue. As the temperature increased, heating at 100 °C, 150 °C, and 200 °C led to a gradual deepening of the color, transitioning from a slightly blackened appearance to localized brown and eventually to brown. At temperatures approximately between

65 °C and 120 °C, the glass fiber reinforced plastic (GFRP) material undergoes its first glass transition, causing the material's resin to shift from a glassy state to a rubbery state [32]. At 250 °C, the surface turned charcoal-black, and the outer protective layer began to partially detach. Upon reaching 300 °C, the entire surface became charcoal-black due to the carbonization of the outer protective layer, which completely detached from the structural layer of the GFRP.

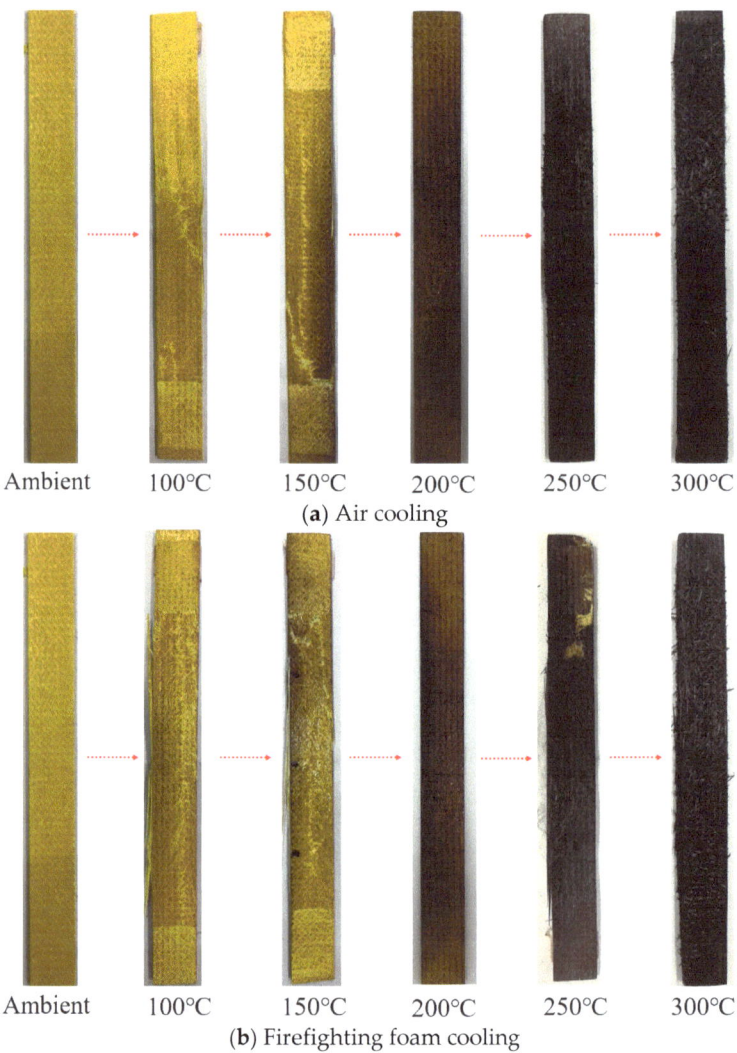

Figure 4. *Cont.*

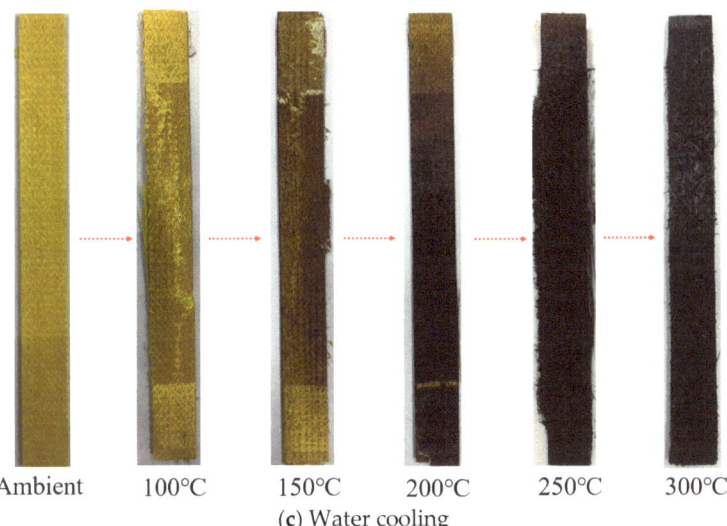

(c) Water cooling

Figure 4. Surface characteristics.

3.1.2. Failure Mode

The tensile failure characteristics of Glass Fiber Reinforced Polymer (GFRP) after high-temperature cooling at various target temperatures are presented in Figures 5–7. Observing the figures reveals distinct stage differences in failure characteristics corresponding to changes in temperature. As the temperature increased from ambient to 100 °C, 150 °C, and 200 °C, the fracture patterns exhibited notable variations. Initially, there was resin adhesion between the fibers, indicating integral destruction of the glass fibers and resin with a concentrated fiber distribution. As the temperature continued to rise, the fibers transitioned gradually into a diffuse filamentous state.

At ambient temperature and 100 °C, GFRP experienced cracking along axial extension, followed by destruction after the fiber bundles burst out. At 100 °C, the three cooling methods showed no significant impact on GFRP failure characteristics, with fiber bonding similar to that at ambient temperature, suggesting that the coordinated working ability of the fibers and resin remained relatively unchanged. At 150 °C and 200 °C, GFRP exhibited cracks extending from the axial center, while the final fracture leaned toward the end of the fixture. In comparison, natural cooling resulted in more dispersed fibers than water cooling and foam cooling, indicating that the latter two methods played a role in the recovery of resin after thermal decomposition.

Simultaneously, with the temperature increase, the flocculent material at the GFRP fracture gradually increased. At 250 °C, the fibers dispersed after failure, indicating that a large portion of the resin in the specimen could not be recovered after cooling. Notably, fiber bundle dispersion was relatively low under natural cooling. When the temperature reached 300 °C, fibers at the fracture of GFRP, cooled by all three methods, exhibited a diffuse filamentous state, indicating de-bonding and complete carbonization of the GFRP. Consequently, recovery through cooling was not possible at this temperature.

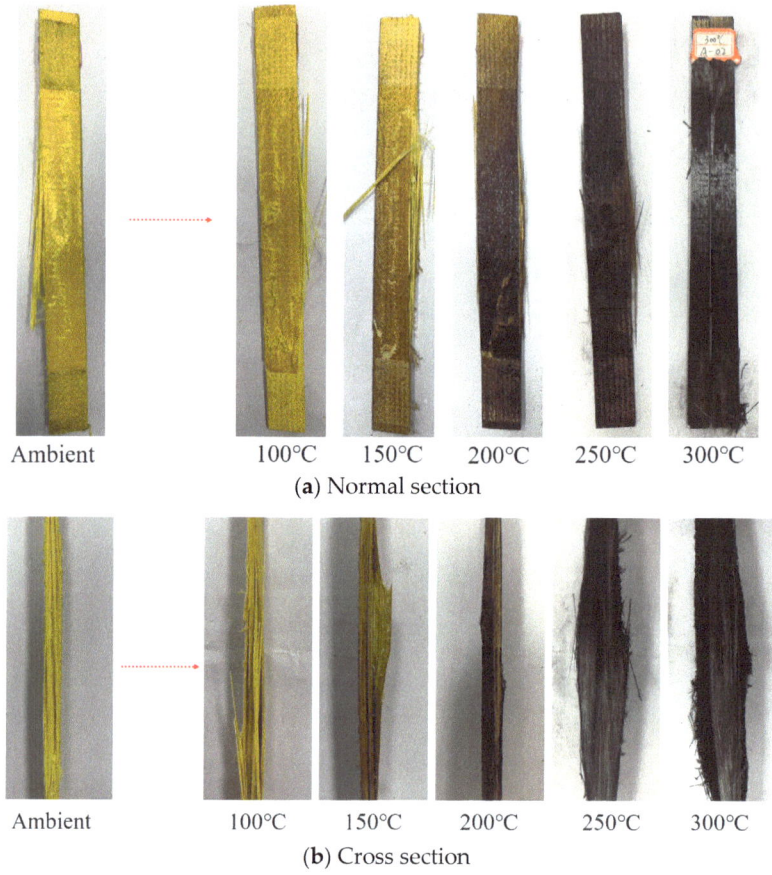

(**a**) Normal section

(**b**) Cross section

Figure 5. Air cooling.

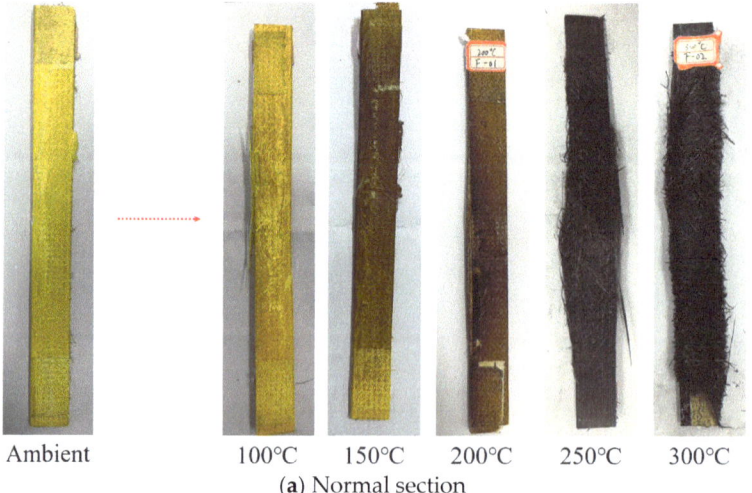

(**a**) Normal section

Figure 6. *Cont.*

(**b**) Cross section

Figure 6. Firefighting foam cooling.

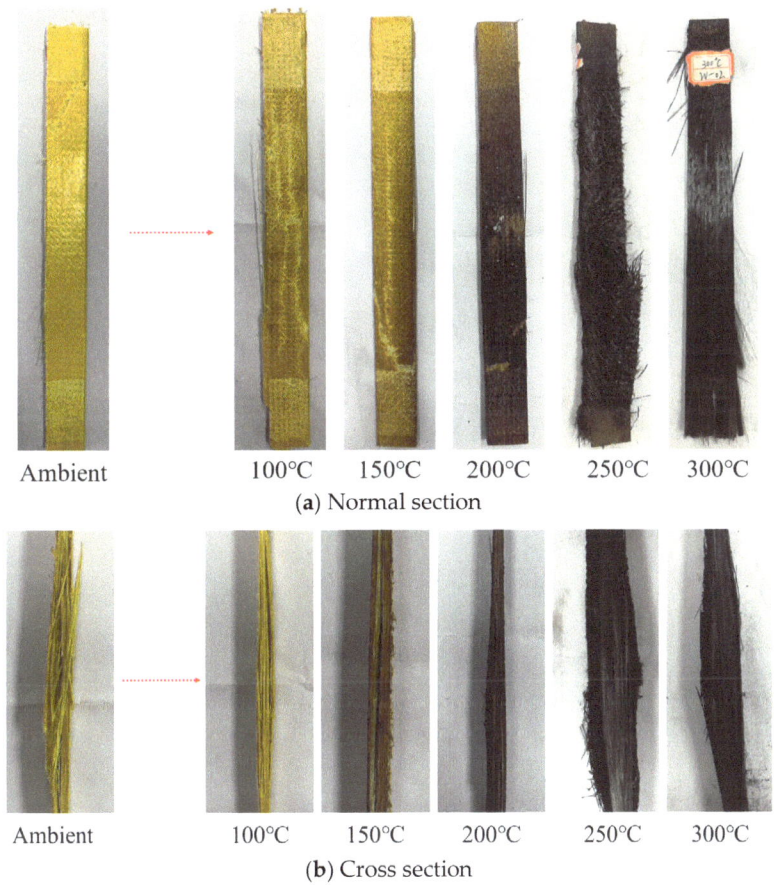

(**b**) Cross section

Figure 7. Water cooling.

3.2. Mechanical Properties

3.2.1. Tensile Strength

The maximum stress value reached when materials undergo damage under the action of force is termed tensile strength. Table 2 illustrates the tensile strength and residual factors of materials after high-temperature cooling at various target temperatures.

At 100 °C, compared to ambient temperature (σ_{max} = 486.38 MPa), the tensile strength under air cooling and foam cooling specimens increased, while water-cooled specimens (σ_{max} = 457.30 MPa) exhibited a decrease. At 100 °C, the resin matrix is undergoing a transition from a glassy state to a rubbery state [32]. This suggests that a reduction reaction enhanced the strength of materials after high-temperature cooling at the target temperature, and air cooling was more conducive to preserving material strength by allowing the reduction reaction to proceed more effectively.

At 150 °C, the residual factor of tensile strength under air cooling was 1.0, indicating effective restoration of material strength. Water and foam cooling exhibited residual factors of 0.90 and 0.83, respectively, indicating a decrease in strength when the temperature exceeded 150 °C. However, under air cooling, the strength exhibited a decreasing trend when the temperature surpassed 200 °C, suggesting that air cooling remained more beneficial for preserving material strength.

At 250 °C, tensile strength significantly decreased, signifying a sharp decline in the resin's bearing capacity due to thermal decomposition, resulting in permanent damage even after cooling. At 300 °C, a change in the damage mode occurred, with the stress slowly decreasing after reaching tensile strength, revealing resin carbonization. Despite cooling, the material lost load-bearing capacity after 300 °C. The strength retention rates for the three cooling methods at the same target temperature followed the order: air cooling > foam cooling > water cooling. After reaching 250 °C, none of the cooling methods could restore the strength.

It indicates that at temperatures above 250 °C, the degree of glass transition of the resin is more pronounced at high temperature. The destruction of the fiber/resin interface is aggravated by the combination of a higher exposure temperature and water cooling. Eventually, the glass fibers debond from the resin, resulting in a significant decrease in the tensile strength of GFRP [33].

Figure 8 illustrates the fitted line between tensile strength and high temperature, showcasing a decrease in the tensile strength of GFRP tubes as temperatures increased. The response of the tensile strength–temperature ratio presented a curvature. The curvatures of air cooling and foam cooling were similar, while water cooling exhibited a gentler change. Regression analysis of the experimental results yielded the following expressions for GFRP tubes:

Air cooling:
$$y_a = -1.07 \times 10^{-2}x^2 + 1.82x + 451.82 \tag{1}$$

Fire foam cooling:
$$y_f = -1.03 \times 10^{-2}x^2 + 1.57x + 463.47 \tag{2}$$

Water cooling:
$$y_w = -6.20 \times 10^{-3}x^2 + 0.30x + 488.37 \tag{3}$$

The reliability coefficients (R^2) for the fit curve under air cooling, foam cooling, and water cooling are 0.941, 0.917, and 0.921, respectively, indicating a favorable matching effect. This implies that the experiment exhibits minimal data dispersion, validating the credibility and accuracy of the collected data.

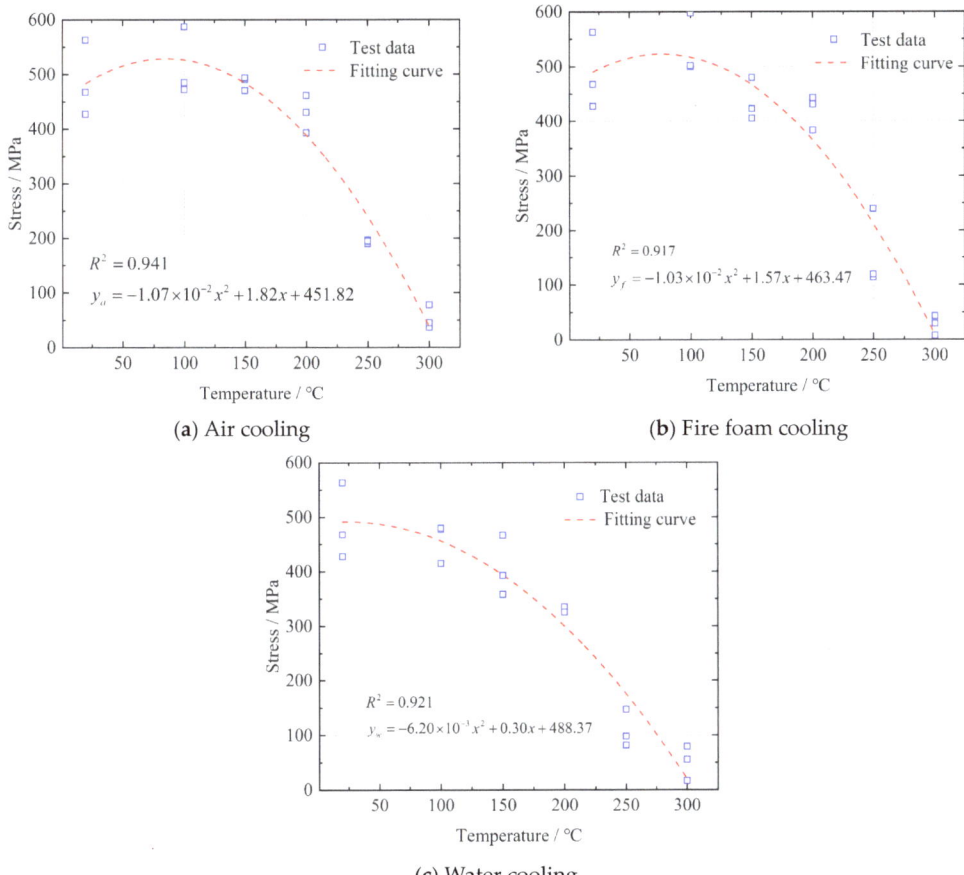

Figure 8. Ultimate strength of GFRP composites after high-temperature cooling.

Table 2. Tensile strength and residual factors of GFRP composites after high-temperature cooling.

Cooling Methods	Temperature/°C	Tensile Strength (MPa)				Residual Factor (σ_T/σ_{20})			
		Group-1	Group-2	Group-3	Average	Group-1	Group-2	Group-3	Average
Air Cooling	Ambient	427.89	563.35	467.89	486.38	0.88	1.16	0.96	1.00
	100	587.68	485.30	472.76	515.24	1.21	1.00	0.97	1.06
	150	490.92	470.92	493.95	485.26	1.01	0.97	1.02	1.00
	200	461.95	431.03	392.76	428.58	0.95	0.89	0.81	0.88
	250	197.30	190.38	195.24	194.31	0.41	0.39	0.40	0.40
	300	77.73	36.54	45.19	53.15	0.16	0.08	0.09	0.11
Fire Foam Cooling	Ambient	427.89	563.35	467.89	486.38	0.88	1.16	0.96	1.00
	100	598.05	499.89	502.16	533.37	1.23	1.03	1.03	1.10
	150	405.41	423.35	480.32	436.36	0.83	0.87	0.99	0.90
	200	443.03	383.24	431.14	419.14	0.91	0.79	0.89	0.86
	250	113.84	119.46	239.89	157.73	0.23	0.25	0.49	0.32
	300	7.14	42.92	29.30	26.45	0.01	0.09	0.06	0.05
Water Cooling	Ambient	427.89	563.35	467.89	486.38	0.88	1.16	0.96	1.00
	100	415.03	477.08	479.78	457.30	0.85	0.98	0.99	0.94
	150	466.38	358.38	392.86	405.87	0.96	0.74	0.81	0.83
	200	326.05	335.03	325.51	328.86	0.67	0.69	0.67	0.68
	250	97.62	81.84	147.24	108.90	0.20	0.17	0.30	0.22
	300	16.43	55.78	79.46	50.56	0.03	0.11	0.16	0.10

3.2.2. Modulus of Elasticity

The experimental findings presented in Table 3 reveal that the elastic modulus of the material under natural and foam cooling at the target temperature of 100 °C experienced a noticeable increase. Specifically, the elastic modulus showed a 20% enhancement under air cooling, indicating a strengthening process. There was no significant difference in the elastic modulus at the target temperatures of 100 °C and 150 °C, and the change in the residual factor of the elastic modulus was within 5%. This suggests that the alteration in the elastic modulus after air cooling within this temperature range is relatively minor, implying no distinct impact on the collaborative working performance of the resin and glass fibers.

Table 3. Elastic moduli and residual factors of GFRP composites after high-temperature cooling.

Cooling Methods	Temperature/°C	Elastic Modulus (GPa)				Residual Factor (E_T/E_{20})			
		Group-1	Group-2	Group-3	Average	Group-1	Group-2	Group-3	Average
Air Cooling	Ambient	37.05	39.16	38.15	38.12	0.97	1.03	1.00	1.00
	100	42.37	54.33	39.96	40.06	1.11	1.42	1.04	1.19
	150	41.78	38.40	39.61	39.93	1.09	1.01	1.03	1.04
	200	43.14	39.39	34.44	38.99	1.13	1.03	0.90	1.02
	250	32.53	28.20	28.97	29.90	0.85	0.74	0.75	0.78
	300	11.26	3.26	10.29	8.27	0.30	0.08	0.26	0.21
Fire Foam Cooling	Ambient	37.05	39.16	38.15	38.12	0.97	1.03	1.00	1.00
	100	41.22	41.74	37.97	40.31	1.08	1.09	0.99	1.05
	150	38.12	36.83	36.21	37.05	0.99	0.97	0.94	0.97
	200	33.70	36.41	38.32	36.15	0.88	0.95	1.00	0.94
	250	17.91	27.87	28.89	24.89	0.46	0.73	0.75	0.65
	300	2.08	9.39	11.40	7.62	0.05	0.24	0.29	0.19
Water Cooling	Ambient	37.05	39.16	38.15	38.12	0.97	1.03	1.00	1.00
	100	36.80	37.98	37.88	37.55	0.96	0.99	0.99	0.98
	150	39.54	39.33	37.12	38.66	1.04	1.03	0.97	1.01
	200	36.17	36.25	32.68	35.03	0.94	0.95	0.85	0.91
	250	13.67	15.05	16.50	15.08	0.35	0.39	0.43	0.39
	300	5.21	1.28	10.46	5.65	0.14	0.03	0.27	0.15

However, after reaching 250 °C, the elastic modulus exhibited a decline, with a notable 60% reduction under water cooling. This decline could be attributed to the fact that the structure of the unsaturated polyester resin experienced significant defects under water cooling. At 300 °C, the residual factor was 20% of the initial elastic modulus under all three cooling methods, and cooling was incapable of restoring the working abilities of the materials. Comparing with the average reduction in about 30% in tensile modulus under high-temperature conditions at 100 °C [34], it can be observed that the tensile modulus of GFRP tends to recover to some extent after cooling.

3.2.3. Ultimate Strain

Table 4 lists the tensile strength and the residual factor of the GFRP composites after high-temperature cooling, which indicated that in the range of 100 °C–300 °C, the tensile strength tends to decrease as the temperature increases. It is worth noting that the tensile strength of foam cooling is generally higher than that of the other two cooling methods, which could be attributed to the fact that foam cooling alters to some extent the bonding effects of the unsaturated polyester and the fiber material, which led to the mass increment of ductility. At 100 °C, the residual factors of the materials in natural and foam cooling are 1.01 and 1.03, respectively, indicating that the material is strengthened to a certain degree. The decrease in tensile strength is greater with water cooling than with other cooling methods, which may be due to defects in the redox process of unsaturated polyester, causing a decrease in the tensile strength of the materials.

Table 4. Ultimate strain and residual factors of GFRP composites after high-temperature cooling.

Cooling Methods	Temperature/°C	Ultimate Strain (%)				Residual Factor ($\varepsilon_T/\varepsilon_{20}$)			
		Group-1	Group-2	Group-3	Average	Group-1	Group-2	Group-3	Average
Air Cooling	Ambient	12.05	14.33	12.51	12.96	0.93	1.11	0.97	1.00
	100	14.31	13.54	11.28	13.04	1.10	1.04	0.87	1.01
	150	11.28	12.01	12.31	11.87	0.87	0.93	0.95	0.92
	200	11.52	11.40	9.80	10.90	0.89	0.88	0.76	0.84
	250	7.69	6.60	7.30	7.20	0.59	0.51	0.56	0.56
	300	6.42	15.35	4.16	8.64	0.50	1.18	0.32	0.67
Fire Foam Cooling	Ambient	12.05	14.33	12.51	12.96	0.93	1.11	0.97	1.00
	100	14.91	13.10	12.01	13.34	1.15	1.01	0.93	1.03
	150	10.14	10.57	11.87	10.86	0.78	0.82	0.92	0.84
	200	11.74	10.46	11.79	11.33	0.91	0.81	0.91	0.87
	250	8.02	6.38	9.80	8.06	0.62	0.49	0.76	0.62
	300	1.85	4.27	14.89	7.03	0.14	0.33	1.15	0.54
Water Cooling	Ambient	12.05	14.33	12.51	12.96	0.93	1.11	0.97	1.00
	100	10.71	12.31	11.81	11.61	0.83	0.95	0.91	0.90
	150	12.07	8.99	10.21	10.42	0.93	0.69	0.79	0.80
	200	8.89	9.31	9.72	9.31	0.69	0.72	0.75	0.72
	250	8.00	5.03	5.88	6.30	0.62	0.39	0.45	0.49
	300	2.81	11.65	5.92	6.79	0.22	0.90	0.46	0.52

3.3. Stress–Strain Curves

Figure 9 depicts the stress–strain curves of GFRP after water, foam, and air cooling at different target temperatures, revealing a brittle damage pattern in the materials. Examining the stress–strain curve of air cooling, it is evident that the tensile strength decreases with rising temperature. Before material destruction, the curve slopes linearly at ambient temperatures of 100 °C, 150 °C, and 200 °C, showing no significant differences. The tensile strength and ultimate strain at the target temperatures of 100 °C and 150 °C are similar, indicating that within the range of 100 °C to 150 °C, there is no notable impact on the coordinated working performance of the resin and glass fibers. However, at 250 °C, the tensile strength significantly drops as only the fiber can withstand the load, leading to thermal decomposition of the resin, which cannot be restored to its working performance after air cooling. At 300 °C, the damage mode changes, with the stress reaching its maximum value and then slowly declining, signifying that the materials have essentially lost their working abilities. Under this temperature condition, the stress–strain curve of the specimen after cooling exhibits significant plastic deformation. Based on the failure characteristics of the specimens, it is concluded that due to the higher exposure temperature, the resin is completely damaged, and only the fibers are in working condition, with the fibers in a divergent state. This results in significant plastic deformation.

Analyzing the strength–strain curve of foam cooling, it is evident that, compared to air cooling at the same temperature, the ultimate strain at 150 °C decreases and is even lower than that at 200 °C. This implies that the internal resin structure of GFRP undergoes changes after air cooling at this temperature, resulting in defects and a reduction in ultimate strain. The damage mode is similar to the air cooling curve under corresponding temperatures at 250 °C and 300 °C, with no significant differences. The cooling mode has a minimal effect on the materials at these temperatures. The strength–strain curves for water cooling show that the tensile strength at each target temperature is lower than those for natural and foam cooling. At 250 °C, the slope of GFRP decreases significantly after water cooling, lower than that of natural and foam cooling under the same target temperature. This is attributed to the weaker retention of strength under water cooling compared to the other two methods. The contact between the materials and water at high temperatures alters the resin structure, leading to defects and a reduction in strength. The materials exhibit plasticity damage at 250 °C and 300 °C after water cooling, resulting in a loss of their bearing capacities.

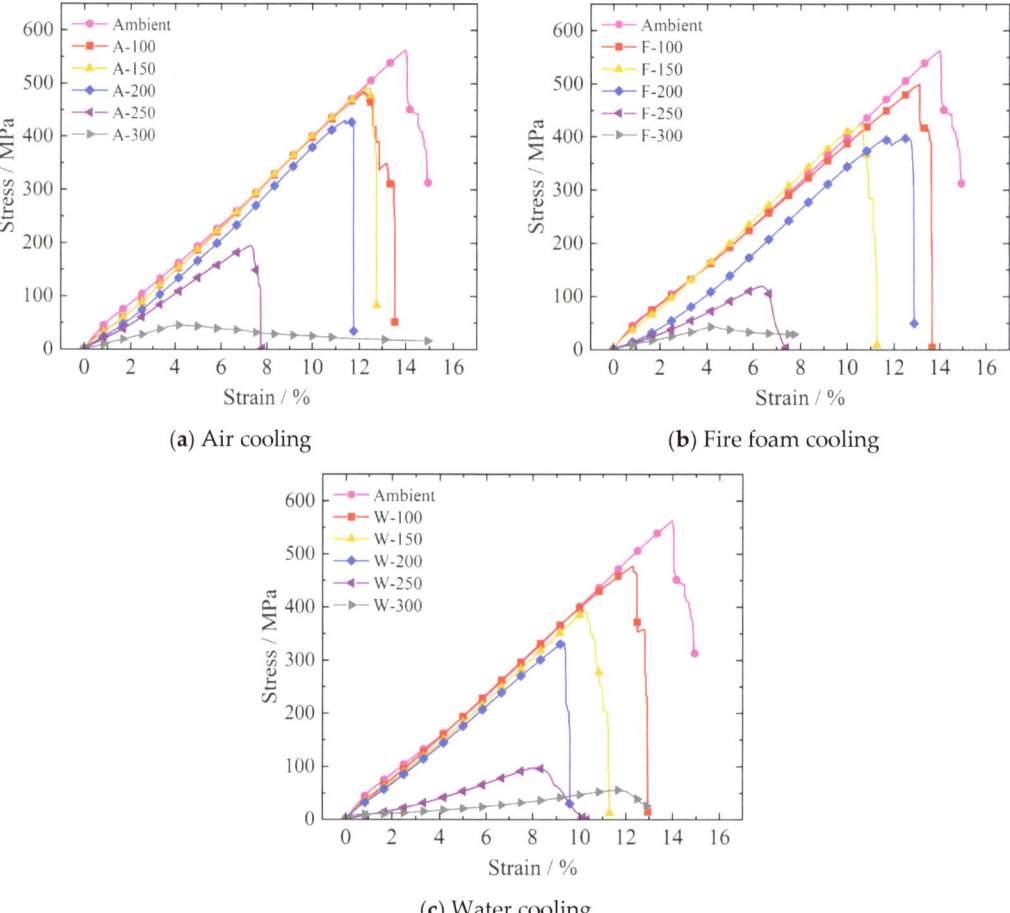

Figure 9. The stress–strain curve of GFRP after high-temperature cooling.

4. Theoretical Analysis

4.1. The T-E Model

To accurately determine the variation in the elastic modulus of GFRP after three cooling methods under different temperatures, the test was conducted using both an extensometer and strain gauges simultaneously. This approach enabled the measurement of strain throughout the tensile process, providing experimental values for the material's elastic modulus. The results of the experimental values are depicted in Figure 7, revealing that the cooling method has a minimal impact on GFRP within the temperature range of 100 °C to 200 °C, with a variation in no more than 7.3%. However, at 250 °C, water cooling displays the most substantial decrease in elastic modulus, showing a reduction in up to 50% compared to the other cooling methods. This observation implies that water cooling induces the softening of the resin matrix from a rubbery state during the cooling process, thereby influencing the stiffness of GFRP.

The values obtained from a nonlinear fit based on the Boltzmann function closely align with those presented in Figure 7. The regression analysis of the relationship between the elastic modulus residual factor and temperature is as follows:

$$y = \frac{A_1 - A_2}{1 + e^{(x-x_0)/k}} + A_2 \qquad (4)$$

Since this paper investigates the relationship between elastic modulus residual factor and temperature, A_1 is the initial value of the elastic modulus discount factor; A_2 is the final value of the elastic modulus discount factor; x_0 is the median temperature of the discount factor; k is the transformation rate of the discount factor; y is the discount factor of the elastic modulus; x is the temperature. To find the coefficient k, divide both sides of Equation (1) by the natural logarithm, Equation (1) becomes:

$$(x - x_0)/k = \ln(A_1 - A_2) - \ln(y - A_2) \tag{5}$$

The coefficients A_1, A_2, x_0 under different cooling methods can be determined from Figure 10, which shows the relationship between the residual factor of elastic modulus and temperature. The results are shown in Table 5. The coefficient k can be obtained by mathematical derivation, which ultimately leads to the equations described by Equations (6)–(8). Air cooling: k_a = 18.84, foam cooling: k_f = 21.17, water cooling: k_w = 15.86.

Air cooling:
$$y_a = \frac{0.86}{1 + e^{(x-266.08)/18.84}} + 0.22 \tag{6}$$

Fire foam cooling:
$$y_f = \frac{0.80}{1 + e^{(x-260.76)/21.17}} + 0.21 \tag{7}$$

Water cooling:
$$y_w = \frac{0.85}{1 + e^{(x-241.42)/15.86}} + 0.15 \tag{8}$$

Figure 10 compares the predicted characteristics of the elastic modulus residual factor temperature with the actual test results. The model's predictions align well with the experimental data, demonstrating the accuracy of the model in forecasting the modulus of elasticity residual factor of GFRP after three cooling methods at various target temperatures.

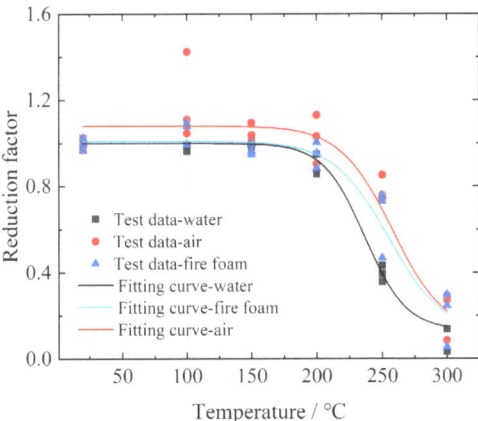

Figure 10. Elastic modulus reduction factors of fire and cooling-affected GFRP.

Table 5. Parameters of Boltzmann model under different cooling modes.

Cooling Method	A_1	A_2	x_0
Air cooling	1.08	0.22	266.08
Fire foam cooling	1.01	0.21	260.76
Water cooling	1.00	0.15	241.42

4.2. Johnson–Cook Model

The original Johnson–Cook (JC) constitutive model [35] can describe the stress–strain relationships under the influence of temperature or strain rate. It is usually considered one of the representative constitutive models due to its simple form and satisfactory performance. Therefore, this model was chosen as the basis for this study. The original model is:

$$\sigma(\varepsilon^p, \dot{\varepsilon}, T) = \left[A + B(\varepsilon^p)^n\right]\left[1 + C\ln\left(\frac{\dot{\varepsilon}}{\dot{\varepsilon}_0}\right)\right]\left[1 - \left(\frac{T - T_R}{T_m - T_R}\right)^m\right] \quad (9)$$

A is the nominal yield stress (MPa) in the tensile process at room temperature, denotes the yield strength of the material, n is the strain hardening parameter, and B is the strain hardening constant (MPa), which can be determined by the fitting method [36]. $\dot{\varepsilon}$, $\dot{\varepsilon}_0$, T, T_R and T_m are the current strain rate, reference strain rate, current temperature, reference temperature, and melting temperature, respectively.

4.2.1. Parametric Analysis

The reference temperature: T_R = 293 K, the reference strain rate: $\dot{\varepsilon}_0$ = 0.005 s^{-1}, A = 12.51 MPa, T_m = 953 K, and T = 373 K. Since strain rate and temperature are not variables in this experiment, Equation (9) can be split into:

$$\sigma(\varepsilon) = A + B\varepsilon^n \quad (10)$$

Converting both sides of the equal value of Equation (10) and dividing them by the natural logarithm gives the following equation:

$$\ln(\sigma(\varepsilon) - A) = n\ln\varepsilon + \ln B \quad (11)$$

After substituting the stress and strain values into Equation (11) for a linear fit, the relationship between $\ln\varepsilon$ and $\ln(\sigma - A)$ under different cooling methods is shown in Figure 11, where n and $\ln B$ denote the slope and initial value of the fitted curve, respectively. From this, the coefficients n, and B are calculated in Table 4.

After obtaining the coefficients n, and B for different cooling methods, Equation (9) is transformed into:

$$\frac{\sigma(\varepsilon)}{(A + B\varepsilon^n)} = 1 + C\ln\left(\frac{\dot{\varepsilon}}{\dot{\varepsilon}_0}\right) \quad (12)$$

As shown in Figure 11, the strain values and strain rates derived from the different cooling methods are brought into Equation (12) for a linear fit, where C denotes the slope of the fitted curve, from which the coefficient C can be calculated.

After the linear fit, the values of C for different cooling methods can be known from Figure 12. To find the parameter m, Equation (13) is changed to:

$$\ln\left[1 - \frac{\sigma(\varepsilon)}{A + B\varepsilon^n}\right] = m\ln\left(\frac{T - T_R}{T_m - T_R}\right) \quad (13)$$

As shown in Figure 13, by bringing the numbers of strains and target temperatures from this test into Equation (13) and then performing a linear fit, m can be yielded, as shown in Table 6.

Table 6. Parameters of Johnson–Cook model under different cooling modes.

Parameters	Air Cooling	Fire Foam Cooling	Water Cooling
n	1.16	1.04	1.10
B	5653.33	4105.16	4865.87
C	0.26	0.30	0.38
m	4.26	3.60	3.49

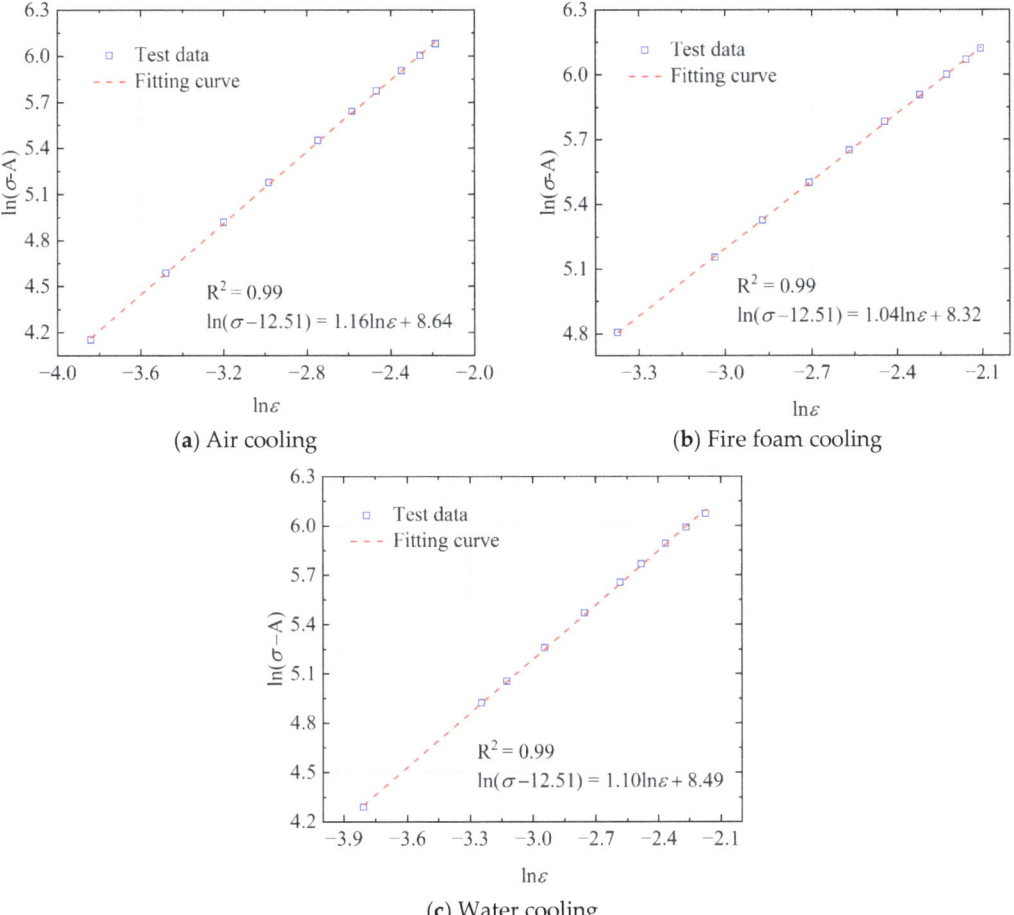

Figure 11. The relationship of $\ln\varepsilon$ and $\ln(\sigma - A)$.

Finally, the relationship between stress (σ), strain (ε), deformation temperature (T), and deformation rate was established based on the JC constitutive model. The modified equation T_0 was obtained through linear fitting, resulting in the predicted model data closely aligning with the actual observations.

Fire foam cooling:

$$T_0 = (0.13T - 12.56)\varepsilon + (6.17 \times 10^{-5}T^2 - 0.018T + 6.18 \times 10^{-3})\varepsilon^2 + 10.33 \quad (14)$$

$$\sigma(\varepsilon^p, \dot{\varepsilon}, T) = \left(12.51 + 4105.16\varepsilon^{1.04}\right)\left[1 + 0.30\ln\left(\frac{\dot{\varepsilon}}{0.005}\right)\right]\left[1 - \left(\frac{T-293}{660}\right)^{3.60}\right] - (0.13T - 12.56)\varepsilon - (6.17 \times 10^{-5}T^2 - 0.018T + 6.18 \times 10^{-3})\varepsilon^2 - 10.33 \quad (15)$$

Air cooling:

$$T_0 = \left(1.08 \times 10^{-3}T^2 - 0.32T + 20.76\right)\varepsilon + 8.65 \quad (16)$$

$$\sigma(\varepsilon^p, \dot{\varepsilon}, T) = \left(12.51 + 5653.33\varepsilon^{1.16}\right)\left[1 + 0.26\ln\left(\frac{\dot{\varepsilon}}{0.005}\right)\right]\left[1 - \left(\frac{T-293}{660}\right)^{4.26}\right] - (1.08 \times 10^{-3}T^2 - 0.32T + 20.76)\varepsilon - 8.65 \quad (17)$$

Water cooling:

$$T_0 = \left(1.07 \times 10^{-3}T^2 - 0.24T + 12.97\right)\varepsilon + \left(-5.12 \times 10^{-4}T^2 + 0.21T - 15.81\right) \quad (18)$$

$$\sigma(\varepsilon^p, \dot{\varepsilon}, T) = \left(12.51 + 4865.87\varepsilon^{1.10}\right)\left[1 + 0.38\ln\left(\frac{\dot{\varepsilon}}{0.005}\right)\right]\left[1 - \left(\frac{T-293}{660}\right)^{3.49}\right] - \left(1.07 \times 10^{-3}T^2 - 0.24T + 12.97\right)\varepsilon - \left(-5.12 \times 10^{-4}T^2 + 0.21T - 15.81\right) \quad (19)$$

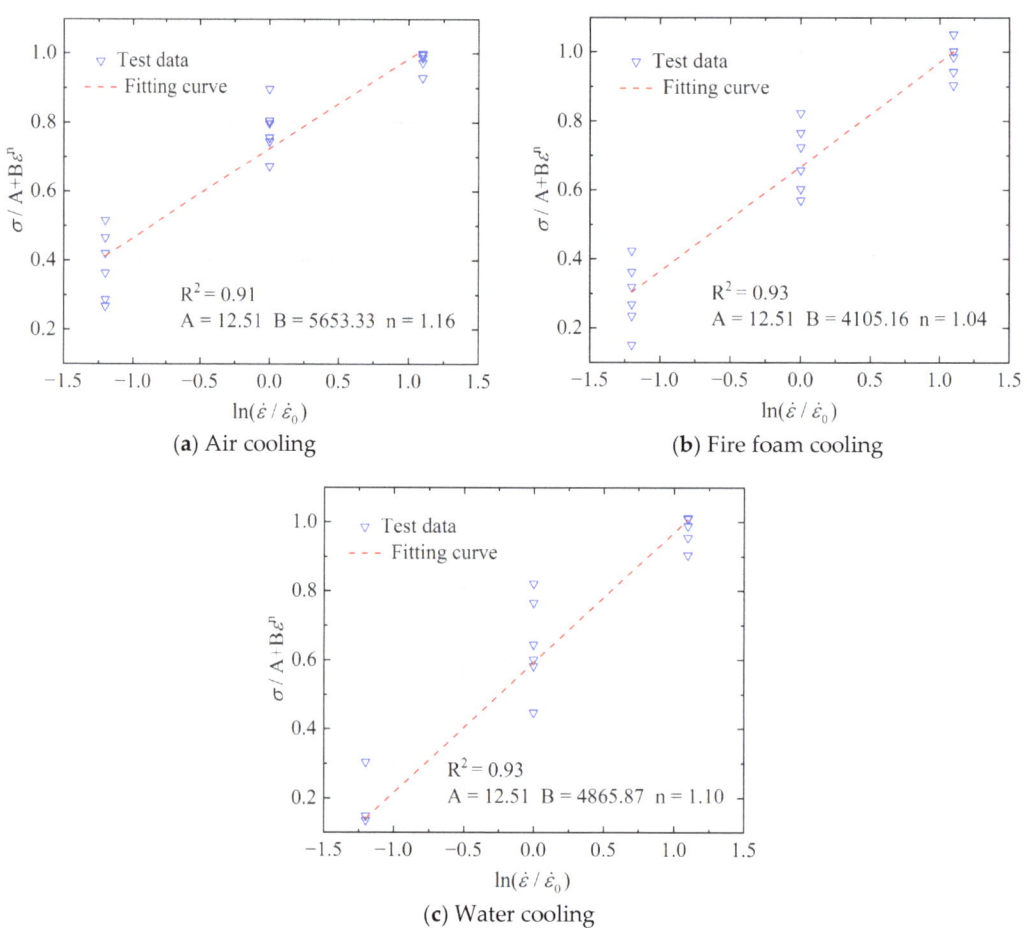

Figure 12. The relationship of $\sigma/A + B\varepsilon^n$ and $\ln(\dot{\varepsilon}/\dot{\varepsilon}_0)$.

4.2.2. Verification of the Constitutive Model

The original JC model demonstrates effective predictions at the reference strain rate and temperature. However, given the consideration of mechanical behavior after high-temperature cooling in this experiment, modifications were made to the JC model, resulting in Equations (15), (17) and (19). Figure 14 compares the stress–strain curves obtained from the tests with the predicted data from the modified constitutive model. It can be observed that the predicted equation accurately describes the effects of the three cooling methods on the mechanical behavior of GFRP at different target temperatures. It is important to note that the equation is specifically applicable to the boundary conditions and materials utilized in this study.

(**a**) Air cooling

(**b**) Fire foam cooling

(**c**) Water cooling

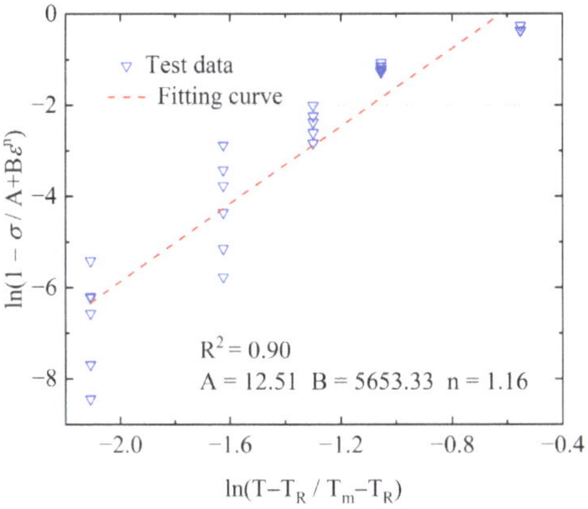

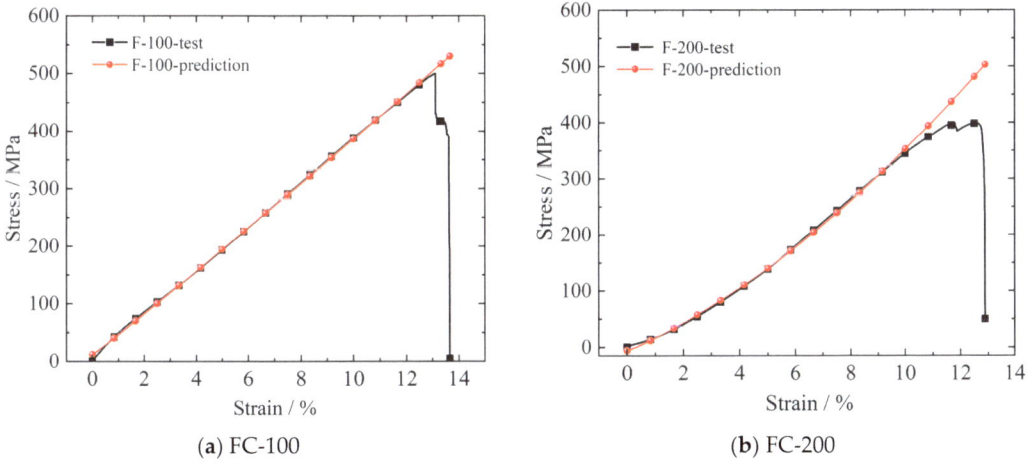

Figure 13. The relationship of $\ln(1 - \sigma/A + B\varepsilon^n)$ and $\ln(T - T_R/T_m - T_R)$.

(**a**) FC-100

(**b**) FC-200

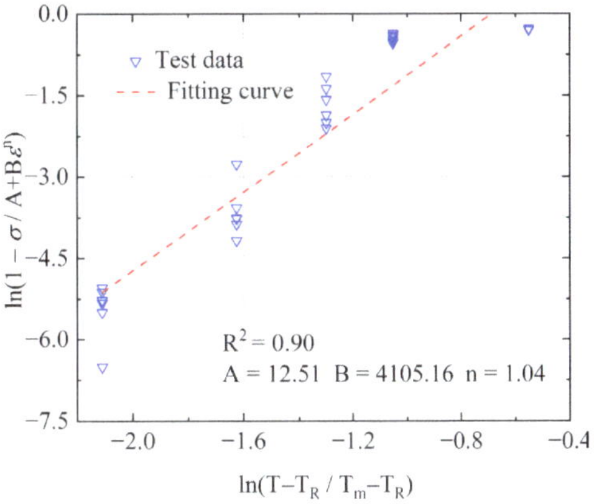

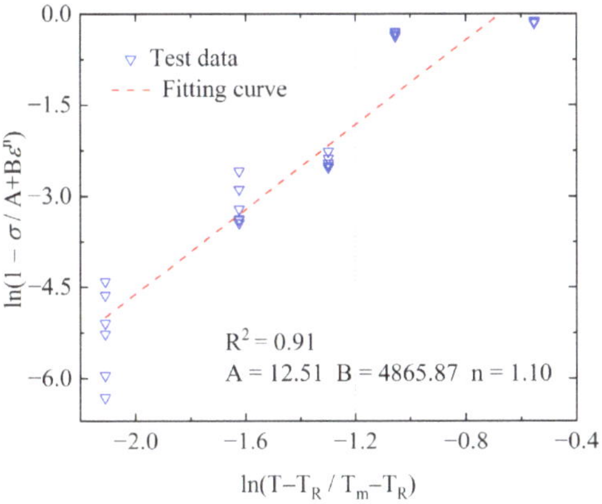

Figure 14. *Cont.*

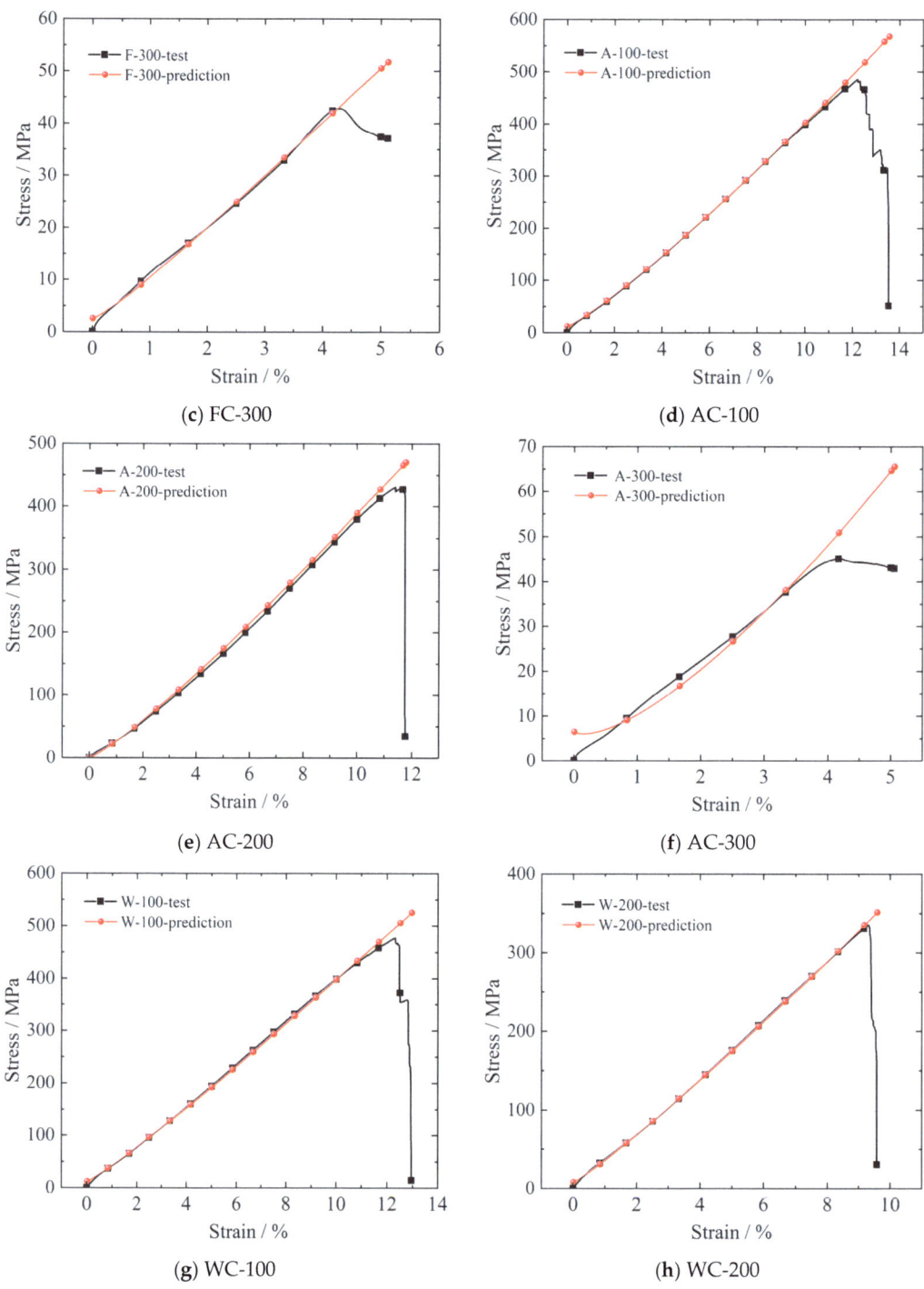

Figure 14. Cont.

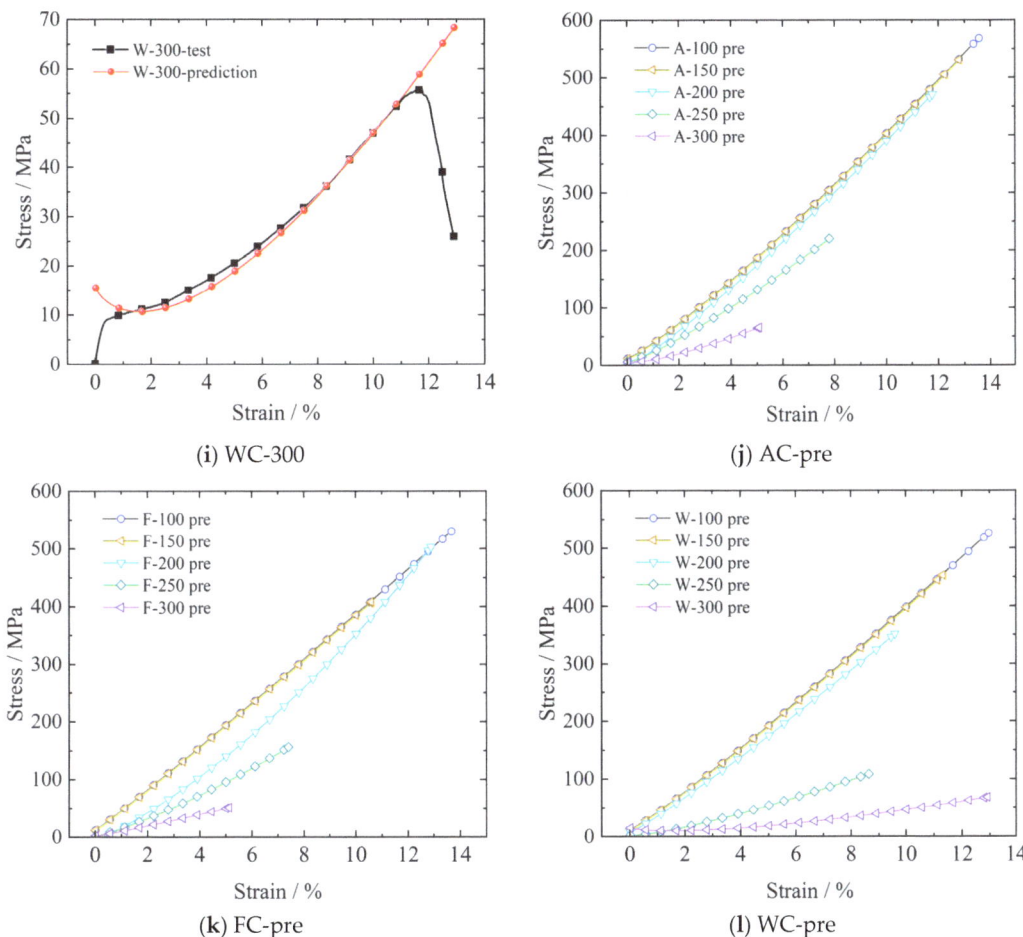

Figure 14. Comparison of predicted and test results.

5. Conclusions

In this paper, the effects of different target temperatures and cooling methods on the tensile properties of GFRP are investigated, and the prediction equations of mechanical properties and theoretical models after high-temperature cooling are derived and verified by experiments. The conclusions of this study are as follows:

1. The elevated temperatures and cooling methods significantly influenced the mechanical properties of GFRP. When exposed to temperatures below 200 °C, the three cooling methods can recover the mechanical properties of GFRP to a greater extent. At temperatures ranging from 20 to 200 °C, the recovery ability of water cooling is the weakest, with tensile strength values of 486.38, 457.30, 405.87, and 328.86. Conversely, natural cooling exhibits the strongest recovery ability, with corresponding tensile strength values of 486.38, 515.24, 485.26, and 428.58; while in 200 °C–300 °C, mechanical properties of GFRP decreased substantially, the ability of the cooling methods to restore the mechanical properties gradually decreased; when at 300 °C, GFRP has basically lost work abilities.

2. The best fire extinguishing method for GFRP materials after exposure to fire is firefighting foam cooling. At 100 °C, the residual factors of elastic modulus (1.05, 1.19) and tensile strength (1.10, 1.06) for GFRP in fire foam cooling and air cooling are greater than those at

ambient temperature. The elastic modulus and tensile strength of GFRP under fire foam cooling and air cooling are greater than those under ambient temperature. They decreased gradually with the increase in temperature. For the same target temperature, the strength retention rates of the materials are air cooling > fire foam cooling > water cooling.
3. The prediction equation of the mechanical properties of GFRP is established based on the experimental results, and the JC constitutive model is also modified to obtain the stress–strain curve equation suitable for this experiment, which is specifically applicable to the boundary conditions and materials utilized in this study.
4. This paper investigates the impact of different firefighting methods on the mechanical properties of GFRP in the event of a fire. Three commonly used cooling methods, namely, fire foam cooling, water cooling, and natural cooling, were employed. This study aims to provide a basis for the post-fire structural safety assessment of buildings.

Author Contributions: Conceptualization, C.Z.; Methodology, C.Z.; Investigation, J.W. and C.Z.; Data curation, J.W.; Writing—original draft, J.W.; Writing—review & editing, C.Z. All authors have read and agreed to the published version of the manuscript.

Funding: This study was supported by the National Natural Science Foundation of China (Grant No. 51868013, 51508482), Sichuan Province Science and Technology Support Program (Grant No. 2022NS-FSC0317) and the Natural Science Foundation of Tibet Autonomous (Grant No. XZ202102YD0035C). The authors would like to express their sincere gratitude for their support.

Data Availability Statement: Data are contained within the article.

Conflicts of Interest: The authors declare no conflict of interest.

References

1. Yang, Y.; Fahmy, M.F.M.; Guan, S.; Pan, Z.; Zhan, Y.; Zhao, T. Properties and applications of FRP cable on long-span cable-supported bridges: A review. *Compos. Part B Eng.* **2020**, *190*, 107934. [CrossRef]
2. Chen, Y.; Zhang, C. Stability analysis of pultruded basalt fiber-reinforced polymer (BFRP) tube under axial compression. *Compos. Struct.* **2024**, *327*, 117660. [CrossRef]
3. Zhang, C.; Chen, Y.; Dou, M. Axial Compression Behaviour and Modelling of Pultruded Basalt-Fibre-Reinforced Polymer (BFRP) Tubes. *Buildings* **2023**, *13*, 1397. [CrossRef]
4. Nawaz, W.; Elchalakani, M.; Karrech, A.; Yehia, S.; Yang, B.; Youssf, O. Flexural behavior of all lightweight reinforced concrete beams externally strengthened with CFRP sheets. *Constr. Build. Mater.* **2022**, *327*, 126966. [CrossRef]
5. Li, Y.; Zhang, C. Effects of high-temperature cooling and ultralow cyclic loading on mechanical behavior of Q450 weathering steel. *J. Constr. Steel Res.* **2023**, *202*, 107783. [CrossRef]
6. Zhang, C.; Li, Y.; Wu, J. Mechanical Properties of Fiber-Reinforced Polymer (FRP) Composites at Elevated Temperatures. *Buildings* **2022**, *13*, 67. [CrossRef]
7. Asaro, R.J.; Lattimer, B.; Ramroth, W. Structural response of FRP composites during fire. *Compos. Struct.* **2009**, *87*, 382–393. [CrossRef]
8. Bai, Y.; Keller, T. Modeling of mechanical response of FRP composites in fire. *Compos. Part A Appl. Sci. Manuf.* **2009**, *40*, 731–738. [CrossRef]
9. Wang, Z.; Zhao, X.-L.; Xian, G.; Wu, G.; Raman, R.K.S.; Al-Saadi, S. Effect of sustained load and seawater and sea sand concrete environment on durability of basalt- and glass-fibre reinforced polymer (B/GFRP) bars. *Corros. Sci.* **2018**, *138*, 200–218. [CrossRef]
10. Correia, J.R.; Keller, T.; Garrido, M.; Sá, M.; Firmo, J.P.; Shahid, M.A.; Machado, M. Mechanical properties of FRP materials at elevated temperature—Definition of a temperature conversion factor for design in service conditions. *Constr. Build. Mater.* **2023**, *367*, 130298. [CrossRef]
11. Spagnuolo, S.; Meda, A.; Rinaldi, Z.; Nanni, A. Residual behaviour of glass FRP bars subjected to high temperatures. *Compos. Struct.* **2018**, *203*, 886–893. [CrossRef]
12. Bai, Y.; Keller, T.; Vallée, T. Modeling of stiffness of FRP composites under elevated and high temperatures. *Compos. Sci. Technol.* **2008**, *68*, 3099–3106. [CrossRef]
13. Ashrafi, H.; Bazli, M.; Najafabadi, E.P.; Vatani Oskouei, A. The effect of mechanical and thermal properties of FRP bars on their tensile performance under elevated temperatures. *Constr. Build. Mater.* **2017**, *157*, 1001–1010. [CrossRef]
14. Li, T.; Zhu, H.; Shen, J.; Keller, T. Thermophysical and thermomechanical properties of basalt-phenolic FRP rebars under high temperature. *Constr. Build. Mater.* **2022**, *342*, 127983. [CrossRef]
15. Khorasani, M.; Muciaccia, G.; Consiglio, A.N.; Mostofinejad, D. Evaluating the behavior and bond properties of FRP spike anchors under confined conditions and elevated temperature. *Compos. Struct.* **2023**, *322*, 117407. [CrossRef]

16. Yu, B.; Kodur, V.K.R. Effect of high temperature on bond strength of near-surface mounted FRP reinforcement. *Compos. Struct.* **2014**, *110*, 88–97. [CrossRef]
17. Dong, K.; Hao, J.; Li, P.; Zhong, C. A nonlinear analytical model for predicting bond behavior of FRP-to-concrete/steel substrate joints subjected to temperature variations. *Constr. Build. Mater.* **2022**, *320*, 126225. [CrossRef]
18. Katz, A.; Berman, N. Modeling the effect of high temperature on the bond of FRP reinforcing bars to concrete. *Cem. Concr. Compos.* **2000**, *22*, 433–443. [CrossRef]
19. Faysal, R.M.; Bhuiyan, M.M.H.; Momin, K.A.; Tafsirojjaman, T.; Liu, Y. A review on the advances of the study on FRP-Concrete bond under hygrothermal exposure. *Constr. Build. Mater.* **2023**, *363*, 129818. [CrossRef]
20. Rosa, I.C.; Firmo, J.P.; Correia, J.R.; Mazzuca, P. Influence of elevated temperatures on the bond behaviour of ribbed GFRP bars in concrete. *Cem. Concr. Compos.* **2021**, *122*, 104119. [CrossRef]
21. Xue, Y.-J.; Wang, W.-W.; Wu, Z.-H.; Hu, S.; Tian, J. Experimental study on flexural behavior of RC beams strengthened with FRP/SMA composites. *Eng. Struct.* **2023**, *289*, 116288. [CrossRef]
22. Shayanfar, J.; Kafshgarkolaei, H.J.; Barros, J.A.O.; Rezazadeh, M. Unified strength model for FRP confined heat-damaged circular and square concrete columns. *Compos. Struct.* **2023**, *307*, 116647. [CrossRef]
23. Wang, S.; Stratford, T.; Reynolds, T.P.S. Linear creep of bonded FRP-strengthened metallic structures at warm service temperatures. *Constr. Build. Mater.* **2021**, *283*, 122699. [CrossRef]
24. Cao, V.V.; Vo, H.B.; Dinh, L.H.; Doan, D.V. Experimental behavior of fire-exposed reinforced concrete slabs without and with FRP retrofitting. *J. Build. Eng.* **2022**, *51*, 104315. [CrossRef]
25. Guo, D.; Gao, W.-Y.; Dai, J.-G. Effects of temperature variation on intermediate crack-induced debonding and stress intensity factor in FRP-retrofitted cracked steel beams: An analytical study. *Compos. Struct.* **2022**, *279*, 114776. [CrossRef]
26. Wu, G.; Dong, Z.-Q.; Wang, X.; Zhu, Y.; Wu, Z.-S. Prediction of Long-Term Performance and Durability of BFRP Bars under the Combined Effect of Sustained Load and Corrosive Solutions. *J. Compos. Constr.* **2015**, *19*, 04014058. [CrossRef]
27. Wang, H.; Gong, Z.; Hao, Y.; Deng, Y.; Zhang, C. Evolution of strain-rate dependent compressive failure behavior of ceramifiable FRP composites at high temperature conditions. *Compos. Sci. Technol.* **2023**, *241*, 110145. [CrossRef]
28. Khaneghahi, M.H.; Najafabadi, E.P.; Shoaei, P.; Oskouei, A.V. Effect of intumescent paint coating on mechanical properties of FRP bars at elevated temperature. *Polym. Test.* **2018**, *71*, 72–86. [CrossRef]
29. Wang, Y.C.; Wong, P.M.H.; Kodur, V. An experimental study of the mechanical properties of fibre reinforced polymer (FRP) and steel reinforcing bars at elevated temperatures. *Compos. Struct.* **2007**, *80*, 131–140. [CrossRef]
30. *GB/T 1447-2005*; Fiber-Reinforced Plastics Composites—Determination of Tensile Properties. General Administration of Quality Supervision, and Standardization Administration of the People's Republic of China: Beijing, China, 2005.
31. *GB/T 38515-2020*; The Method for High Temperature Mechanical Properties of Quartz Fabric-Reinforced Resin Matrix Composite Materials. General Administration of Quality Supervision, and Standardization Administration of the People's Republic of China: Beijing, China, 2020.
32. Robert, M.; Benmokrane, B. Behavior of GFRP Reinforcing Bars Subjected to Extreme Temperatures. *J. Compos. Constr.* **2010**, *14*, 353–360. [CrossRef]
33. Li, C.; Yin, X.; Wang, Y.; Zhang, L.; Zhang, Z.; Liu, Y.; Xian, G. Mechanical property evolution and service life prediction of pultruded carbon/glass hybrid rod exposed in harsh oil-well condition. *Compos. Struct.* **2020**, *246*, 112418. [CrossRef]
34. Mazzuca, P.; Firmo, J.P.; Correia, J.R.; Castilho, E. Influence of elevated temperatures on the mechanical properties of glass fibre reinforced polymer laminates produced by vacuum infusion. *Constr. Build. Mater.* **2022**, *345*, 128340. [CrossRef]
35. Johnson, G.R.; Cook, W.H. Fracture characteristics of three metals subjected to various strains, strain rates, temperatures and pressures. *Eng. Fract. Mech.* **1985**, *21*, 31–48. [CrossRef]
36. He, A.; Xie, G.; Zhang, H.; Wang, X. A comparative study on Johnson–Cook, modified Johnson–Cook and Arrhenius-type constitutive models to predict the high temperature flow stress in 20CrMo alloy steel. *Mater. Des.* **2013**, *52*, 677–685. [CrossRef]

Disclaimer/Publisher's Note: The statements, opinions and data contained in all publications are solely those of the individual author(s) and contributor(s) and not of MDPI and/or the editor(s). MDPI and/or the editor(s) disclaim responsibility for any injury to people or property resulting from any ideas, methods, instructions or products referred to in the content.

Article

Study on Bonding Behavior between High Toughness Resin Concrete with Steel Wire Mesh and Concrete

Qu Yu [1,2], Yu Ren [1,*], Anhang Liu [1] and Yongqing Yang [1,2]

[1] School of Civil Engineering, Southwest Jiaotong University, Chengdu 610031, China; yuqu2014swjtu@163.com (Q.Y.); lah@my.swjtu.edu.cn (A.L.); yangyongqingx@163.com (Y.Y.)
[2] Sichuan Jiaoda Engineering Detection & Consulting Co., Ltd., Chengdu 610031, China
* Correspondence: renyu@my.swjtu.edu.cn

Abstract: This paper investigates the interfacial bonding behavior between high toughness resin concrete with steel wire mesh (HTRCS) and concrete. A total of five sets of fifteen double shear specimens were tested for parameters including concrete strength and material properties of HTRCS composites. The test results showed that the failure mode of DS1 specimens was partial debonding and fracture, and the rest of the specimens were the fracture of HTRCS. The concrete strength and reinforcement ratios of HTRCS composites were positively correlated with interfacial adhesion properties. When the concrete strength was increased from C30 to C40 and C50, the ultimate load increased by 43.4% and 43.2%, respectively. The ultimate load capacity increased by 32.1%, with the reinforcement ratio of HTRCS composites increasing from 1.05% to 1.83%. Moreover, the bonding slip model and the bearing capacity formula for the interface between HTRCS composites and concrete were proposed, and the calculation values were in good agreement with the test values, with an average value of 0.978.

Keywords: high toughness resin concrete with steel wire mesh; concrete; interface bonding behavior; double shear test

1. Introduction

Improving the normal performance and extending the service life of RC structures through simple and economical reinforcement measures is an inevitable requirement for the research and application of RC structures [1–3]. Traditional reinforcement methods include pasting fiber composite fabric [4], pasting steel plates [5], section enlargement methods [6,7], and using the NSM technique with steel bars or CFRP bars [8,9]. The pasting fiber composite reinforcement method considers the characteristics of lightweight and high strength FRP, and at the same time it has the advantage of low thickness of reinforcement; however, there are problems with the hollowing of the bonded interface, and the quality of the interface bond cannot be guaranteed [10]. The steel plate external bonding method has the advantage of convenient construction, but it has the problem that the steel plate cannot fit the concrete concave and convex surface well [5]. The section enlargement method can effectively enhance the structural stiffness and bearing capacity, but there are defects that increase the structural deadweight, prolong the construction period, and reduce the vertical clearance under the bridge [11]. The NSM technique embeds CFRP bars, steel bars, and adhesives into the grooves of the concrete members, but it requires extensive interface treatment [12]. In addition, fewer studies have been conducted on the effects that factors such as mandatory protective layer thickness and temperature variations have on it [13].

In recent years, the strengthening method of combining cement mortar and steel wire mesh has gradually gained the attention of scholars because of its convenient construction and excellent strengthening performance [14,15]. Marthong [16] used galvanized wire mesh mortar layers to reinforce concrete columns of different cross-sectional shapes and found that axial loads increased by 20% and 19% when circular and square columns were reinforced with one turn of GSWM, respectively. Zhang et al. [17] investigated the flexural performance of reinforced concrete (RC) T-beams reinforced with HSSWM-PUC composites and found that the reinforced beams exhibited a 34% improvement with an increase in PUC thickness from 20 to 30 mm, and a 31.7% increase in ultimate and yield loads. However, this method is limited by the tensile properties of ordinary cement mortar, resulting in limited overall crack resistance of the reinforcement and premature failure of the reinforcement [18]. To solve the problem of early cracking of the reinforcement layer, related studies have proposed using engineering cementitious composites (ECCs) as binders and anchors [19]. With multicrack development and excellent ductility, ECC has shown it has a good effect on inhibiting cracks. However, ECC, as an inorganic material, has weak bonding and usually suffers from interfacial bond failure damage [20]. In addition, by drilling holes on the concrete surface, planting rebar and other methods can improve the bonding performance between the interface of ECC and concrete; however, there are also problems that affect its structural integrity and long construction period [21,22].

Due to the above reasons, this paper proposes high toughness resin concrete with steel wire mesh (HTRCS) composite material. High toughness epoxy resin concrete has the advantages of strong adhesion, high tensile strength, high toughness, high fluidity, and short curing time, which is conducive to rapid construction, shortening the reinforcement construction period [23]. Meanwhile, steel wire mesh embedded in the resin concrete enhances the strength of the composite material. Furthermore, the basis of the HTRCS composite reinforcement of RC members depends on whether HTRCS composite and RC matrix can work together, that is, the bonding behavior at the interface between HTRCS composite and concrete. Similar to other cement mortar composites for reinforcing RC structures, the bond strength of HTRCS composites is mainly related to the cross-sectional area of the wire mesh in the composite, the cross-sectional area of the resin concrete, and the strength of the concrete surface [24].

This paper investigates the interfacial bonding behavior between HTRCS composite material and concrete, which is investigated using a double shear test. Furthermore, the failure mode, force-displacement response, and strain of the specimens are analyzed. Moreover, we propose a bond-slip model and a bearing capacity formula for the interface between the HTRCS composite and the concrete.

2. Experimental Investigation
2.1. Details of the Specimens

To study the bond performance of HTRCS composite reinforced concrete members, double shear specimens were designed and fabricated to carry out interfacial bond tests. The dimensions of the specimens are shown in Figure 1. The test section was 300 mm × 150 mm × 150 mm, and the fixed section was 150 mm × 150 mm × 150 mm. The thickness of resin concrete was 20 mm and 40 mm, and the dimensions of steel wire mesh (transverse × longitudinal) were 13 mm × 25 mm, 15 mm × 25 mm, and 30 mm × 25 mm, respectively. The interfacial test section was bonded at a length of 300 mm and a width of 120 mm to ensure sufficient and effective bond length to observe strain transfer. The bond length of the fixed section was 150 mm, while the clip was used to apply pressure in the normal direction to enhance the interfacial bond. Before casting the HTRCS composite, the concrete interface needs to be cleaned to keep the interface clean.

A total of five types of specimens were designed for this test, and the main parameters were concrete strength and HTRCS material properties. Three specimens of each type were cast, and a total of fifteen double shear specimens were cast. For specimen numbers, "DS" represents the abbreviation "Double Shear". DS1, DS2, and DS3 specimens were used to study the effect of concrete strength on bond performance. DS3 and DS4 specimens were used to study the effect of the cross-sectional area of the steel wire mesh on bond behavior when reinforcement ratios were approximately equal. DS4 and DS5 specimens mainly show the effect of reinforcement ratios in HTRCS on the bond behavior. The specimen parameters are shown in Table 1.

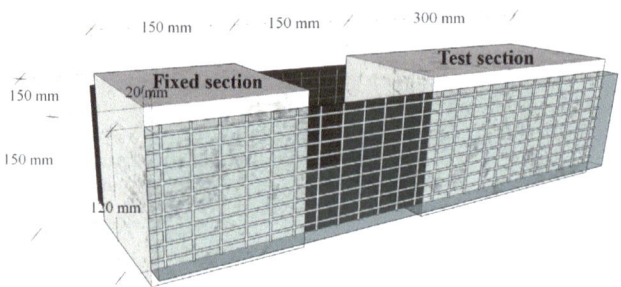

Figure 1. Dimensions of the specimens.

Table 1. Types of specimens.

Specimen	Concrete	Resin Concrete			Steel Wire Mesh			
		Thickness (mm)	Width (mm)	Diameter of Single Bar (mm)	Interval (mm)	A_s (mm^2)	ρ/%
DS1	C30	20		2	13 × 25	28.3	1.18
DS2	C40	20		2	13 × 25	28.3	1.18
DS3	C50	20	120	2	13 × 25	28.3	1.18
DS4	C50	40		4	30 × 25	50.2	1.05
DS5	C50	40		4	15 × 25	87.9	1.83

Note: A_s is the cross-sectional area of the steel wire mesh, ρ is the reinforcement ratio of HTRCS.

2.2. Fabrication of the Specimens

Figure 2 shows the specimen fabrication process according to Chinese codes [25] with the following procedure:

(1) Casting concrete: Standard molds were used to cast the concrete.
(2) Preparing wire mesh: Trim the wire mesh to the required dimensions and paste the strain gauges according to Section 2.4.
(3) Preparing molds: Wooden formwork was made, placing the test and fixed sections of concrete.
(4) Casting resin concrete: Pouring resin concrete in wooden formwork.
(5) Removing molds: After 24 h, remove the wooden formwork and remove the foam boards from the loading section.
(6) Specimen curing: The fabricated double shear specimens were cured under standard conditions for 5 days.

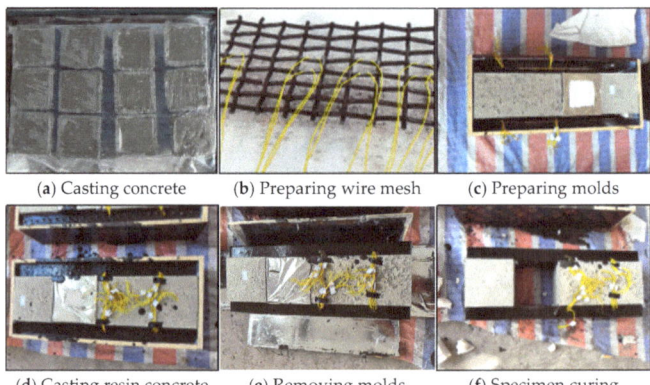

Figure 2. Fabrication of the specimens.

2.3. Materials Properties

The mix ratio of resin concrete used for the specimens [23] was resin colloid:hardener: continuous graded aggregate = 4:1.16:24.84. The material properties of resin concrete and ordinary concrete were measured according to Chinese codes [25]. The compressive strengths of C30, C40, and C50 ordinary concrete were measured to be 30.1, 42.3, and 54.6 MPa, and the tensile strengths were 2.35, 2.77, and 2.98 MPa, respectively. The tensile strength was 7.8 MPa, the modulus of elasticity was 12,500 MPa, and the compressive strength of resin concrete was 102.3 MPa [23], as shown in Table 2. The yield strength of steel wire mesh was 412 MPa.

Table 2. Properties of resin concrete.

Materials	Tensile Strength (MPa)	Elastic Modulus (MPa)	Compressive Strength (MPa)
Resin concrete	7.8	1.25×10^4	102.3

2.4. Testing Procedure

The test setup used in this test is shown in Figure 3. The test setup consists of fixed, loading, and testing sections, in which the loading section includes load cells and hydraulic jacks. In the fixing section, normal pressure is applied using a clamp to improve the interfacial bond between the HTRCS composites and the concrete, thus ensuring that the damage occurs in the test section. In addition, the specimen should be kept parallel to the slide throughout the test to ensure that the specimen can slide freely with the rail. The loading level difference was about 1 kN, and the loading interval between two adjacent levels was about 3 s. When the displacement suddenly increased or the load increase rate slowed down, displacement-controlled loading was used, with a loading rate of 0.02 mm/s.

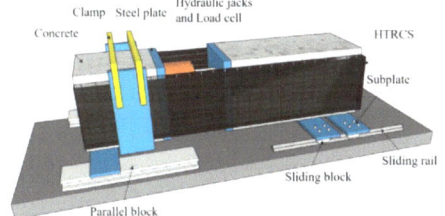
(**a**) Description of the device

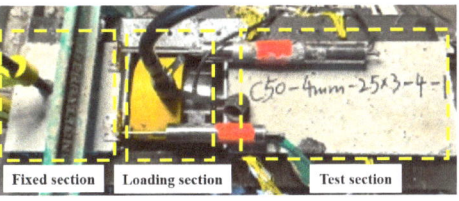
(**b**) On-site test

Figure 3. Test setup.

Figure 4 shows the arrangement of the strain measurement points of the wire mesh, and the specimen displacement measurement points. The first strain gauge, F1, is used to simulate the strain at the free end and is positioned in the middle of the loading section. Strain gauges F2–F8 are set on the test section to measure the strain in the test section. In addition, two Linear Variable Differential Transformers (LVDTs) were installed at the front of the test section to measure the relative displacement of the specimens.

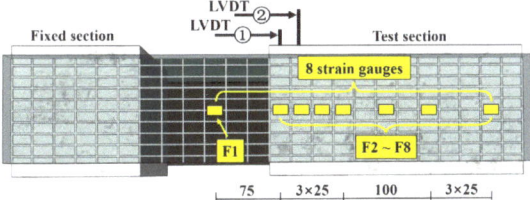

Figure 4. Measurement points.

3. Test Results

3.1. Crack Patterns and Failure Modes

The failure modes of the double shear specimens are shown in Figure 5. Partial debonding and fracture of the HTRCS occurred in the DS1 specimen. The failure mode of the DS1 specimen was mainly due to the low interfacial bond between the HTRCS and the concrete when the concrete strength was low. At the same time, with the increase of load, the local stress concentration at the front end of the test section led to interfacial debonding and the subsequent fracture of the HTRCS material, as shown in Figure 5a. The fracture of the HTRCS in the loaded section occurred in the DS2, DS3, DS4, and DS5 double shear specimens, as shown in Figure 5b. This is because when the concrete strength increases, the interfacial bond strength increases, and therefore no interfacial debonding occurs. Meanwhile, the deformation of the HTRCS composites was less pronounced before the specimen fracture because the resin concrete in this study is a linear elastic material with an ultimate tensile strain of 0.000624. When the specimen was damaged, the resin concrete was pulled out and the wire mesh did not yield. A crisp sound was emitted at the time of destruction, and the failure process was rapid.

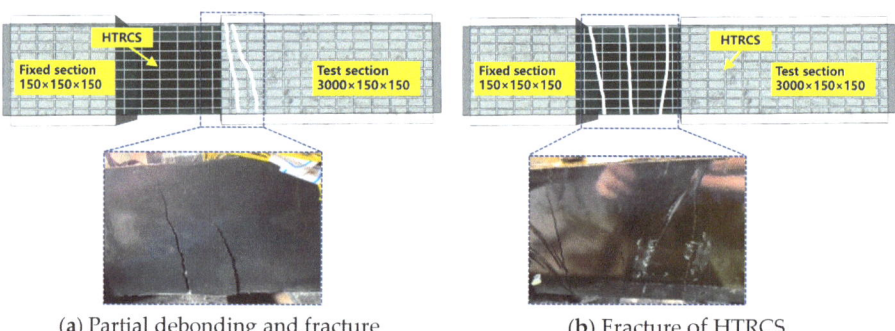

(a) Partial debonding and fracture   (b) Fracture of HTRCS

Figure 5. Failure modes.

The specimen extended interlayer cracks at the vertical cracks, as shown in Figure 6. Due to that, the resin concrete in the HTRCS composites pulled off when the specimen was damaged, and the resin concrete was squeezed by the wire mesh; thus, the interlayer cracks appeared. Furthermore, the bond between the wire mesh and the resin concrete was good during the whole test, and no significant slippage occurred.

Figure 6. Interlayer cracks.

3.2. Maximum Load and Displacement

Figure 7 shows the force-displacement curves of the double shear specimens. The results of maximum load P_{max}, average maximum load $\overline{P}_{max}$, displacement S, and average displacement $\overline{S}$ of the specimens are shown in Table 3. The relationship between force and displacement was basically linear for all specimens. This is due to the fact that the resin concrete in HTRCS composites is a linear elastic material, and the wire mesh did not yield when the resin concrete reached its ultimate strength. Therefore, HTRCS composites can be treated as linear elastic materials, the interfacial bonding is sufficient, and no interfacial peeling occurs, so the load-displacement curves are basically linear.

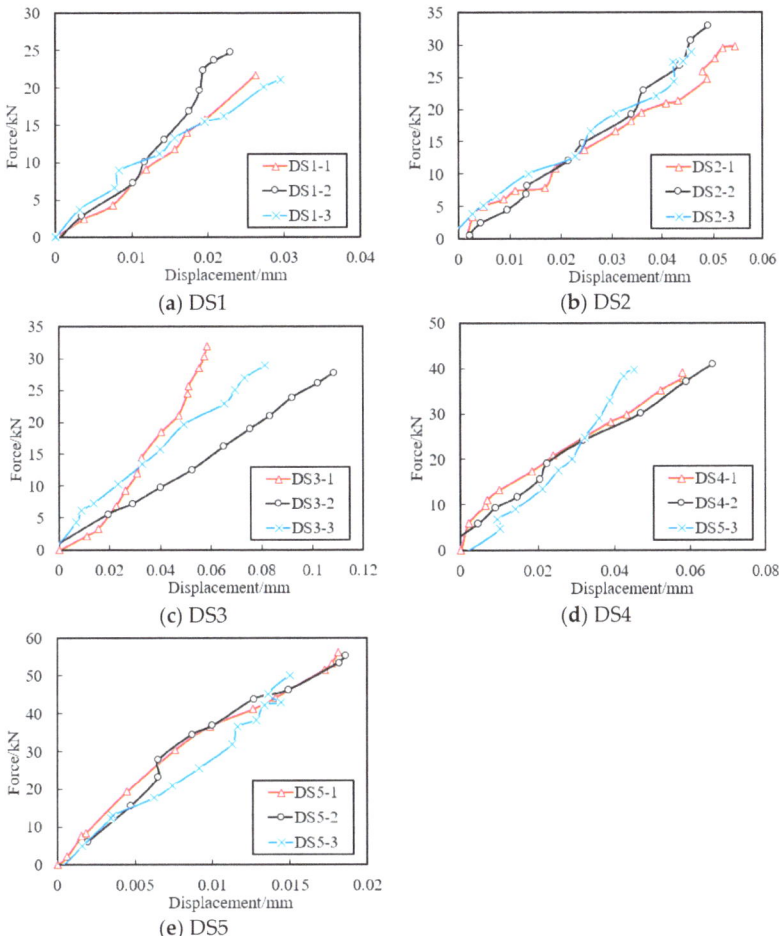

Figure 7. Force-displacement curves.

A comparison of the DS1, DS2, and DS3 specimens shows that the load carrying capacity is affected by the strength of the concrete. The ultimate loads of the DS2 and DS3 specimens are 43.4% and 43.2% higher than that of the DS1 specimen, respectively. This is because when the concrete is lower, the bond between HTRCS composites and concrete is weaker. After debonding at the front end of the test section, the partial HTRCS material fractured, so the bearing capacity of the DS1 specimen was lower. From specimens DS2 and DS3, it can be seen that their ultimate loads are basically the same. This is because when the concrete strength is increased, the interfacial bond is enhanced, and the ultimate load of the specimens is mainly related to the reinforcement ratio of HTRCS, which is the same for both groups of specimens.

Table 3. Maximum load and displacement.

Specimen	P_{max} (kN)	$\overline{P}_{max}$ (kN)	s/mm	$\overline{S}$ (mm)	Failure Modes
DS1-1	21.18		0.026		Partial debonding and fracture
DS1-2	24.75	22.37	0.023	0.026	Partial debonding and fracture
DS1-3	21.16		0.029		Partial debonding and fracture
DS2-1	32.92		0.054		Fracture of HTRCS
DS2-2	32.31	32.07	0.048	0.049	Fracture of HTRCS
DS2-3	30.98		0.046		Fracture of HTRCS
DS3-1	32.05		0.061		Fracture of HTRCS
DS3-2	33.86	32.02	0.108	0.083	Fracture of HTRCS
DS3-3	30.16		0.081		Fracture of HTRCS
DS4-1	40.05		0.058		Fracture of HTRCS
DS4-2	42.14	40.69	0.066	0.056	Fracture of HTRCS
DS4-3	39.88		0.045		Fracture of HTRCS
DS5-1	56.89		0.018		Fracture of HTRCS
DS5-2	54.73	53.75	0.019	0.017	Fracture of HTRCS
DS5-3	49.64		0.015		Fracture of HTRCS

A comparison of the DS3 and DS4 specimens shows that the ultimate load of the specimens is positively correlated with the cross-sectional area of the reinforcement when the difference in the reinforcement ratio in the HTRCS composites is small. The ultimate load of the DS4 specimen is increased by 27.1% compared to the DS3 specimen. A comparison of the DS4 and DS5 specimens shows that when the cross-sectional area of resin concrete is determined, the reinforcement ratio of HTRCS composites is positively related to the ultimate load capacity of the specimens, and the ultimate load capacity increased by 32.1% with the reinforcement ratio of HTRCS composites increasing from 1.05% to 1.83%.

3.3. Strain-Distance Analysis

The strain distribution of the specimen is shown in Figure 8. Strain gauge F1 simulates the strain at the free end, with the distance of 0 mm in the test section. Strain gauges F2–F8 draw the strain distribution curve according to the actual distance from the end of the test section. Because the strain distribution curves of three specimens in each type are basically the same, one specimen in each group is selected for display.

The DS1 specimen was damaged when the strain reached 400 $\mu\varepsilon$, while the strains of the DS2 and DS3 specimens reached 600 $\mu\varepsilon$, which is close to the ultimate tensile strain of resin concrete. Meanwhile, the stress transfer distance is about 140 mm for the DS1 specimen and 190 mm for the DS2 and DS3 specimens. A comparison of the DS1, DS2, and DS3 specimens shows that as the strength of the concrete increases, the bonding effect between the interfaces is enhanced, the HTRCS composite is more fully utilized, and the strain transfer range is increased.

The strain transfer distance of the DS4 specimen is about 190 mm, and that of the DS5 specimen is about 260 mm. By comparing the DS4 and DS5 specimens, it can be seen that with the increase of the reinforcement rate in HTRCS, the strain transfer distance increases and the stress transfer effect is more significant.

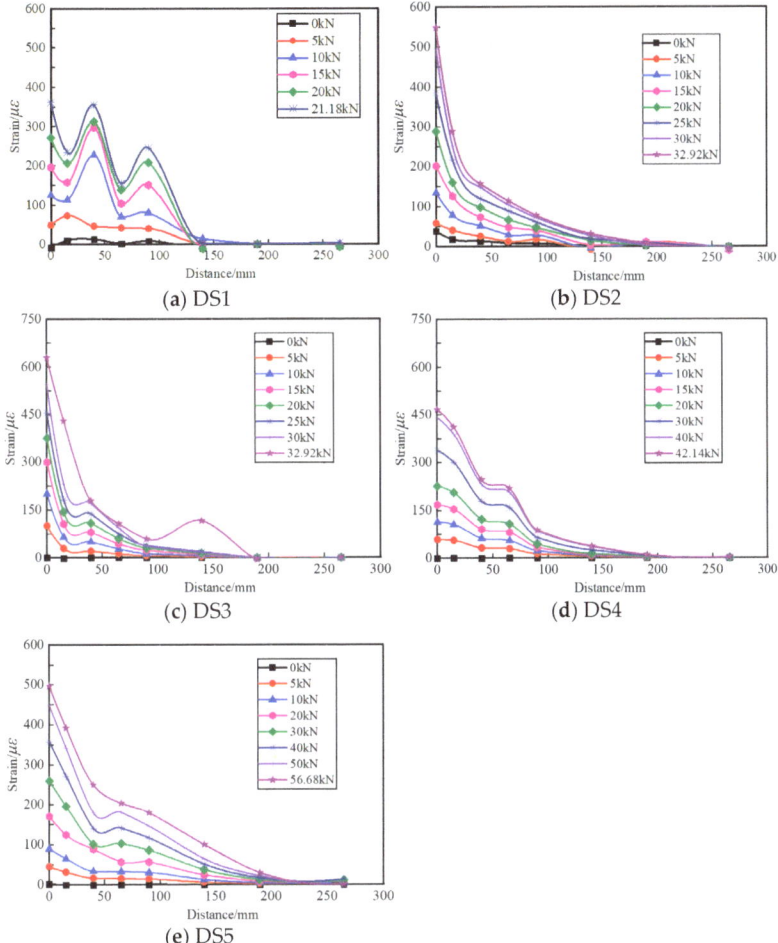

Figure 8. Strain distance curve.

4. Bond-Slip Model and Bearing Capacity Formula

4.1. Bond-Slip Model

In order to investigate the material properties of HTRCS composites and the effect of concrete strength on the bonding behavior at the interface between HTRCS composites and concrete, the local stress-slip curves need to be obtained first. The strain distance curves (Figure 8) from the double shear experiments were converted by the conversion formula proposed in [26], and the expressions are given as follows:

$$\tau_i = \frac{E_H t_H (\varepsilon_i - \varepsilon_{i-1})}{\Delta d} \quad (1)$$

$$S_i = \frac{\Delta d}{2}(\varepsilon_0 + 2\sum_{j}^{i-1} \varepsilon_j - \varepsilon_i) \quad (2)$$

where τ_i and S_i are the interfacial bonding stress and slip at strain gauge i; ε_0 is the strain at the free end, which is the strain of F1; $\varepsilon_j (j = 1, \ldots, i)$ is the strain at strain gauge j; E_H is the elastic modulus of the HTRCS composite; t_H is the thickness of the HTRCS composite; and Δd is the corresponding distance between each strain gauge.

In exploring the interfacial bonding behavior between HTRCS and concrete, we assumed that the deformation of resin concrete and steel wire mesh before specimen damage are the same. Therefore, the expressions for the ultimate stress and modulus of elasticity of HTRCS are as follows:

$$f_H = (f_{rc}A_{rc} + \varepsilon_{rc}E_s A_s)/A_H \tag{3}$$

$$E_H = (E_{rc}A_{rc} + E_s A_s)/A_H \tag{4}$$

where f_H is the tensile stress of the HTRCS; f_{rc} is the tensile strength of the resin concrete; A_H, A_{rc}, and A_s are the cross-sectional area of the HTRCS, resin concrete, and steel wire mesh, respectively; E_{rc} and E_s are the elasticity moduli of the resin concrete and steel wire mesh, respectively; ε_{rc} is the tensile strain of the resin concrete; and $\varepsilon_{rc} = 0.000624$.

After plotting the local stress-slip curves, the local stress-slip relationship closer to the free end is selected for fitting the overall stress-slip curve. In addition, the stress-slip model is compared to the formulae presented in [27,28], in which the model shape is controlled by three parameters with the following expression:

$$\tau = \tau_{max} \frac{S}{S_0} \frac{n}{(n-1) + (S/S_0)^n} \tag{5}$$

where τ_{max} is the maximum shear stress, S_0 is the slip at peak, and n is the coefficient of the softening branch.

The fitting diagram of the interface bond-slip curve is shown in Figure 9. The fitting results are somewhat discrete, but basically reflect the bond-slip relationship at the interface. Similarly, bond-slip curves were plotted for all specimens, and the obtained curve control parameters, τ_{max}, S_0, the tensile strength of concrete (f_t), and the elasticity modulus of the HTRCS (E_H), are listed in Table 4. According to the experimental data, the softening coefficient n is uniformly taken as 3.6.

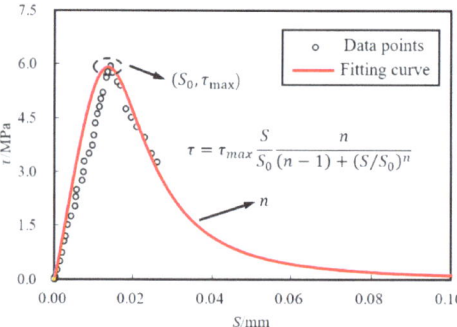

Figure 9. Interface bond-slip curve.

From Table 4, it can be seen that the bond-slip curve parameters are more significantly affected by the HTRCS material properties and concrete strength. Considering the dimensionless design, $\sqrt{f_t E_H}$ is the independent variable, and τ_{max} and S_0 are the dependent variables. Figure 10 illustrates the regression analysis and gives Equations (5) and (6) with their R^2 as 0.859 and 0.791, respectively.

$$\tau_{max} = 0.577 e^{0.001\sqrt{f_t E_H}} \tag{6}$$

$$S_0 = 0.0008 e^{0.013\sqrt{f_t E_H}} \tag{7}$$

Table 4. Parameters of bond-slip curve.

Specimens	τ_{max}	S_0	f_t/MPa	E_H/MPa
DS1-1	5.42	0.0125	2.35	2.0×10^4
DS1-2	5.35	0.0110	2.35	2.0×10^4
DS1-3	4.72	0.0120	2.35	2.0×10^4
DS2-1	5.10	0.0201	2.77	2.0×10^4
DS2-2	5.91	0.0134	2.77	2.0×10^4
DS2-3	5.8	0.0200	2.77	2.0×10^4
DS3-1	6.65	0.0172	2.98	2.0×10^4
DS3-2	6.22	0.0175	2.98	2.0×10^4
DS3-3	6.21	0.0190	2.98	2.0×10^4
DS4-1	8.50	0.0260	2.98	2.4×10^4
DS4-2	9.33	0.0197	2.98	2.4×10^4
DS4-3	7.64	0.0250	2.98	2.4×10^4
DS5-1	5.98	0.0195	2.98	1.9×10^4
DS5-2	6.09	0.0287	2.98	1.9×10^4
DS5-3	6.10	0.0190	2.98	1.9×10^4

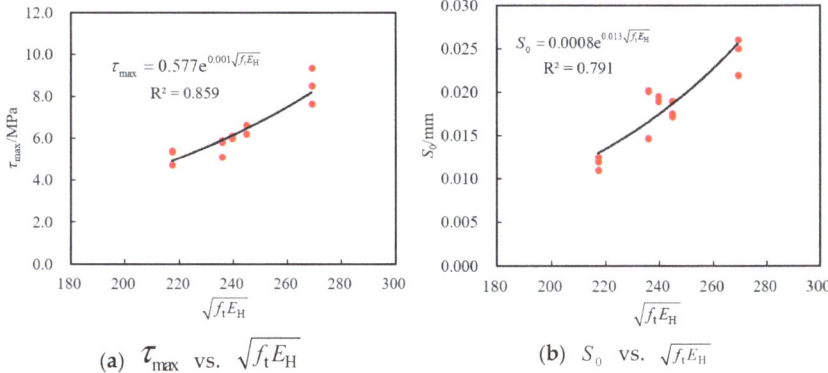

Figure 10. Regression analysis.

As can be seen from Figure 10, τ_{max} and S_0 are shown to be positively correlated with $\sqrt{f_t E_H}$, indicating that the higher the concrete strength and the elastic modulus of HTRCS, the better the bond behavior between the interface of HTRCS and concrete. The high R^2 values of the equation established by regression analysis indicate that the bond-slip model of HTRCS and concrete established in this paper has good accuracy. Furthermore, an accurate bond-slip model is also required for the establishment of an interfacial bearing capacity formula.

1.2. Bearing Capacity Formula

There are few studies on the load bearing capacity formulae at the interface between HTRCS and concrete. However, HTRCS composite material can be regarded as a sheet, and its bond to concrete can be referred to as the bond between an FRP sheet and concrete. Lu [29] and Neubauer [30] have conducted a series of tests and an extensive finite element study on the bond between the sheet and concrete, and they concluded that the load bearing capacity equation is mainly affected by factors such as the elastic modulus of the bonding material, thickness, bond length, and concrete strength. With reference to the relevant literature, for the interface bonding between HTRCS and concrete, the following bearing capacity calculation formula is proposed:

$$P_u = k\beta_w \beta_l b_H \sqrt{2 E_H t_H f_t} \qquad (8)$$

$$\beta_w = \sqrt{\frac{2.25 - b_f/b_c}{1.25 + b_f/b_c}} \tag{9}$$

$$\beta_l = \frac{L}{L_e}(2 - \frac{L}{L_e}), \ L < L_e \tag{10}$$

$$\beta_l = 1, \ L \geq L_e \tag{11}$$

$$L_e = 1.33 \frac{\sqrt{E_f t_f}}{f_t} \tag{12}$$

where P_u is the interface bearing capacity; k is the interface bonding coefficient; β_w is the width influence coefficient; β_l is the anchorage length coefficient; b_H is the width of the HTRCS; b_c is the width of the concrete; L is the actual adhesive length; L_e is the effective bonding length; and f_t is the tensile strength of the concrete.

In the above established formula for calculating the interfacial bond bearing capacity of HTRCS composite material and concrete, the interfacial bond coefficient k is unknown. Through regression analysis, $k = 0.286$ is obtained, and the square of the correlation coefficient is found to be 0.857, as shown in Figure 11.

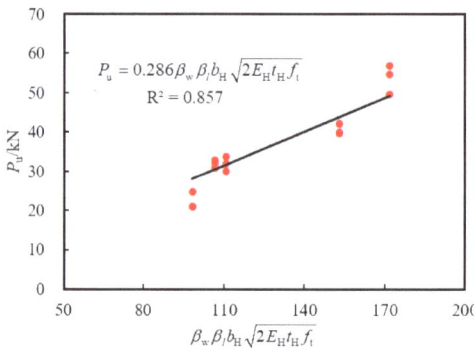

Figure 11. Regression analysis of k.

Currently, there are fewer studies on HTRCS composites, and even fewer on the interfacial bonding properties between HTRCS and concrete, so this paper is only based on experimental data to validate the bearing capacity equations established above. Table 5 shows the comparison between the calculated and test values for each specimen, SD is standard deviation, and COV is coefficient of variation. The prediction accuracy of the proposed load-bearing capacity formula is good, and the average value of the ratio of test values to calculated values is 0.978; the SD is 0.115 and the COV is 0.118. Therefore, the proposed formulae can accurately predict the interfacial bearing capacity between HTRCS and concrete. However, the applicability of the proposed prediction model is limited. The effects of factors such as interfacial bond length, the width ratio of concrete to composite layers, and interfacial cleanliness on interfacial bond behavior have not been tested and analyzed in depth. Therefore, in future studies, it is necessary to further explore the modification of the formulae by the key parameters and provide suggestions for engineers in practical applications.

Table 5. Comparison between test values and calculation values.

Specimens	$P_{u,exp}$/kN	$P_{u,cal}$/kN	$P_{u,exp}/P_{u,cal}$
DS1-1	21.18	28.03	0.76
DS1-2	24.75	28.03	0.88
DS1-3	21.16	28.03	0.75
DS2-1	32.92	30.42	1.08
DS2-2	32.31	30.42	1.06
DS2-3	30.98	30.42	1.02
DS3-1	32.05	31.59	1.01
DS3-2	33.86	31.59	1.07
DS3-3	30.16	31.59	0.95
DS4-1	56.89	49.12	1.16
DS4-2	54.73	49.12	1.11
DS4-3	49.64	49.12	1.01
DS5-1	40.05	43.72	0.92
DS5-2	42.14	43.72	0.96
DS5-3	39.88	43.72	0.91
	Average		0.978
	SD		0.115
	COV		0.118

5. Conclusions

In this paper, the interfacial bond behavior of HTRCS material and concrete is discussed through a double shear test. The main parameters involve concrete strength and HTRCS material properties. The failure modes, force-displacement relationship, and strain distribution of specimens are analyzed. The bond-slip model and bearing capacity formula are proposed. The following conclusions are drawn:

(1) The failure mode of DS1 specimens was partial debonding and fracture, and the rest of the specimens were the fracture of HTRCS. The bond between the wire mesh and the resin concrete was good, and there was no obvious slip.

(2) The concrete strength and reinforcement ratios of HTRCS composites were positively correlated with interfacial adhesion properties. When the concrete strength was increased from C30 to C40 and C50, the ultimate load increased by 43.4% and 43.2%, respectively. The ultimate load capacity increased by 32.1% with the reinforcement ratio of HTRCS composites increasing from 1.05% to 1.83%.

(3) The stress transfer effect is positively correlated with the concrete strength and reinforcement ratio of HTRCS; the higher the concrete strength and reinforcement ratio of HTRCS, the further the stress transfer distance.

(4) A bond-slip model for the interface between HTRCS and concrete is proposed and corrected according to the test parameters. The calculation formula for the interface bearing capacity is also proposed, and the test values are in good agreement with the calculated values, with an average value of 0.978.

Author Contributions: Conceptualization, Q.Y. and Y.Y.; investigation, A.L.; data curation, A.L.; writing—original draft preparation, Q.Y. and Y.R.; writing—review and editing, Q.Y. and Y.R.; visualization, A.L.; supervision, Y.Y.; project administration, Y.Y.; funding acquisition, Y.Y. All authors have read and agreed to the published version of the manuscript.

Funding: This work was supported by the Science and Technology Project of Sichuan Province (Grant No. 2018RZ0102).

Data Availability Statement: Data are contained within the article.

Conflicts of Interest: Qu Yu and Yongqing Yang were employed by the company Sichuan Jiaoda Engineering Detection & Consulting Co., Ltd. The remaining authors declare that the research was conducted in the absence of any commercial or financial relationships that could be construed as a potential conflict of interest.

References

1. Guo, R.; Ren, Y.; Li, M.; Hu, P.; Du, M.; Zhang, R. Experimental study on flexural shear strengthening effect on low-strength RC beams by using FRP grid and ECC. *Eng. Struct.* **2021**, *227*, 111434. [CrossRef]
2. Yang, X.; Gao, W.-Y.; Dai, J.-G.; Lu, Z.-D.; Yu, K.-Q. Flexural strengthening of RC beams with CFRP grid-reinforced ECC matrix. *Compos. Struct.* **2018**, *189*, 9–26. [CrossRef]
3. Yu, R.; Xia, M.; Yi, P.; Guo, S.; Guo, R. Study on the uniaxial tensile behavior of an FRP grid-ECC composite layer. *Case Stud. Constr. Mater.* **2024**, *20*, e02909.
4. Zhou, Y.W.; Guo, M.H.; Sui, L.L.; Xiong, F.; Hu, B.; Huang, Z.; Yun, Y. Shear strength components of adjustable hybrid bonded CFRP shear-strengthened RC beams. *Compos. Part B Eng.* **2018**, *163*, 36–51. [CrossRef]
5. Rakgate, S.M.; Dundu, M. Strength and ductility of simple supported R/C beams retrofitted with steel plates of different width-to-thickness ratios. *Eng. Struct.* **2018**, *40*, 192–202. [CrossRef]
6. Torabian, A.; Isufi, B.; Mostofinejad, D.; Ramos, A.P. Behavior of thin lightly reinforced flat slabs under concentric loading. *Eng. Struct.* **2019**, *196*, 109327. [CrossRef]
7. Huixiang, M.; Zhigang, L.; Zhang, Y.; Haoran, X. Research on the Bearing Capacity of the Cross Section of Reinforced Concrete Beams Strengthened with Increasing Section Method Considering the Effect of the Second Loading. *Adv. Res.* **2015**, *5*, 1–9. [CrossRef]
8. Shakir, Q.M.; Abdlsaheb, S.D. Rehabilitation of partially damaged high strength RC corbels by EB FRP composites and NSM steel bars. *Structures* **2022**, *38*, 652–671. [CrossRef]
9. Shakir, Q.M.; Abdlsaheb, S.D. Strengthening of the self-compacted reinforced concrete corbels using NSM steel bars and CFRP sheets techniques. *J. Eng. Sci. Technol.* **2022**, *17*, 1764–1780.
10. Singh, S.; Vummadisetti, S.; Chawla, H. Influence of curing on the mechanical performance of FRP laminates. *J. Build. Eng.* **2018**, *16*, 1–19. [CrossRef]
11. Li, Y.Y.; Guo, B.; Liu, J. Research on reinforced concrete beam enlarged cross section method experiment and finite element simulation. *Appl. Mech. Mater* **2014**, *638*, 208–213. [CrossRef]
12. De Lorenzis, L.; Teng, J.G. Near-surface mounted FRP reinforcement: An emerging technique for strengthening structures. *Compos. Part B Eng.* **2007**, *38*, 119–143. [CrossRef]
13. Zhang, S.S.; Yu, T.; Chen, G.M. Reinforced concrete beams strengthened in flexure with near-surface mounted (NSM) CFRP strips: Current status and research needs. *Compos. Part B Eng.* **2017**, *131*, 30–42. [CrossRef]
14. Liu, Z.-Q.; Guo, Z.-X.; Ye, Y. Flexural behaviour of RC beams strengthened with prestressed steel wire ropes polymer mortar composite. *J. Asian Arch. Build. Eng.* **2021**, *21*, 48–65. [CrossRef]
15. Emara, M.; Rizk, M.; Mohamed, H.; Zaghlal, M. Enhancement of circular RC columns using steel mesh as internal or external confinement under the influence of axial compression loading. *Frat. Integrità Strutt.* **2021**, *15*, 86–104. [CrossRef]
16. Marthong, C. Compressive behavior of galvanized steel wire mesh (GSWM) strengthened RC short column of varying shapes. *Struct. Monit. Maint.* **2020**, *7*, 215–231.
17. Zhang, K.; Xuan, J.; Shen, X.; Xue, X. Investigating the Flexural Properties of Reinforced Concrete T-Beams Strengthened with High-Strength Steel Wire Mesh and Polyurethane Cement. *J. Bridge Eng.* **2024**, *29*, 86–104. [CrossRef]
18. Zhao, H. Axial compressive behaviour of concrete strengthened with steel rings, wire mesh and modified high strength mortar (MHSM). *Constr. Build. Mater.* **2020**, *250*, 118938. [CrossRef]
19. Yuan, F.; Chen, M.; Pan, J. Flexural strengthening of reinforced concrete beams with high-strength steel wire and engineered cementitious composites. *Constr. Build. Mater.* **2020**, *254*, 119284. [CrossRef]
20. Wang, X.; Yang, G.; Qian, W.; Li, K.; Zhu, J. Tensile behavior of high-strength stainless steel wire rope (HSSSWR)-reinforced ECC. *Int. J. Concr. Struct. Mater.* **2021**, *15*, 43. [CrossRef]
21. Zheng, Y.-Z.; Wang, W.-W.; Brigham, J.C. Flexural behaviour of reinforced concrete beams strengthened with a composite reinforcement layer: BFRP grid and ECC. *Constr. Build. Mater.* **2016**, *115*, 424–437. [CrossRef]
22. Zheng, Y.-Z.; Wang, W.-W.; Mosalam, K.M.; Fang, Q.; Chen, L.; Zhu, Z.-F. Experimental investigation and numerical analysis of RC beams shear strengthened with FRP/ECC composite layer. *Compos. Struct.* **2020**, *246*, 112436. [CrossRef]
23. Yan, M. *Concrete with Steel High Toughness Resin Wire Mesh and the Reinforcement Theoretical Research on Prestressed Concrete Simply Supported Plate Beam Bridge*; Southwest Jiaotong University: Chengdu, China, 2015.
24. Ren, Y.; Gao, Z.; Xia, M.; Guo, R. Study on bonding behavior of FRP grid-ECC composite layer and concrete with quantitative interface treatment. *Eng. Struct.* **2023**, *294*, 116768. [CrossRef]
25. *GB 50010-2010*; Code for Design of Concrete Structures. China Architecture & Building Press: Beijing, China, 2010.
26. Dai, J.G.; Ueda, T.; Sato, Y. Development of the nonlinear bond stress-slip model of fiber reinforced plastics sheet-concrete interfaces with a simple method. *J. Compos. Constr.* **2005**, *9*, 52–62. [CrossRef]
27. Ferracuti, B.; Savoia, M.; Mazzotti, C. Interface law for FRP—Concrete delamination. *Compos. Struct.* **2007**, *80*, 523–531. [CrossRef]
28. Lu, X.Z.; Teng, J.G.; Ye, L.P.; Jiang, J.J. Bond–slip models for FRP sheets/plates bonded to concrete. *Eng. Struct.* **2005**, *27*, 920–937. [CrossRef]

29. Lu, X.Z. *Studies on FRP-Concrete Interface*; Tsinghua University: Beijing, China, 2005.
30. Neubauer, U.; Rostasy, F.S. Design aspects of concrete structures strengthened with externally bonded CFRP plates. In Proceedings of the 7th International Conference on Structural Faults and Repair, Edinburgh, UK, 8 July 1997; ECS Publications: Edinburgh, UK, 1997; Volume 2, pp. 109–118.

Disclaimer/Publisher's Note: The statements, opinions and data contained in all publications are solely those of the individual author(s) and contributor(s) and not of MDPI and/or the editor(s). MDPI and/or the editor(s) disclaim responsibility for any injury to people or property resulting from any ideas, methods, instructions or products referred to in the content.

Article

Shear Transfer in Concrete Joints with Non-Metallic Reinforcement

Lore Zierul *, Enrico Baumgärtel, David Sandmann and Steffen Marx

Institute of Concrete Structures, TUD Dresden University of Technology, 01062 Dresden, Germany
* Correspondence: lore.zierul@tu-dresden.de

Abstract: The use of non-metallic reinforcement can significantly reduce the carbon footprint of the construction sector. Mixed structures made out of steel and non-metallic reinforcement should be avoided due to the risk of galvanic corrosion. So far, researchers have been focusing on the load-bearing behavior in the longitudinal direction of the fibers. In this study, the behavior of the fibers in the non-metallic reinforcements is analyzed perpendicular to the fiber orientation. Therefore, a uniaxial shear test on a single bar (uniaxial shear test), as well as a series of push-off tests with reinforcements embedded in the concrete, was carried out. For both experiments, bars made of carbon fiber-reinforced polymers (CFRPs) and glass fiber-reinforced polymers (GFRPs) were investigated. In order to analyze the influence of non-metallic reinforcement in the joint, specimens without reinforcement have been tested as well. Also, the joint roughness and reinforcement ratio of the concrete joint was varied in the tests. The determined transverse shear strengths for the single bar exceed the values of the producer. For the push-off test, high standard deviations occurred, making it difficult to draw firm conclusions. Nevertheless, it is shown that increasing the amount of reinforcement leads to higher ultimate forces. The presented study emphasizes the necessity of further studies of the shear transfer in concrete joints.

Keywords: shear transfer; joints; carbon fiber-reinforced polymers (CFRPs); glass fiber-reinforced polymers (GFRPs); non-metallic reinforcement; carbon reinforced concrete; adhesion; friction; dowel action

Citation: Zierul, L.; Baumgärtel, E.; Sandmann, D.; Marx, S. Shear Transfer in Concrete Joints with Non-Metallic Reinforcement. *Buildings* **2024**, *14*, 1975. https://doi.org/10.3390/buildings14071975

Academic Editors: Bo Wang, Rui Guo, Muye Yang, Weidong He and Chuntao Zhang

Received: 24 May 2024
Revised: 12 June 2024
Accepted: 24 June 2024
Published: 30 June 2024

Copyright: © 2024 by the authors. Licensee MDPI, Basel, Switzerland. This article is an open access article distributed under the terms and conditions of the Creative Commons Attribution (CC BY) license (https://creativecommons.org/licenses/by/4.0/).

1. Introduction

1.1. Motivation

The cement production is responsible for 8% of global annual CO_2 emissions [1]. To protect the steel reinforcement in order to avoid corrosion in steel-reinforced concrete, a minimum concrete cover must be maintained. The development of carbon and glass fiber-reinforced polymers is opening up new possibilities for lightweight structures. Due to its inherent corrosion resistance, high tensile strength and small weight, non-metallic reinforcement has become a promising substitute for conventional steel bars [2]. Broad research has been conducted concerning the behavior of non-metallic reinforcement in pure tensile tests, as well as experiments for non-metallic reinforcement embedded in concrete. The used reinforcement shapes varied from fibers to single yarns over to grids and bars [3–5]. According to the German DAfStb-guideline for non-metallic reinforcement published in March 2024 [6], the cross-sectional area of a non-metallic reinforcement crossing a joint must not yet be accounted for in the shear resistance. Thus, the load-bearing capacity is underestimated.

1.2. Load-Bearing Behavior in Joints

The ultimate load of a joint is reached after the mechanisms of adhesion, friction and the load-carrying action of the reinforcement are exceeded. The latter can be divided into dowel action and clamping. An overview of the described mechanisms is shown in Figure 1.

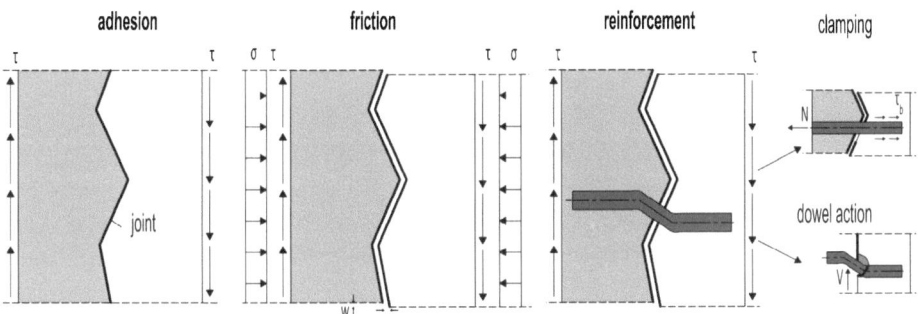

Figure 1. Load-carrying mechanisms in a joint exposed to shear [7].

For adhesion, physical bonds such as van der Waals forces in the form of hydrogen bonds and dipole-dipole bonds are involved. However, chemical bonds contribute to only a small part of the load transfer through ionic and covalent bonds compared to physical mechanisms such as mechanical interlock [8].

After adhesion is exceeded, a surface displacement occurs, which leads to interlocking due to surface roughness and the interaction of the aggregate grains with each other. In addition to the aggregate, this interlocking is also dependent on the reinforcement crossing the joint. The reinforcement resists the displacement of the crack surface, resulting in a clamping and dowel effect. In the case of the clamping effect, the forces are transferred from the joint to the interior of the component along the length of the reinforcement. In the case of the dowel effect, a combined load transfer takes place via the bending stress and the diagonal tension of the reinforcement [9].

The behavior of concrete joints has been extensively investigated in the past using steel reinforcement, e.g., [7,9–12]. A variation in roughness for unreinforced specimens was investigated by differentiating a surface that had been water-blasted or sand-blasted, as well as a surface that had been casted against formwork or casted without surface treatment [13]. According to the test results, the load bearing is dominated by friction and cohesion. Based on experiments using round steel dowels clamped in concrete, the embedment length upon which no changing in the ultimate load was stated equals 5·Ø [11]. A model of the pressure distribution along a reinforcement, assuming it behaves like a pole embedded in the ground, was provided first by [10]. For an embedment length of >6–8·Ø, concrete splintering instead of spalling occurs, and, additionally, a yielding of the reinforcement is observed. The model was modified by determining that the concrete section directly at the surface until the yielding zone plasticizes. Below this yielding zone the concrete does not form cracks [12]. The yielding zone can be found at 1·Ø [12].

The development of [14] was based on the discovery that reinforcements crossing a joint are only involved after a certain opening or displacement of the joint. The mechanisms for a slip smaller and greater than 0.05 mm, as well as a reinforcement ratio smaller or greater 0.05%, are distinguished by [14]. The case of a slip smaller than 0.05 mm, as well as a reinforcement smaller than 0.05%, is depicted in (1), consisting of the load-carrying behavior through adhesion and friction.

$$\tau_{Rd} = c_a \cdot f_{ctd} + \cdot \sigma_n \leq 0.5 \cdot \nu \cdot f_{cd} \tag{1}$$

With

c_a	Coefficient for adhesive bond;
f_{ctd}	Design tensile strength of concrete;
	Coefficient of friction;
σ_n	Stress due to external normal force;
ν	Reduction factor for shear strength of diagonal concrete strut;
f_{cd}	Design compressive strength of concrete.

When the slip exceeds 0.05 mm, the reinforcement participates in the load-bearing, and the resisting shear strength is depicted in (2) as in [14].

$$\tau_{Rd} = c_r \cdot f_{ck}^{\frac{1}{3}} + \cdot (\sigma_n + \rho \cdot \kappa_1 \cdot f_{yd}) + \kappa_2 \cdot \rho \cdot \sqrt{f_{yd} \cdot f_{cd}} \leq \beta_c \cdot \nu \cdot f_{cd} \quad (2)$$

With the formula symbols, in addition to the listing beforehand

c_r	Coefficient for adhesive bond for slip >0.05 mm;
f_{ck}	Characteristic compressive strength of concrete;
κ_1	Coefficient of efficiency for tensile force that can be activated in the reinforcement;
f_{yd}	f_{yd} Design tensile strength of steel;
κ_2	κ_2 Coefficient for flexural resistance of reinforcement (dowel action);
β_c	β_c Coefficient allowing for angle of diagonal concrete strut.

This article focuses on the strength of non-metallic bars perpendicular to their fiber orientation. Since non-metallic bars consist of pure fibers and a surrounding polymer impregnation, the material properties differ widely depending on the direction of loading. The load-carrying capacity perpendicular to the fiber orientation is examined in an experiment for a single bar in a uniaxial shear test and for the reinforcements embedded in concrete in a push-off test. The test series for the concrete joints analyzes the influence of the roughness of the joint, as well as the amount and type of reinforcement.

1.3. Novelty of the Test Setup

The presented test setups had not yet been used in the context of non-metallic reinforcement. The uniaxial shear test of a single bar was conducted inspired by [15] performing a ASTM D7617/D7617M [16] shear test on glass fiber-reinforced polymers with varying profiles. This test-setup was not adopted entirely since the blade pushing down on the reinforcement bar is surrounding it, which translates to stiff behavior from the steel compared to the soft behavior of concrete that exists in reality [17]. A reinforcing bar with a spiral sheathing as a profile had a greater load-bearing capacity than milling off material [15]; thus, both types of profiles were used in the presented test setup. In addition, carbon fiber-reinforced polymer bars (CFRPs) exhibited no recorded information on shear behavior; hence, this study guarantees new insights into the behavior of non-metallic reinforcements. Regarding the push-off test of reinforcement crossing a joint, first insights in the role of dowel action using CFRP textiles have been given. It was stated that dowel action in the shear transfer of FRP textile-reinforced members should not simply be negated [18]. The test setup used plates of two thicknesses, predefining a crack specified with and inducing a normal force. The predefinition of a crack in the presented test setup was executed by creating two joints (see Section 2.2), and the introduction of a normal force was avoided in order to exclude positive effects on the shear capacity (see Formulas (1) and (2)).

2. Materials and Methods
2.1. Materials

As for the reinforcement, two different types of bars were used. A yarn made of carbon fibers is helically twisted around the carbon core of the bar (CFRP). The glass bar profile (GFRP) is twisted as well, but it is achieved by milling out material (see Figure 2). The nominal diameter is defined as the diameter where there is a completely circular cross section. Further properties from the technical data sheet can be found in Table 1.

Figure 2. Detail of investigated reinforcement: carbon reinforcement—black: solidian REBAR D8-CCE; white: Schöck ComBAR, d = 8 mm.

Table 1. Properties from technical data sheet [19,20].

Properties	Unit	Carbon	Glass
Nominal diameter	mm	8	8
Ultimate tensile strength	N/mm^2	2100 [1]	1000
Ultimate strain	‰	15 [2]	7.4 [3]
Modulus of elasticity	N/mm^2	140,000 [4]	60,000 [5]
Fiber volume content	%	64	75
Nominal cross-sectional area	mm^2	50.3	50.3

[1] Characteristic short-time tensile strength regarding the nominal cross-sectional area. [2] Characteristic elongation at break. [3] Strain at Ultimate Limit State. [4] Average modulus of elasticity regarding nominal cross-sectional area. [5] Tension modulus of elasticity.

The concrete had a maximum aggregate size of 8 mm and the quantitative amount can be found in Table 2. According to DIN EN 206 [21], aggregate mixtures with an aggregate size smaller than 4 mm are classified as mortar. However, because of the relatively small aggregate size, the testing of the envisaged compressive strength was verified by prisms (160 × 40 × 40 mm^3) according to the standard test method for mortar DIN EN 1015-11 [22]. For each concrete batch, the compressive strength f_c was determined on three prisms, and the flexural tensile strength $f_{ct,fl}$ was determined on the two halves after each compression test on the day of testing (age 28 to 32 days). The average compressive strength of the concrete prisms was scaled to experiments of cube dimensions and resulted in 21.6 N/mm^2 [23].

Table 2. Mix design of the cementitious matrix for 1 m^3.

Component	Volume in dm^3	Density in kg/cm^3	Mass in kg
Water	199.81	1	199.81
Cement	105.37	3	316.1
Aggregate size 0–2 mm	379.87	2.63	999.05
Aggregate size 0–8 mm	297.21	2.63	781.66
Pore	20	-	-
Sum	1002.25 [1]	-	-

[1] The volume differs from 1000 cm^3 due to rounding.

2.2. General Procedure and Experimental Program

The present study differentiates between the uniaxial shear test and the push-off test. One main objective was to investigate the shear capacity of the samples in terms of the load-displacement characteristics. For the push-off test, the following parameters were varied:

- Roughness (smooth, rough: R_t = 1.41 mm);
- Ratio of reinforcement (one or two reinforcements crossing the joint);
- Type of reinforcement (carbon fiber-reinforced polymer, glass fiber-reinforced polymer).

Two reinforcements crossing the joint were only investigated for the smooth joint (geometry as in Figure 3). Three specimens were examined for the reinforced configuration, resulting in 18 reinforced and 6 unreinforced samples. Regarding the uniaxial shear test, 4 samples per reinforcement type, in total 8 samples were tested (Table 3). For each reinforcement type in the uniaxial shear test, only one sample was not tested until failure.

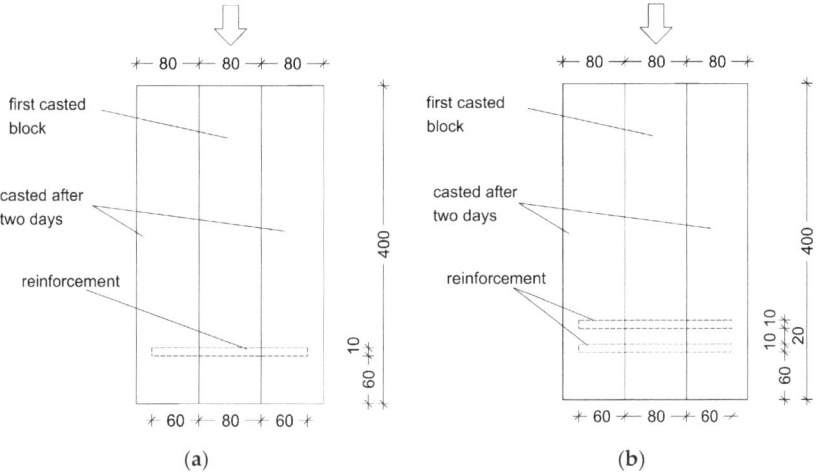

Figure 3. Specimen for the push-off test with: (**a**) one reinforcement crossing the joint; (**b**) two reinforcements crossing the joint.

Table 3. Number of experiments conducted for each test setup.

Uniaxial Shear Test		Push-Off Test			
Reinforcement Type	Specimens	Reinforcement	Surface of the Joint	Reinforcements Crossing the Joint	Specimens
Carbon	4		Smooth	1	3
Glass	4	Carbon	Smooth	2	3
			Rough	1	3
			Smooth	1	3
		Glass	Smooth	2	3
			Rough	1	3
		None	Smooth	-	3
			Rough	-	3
Total: 8 specimens		Total: 24 specimens			

2.3. Sample Preparation

The length of the bar for the uniaxial shear test was determined to 180 mm, respecting the conditions of ISO 10406-1 [24] being not more than 300 mm. The condition of the length being less than five times the shear plane interval could not be satisfied due to the restrictions of the load-charging equipment of the laboratory. The cutting was realized by using a diamond circular saw.

For the experiments of concrete joints, a three-layered specimen was built. The specimens used for this study were produced at the Otto-Mohr-Laboratorium (OML, Dresden, Germany). A formwork of sealed timber (340 × 280 × 90 mm^3) was prepared with three casting boxes. The inner part was casted with a fresh concrete consistency class of F3 according to DIN EN 12350-5 [24,25]. The specimens were roughened one day after casting by loosening the limiting formwork of the inner part. In order to roughen the generated surface evenly, the specimens were extracted from the formwork. The roughness was obtained by using water pressure. Determining the roughness coefficient R_t according to Kaufmann [26,27] as depicted in Figure 4a resulted in a roughness of 1.39 mm for the first batch and 1.42 mm for the second batch. A few hours after the roughening process, the concrete was embedded again in the formwork as in Figure 4b, casting the outer parts one day afterwards.

(a)

(b)

Figure 4. Production of the samples with roughened concrete joints: (**a**) determining roughness depth; (**b**) formwork ready for second casting.

Due to the profile of the bars, especially for the carbon bars, the diameter of the hole in the formwork was not big enough to remove the formwork. Turning the formwork out along the profile resulted in stress being put on the young concrete, thus destroying it. Two specimens in the configuration of two reinforcements crossing the joint version had to be casted again. Probably, microcracks existed for the other specimens as well, albeit without being noted. Until the day of testing, the specimens remained in a climate chamber at a temperature of 20 °C with 65% relative humidity.

2.4. Test Setup and Instrumentation

2.4.1. Uniaxial Shear Test of Single Bar

Since the tests were of tentative nature, the test set-up was designed based on ISO 10406-1 [24]. The setup, as shown in Figure 5a, consisted of an upper blade (in this case, the machine charging equipment) exerting downward pressure on the bar and two lower blades supporting the bar. The demanded distance of the two lower blades of 50 mm could not be guaranteed by the load application of the testing machine at the Otto-Mohr-Laboratorium (OML) with 60 mm.

Figure 5. Uniaxial shear test: (**a**) front view of test-setup; (**b**) side view scheme of test-setup (dimensions in mm).

The upper blade was not adapted to the diameter of the tested bar and a plane surface, whereas the halves of the lower blades were drilled with the diameter according to the tested reinforcement type. In order to prevent the lower blade halves from opening up, four screws assured the position between two halves and the position on the test machine (Figure 5b). The setup was screwed to the frame of the testing machine with a compressive load capacity of 250 kN.

The load was applied in a displacement-controlled manner at a loading rate of 1 mm/min until failure for three specimens and until extreme bending for one specimen. Before starting the test, a force of 100 N was applied in order to align the calotte of the test setup.

2.4.2. Push-off Test of Reinforcement Crossing a Joint

The specimen with three concrete parts was tested while standing on metal blocks, allowing the inner part to move downward without any hindrance. The aim of the test setup was to determine the force that could be transferred by the reinforcement. If the specimen is clamped on both sides, the normal force created while doing so contributes to the resisting force of the joint. In order to avoid this effect, steel profiles were placed on both sides, with distance to the specimen only for the case of falling (Figure 6).

Figure 6. Push-off test setup of reinforcement crossing a joint.

The load was applied in a displacement-controlled manner at a loading rate of 0.5 mm/min until the crushing of the concrete around the reinforcement bar. A force of 200 N was applied in order to align the load from the metal plates to the specimen. During the experiments, the machine force F, as well as the vertical machine displacement w, were measured. In order to monitor the horizontal opening of the joints, one inductive displacement sensor (linear variable differential transformer—LVDT) was installed on the upper part, and one was installed on the lower part beneath the reinforcement bar. The displacement of the joints was determined with two vertical inductive displacement sensors, one in each joint.

3. Results and Discussion

3.1. Uniaxial Shear Test Result of Single Bar

The experiments were conducted in order to deliver a better understanding of the shear failure of non-metallic reinforcement. Figure 7 shows photographs of the failure pattern. All specimens failed on the left lower blade after a displacement of over 6.7 mm regarding the carbon reinforcement and 5.3 mm for the glass reinforcement. Until the failure bending occurred in the middle of the bar (see Figure 5a). Some fibers close to the support are detached and a crack in the surface propagates widely from the joint away further into the middle of the reinforcement bar, as it can be seen in Figure 7b (orange arrows).

Figure 7. Uniaxial shear test after failure: (**a**) carbon reinforcement; (**b**) glass reinforcement.

Regarding the tensile strength, one can note that the characteristic short-time tensile strength of carbon reinforcement is higher than the one of glass reinforcement (2100 N/mm^2 > 1000 N/mm^2 [19,20,28]). This tendency was also observed in the shear force, as shown in Figure 8. In addition, the curve trend varies from parabolic for the carbon bars to parabolic with two drops at a displacement of approximately 1 and 2 mm for the glass bar. One offered explanation is the failure of the surrounding matrix, which is scattered during the test of all the carbon bars and two times significantly for the glass bars.

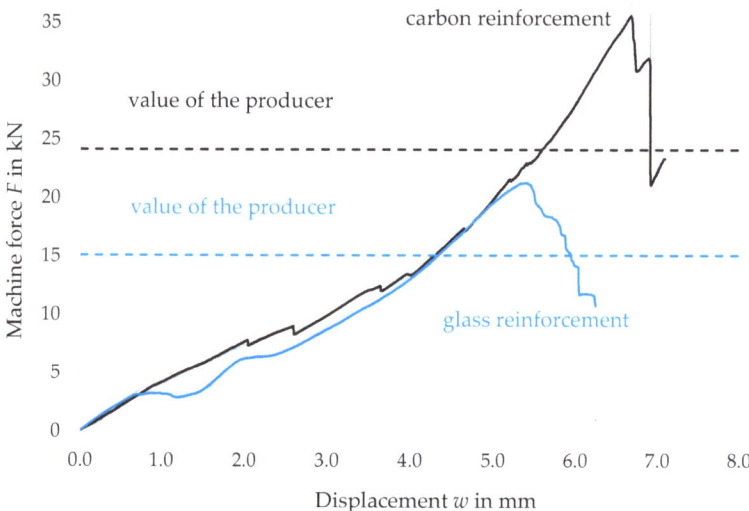

Figure 8. Force-machine displacement diagram for the uniaxial shear test comparing carbon reinforcement with glass reinforcement. The highlighted curve represents the average value.

The shear strength (τ_s) can be calculated with the aid of dividing the ultimate force (P_s) by two times the cross section according (A) to ISO 10406-1 [20] as follows:

$$\tau_s = \frac{P_s}{2 \cdot A}$$

On average, the samples achieved an average shear strength of 366.5 N/mm² regarding the carbon reinforcement and 213.7 N/mm² for the glass reinforcement according to Table 3. The values for the shear strengths of the producers were smaller than the experimentally determined values. The producer of the carbon reinforcement removed the profiling prior to testing. As the nominal cross section stays the same, it translates to a lower ultimate force in the test of the producer compared to our own obtained ultimate force. The profiling, therefore, led to higher ultimate forces and should be taken into account. According to [19], the profile was removed in order to provide replicability, and the values were rounded down, ensuring safety in the dimensioning. The blade in the shear test of the carbon reinforcement producer surrounded the reinforcement, whereas the blade in the here presented test setup pushed only onto the surface of the reinforcement. This reasoning explains the ratio of the producers' value to the average experimentally developed values of 65% for the carbon reinforcement and 70% regarding the glass reinforcement. Further investigation showed that a higher value for the rib inclination angle suggests a greater tendency towards splitting [29]. Comparing the used reinforcements, this translates to a greater tendency towards splitting for the glass fiber-reinforced polymer than for the carbon fiber-reinforced polymer. Figure 7 confirms that GFRPs have greater splitting than CFRPs. The details of the test results are listed in Table 4.

Table 4. Uniaxial shear test results.

	Carbon Reinforcement			Glass Reinforcement		
Sample No.	P_s in kN	w in mm	τ_s in N/mm²	P_s in kN	w in mm	τ_s in N/mm²
1	36.9	6.7	366.8	21.8	5.6	216.7
2	35.9	6.9	356.9	21.8	5.4	217.7
3	37.8	6.9	375.7	20.9	5.3	207.8
Average value	36.9	6.8	366.5	21.5	5.4	213.7
Variation	0.9	0.01	89.3	0.3	0.02	26.7
Standard deviation	1.0	0.1	9.4	0.5	0.2	5.2
Average transversal shear strength (producer value)	240 N/mm²			150 N/mm²		

3.2. Push-off Test Results of Reinforcements Crossing a Joint

With 6 unreinforced specimens and nine specimens for each reinforcement type, the shear behavior of concrete joints was assessed. Figure 9 shows a selection of possible failure patterns of the specimens. The specimens with no reinforcement failed brittle, and the entire specimen tilted towards one side (Figure 9a). The rough surface of the joint created small concrete spalling along the joint. Comparing the specimen where there was one reinforcement crossing the joint to two reinforcements, it was noted that the failure of the concrete was also brittle, but the three parts of the specimens were held together by the reinforcement. In addition, cracks occurred above the reinforcement (Figure 9b,c orange arrows) and propagated from the joint towards the end of the reinforcement. The formation of the crack coincided with the peak load, as in [18]. The experiments for the specimens with two reinforcements crossing the joint were conducted until a machine displacement of at least 5.6 mm was achieved, resulting in crack widths of 2.5 mm and even separating the concrete blocks above the reinforcement from the part below. This enabled an analysis of the reinforcement embedded in the concrete.

In order to analyze the torsion of the reinforcement and the crack in the cross section, the three parts of the specimens were separated from each other. The cracks in the concrete are caused by exceeding the concrete tension force, resulting in a crack. The cracks begin on the surface and disperse toward the reinforcement, as shown in Figure 10a. The formation of the cracks results from the destroyed adhesive bond along the joint. The reinforcement that crosses the joint is therefore the most contributing component for transferring the force of the inner concrete part to the two outer concrete parts. The vertical force on the test specimen is transformed into a horizontal force by the reinforcement. The surrounding concrete, thus, receives a tensile force, which the concrete withstands less effectively than a compressive force. The appearance of longitudinal cracks in the plane of the reinforcement is confirmed by [18].

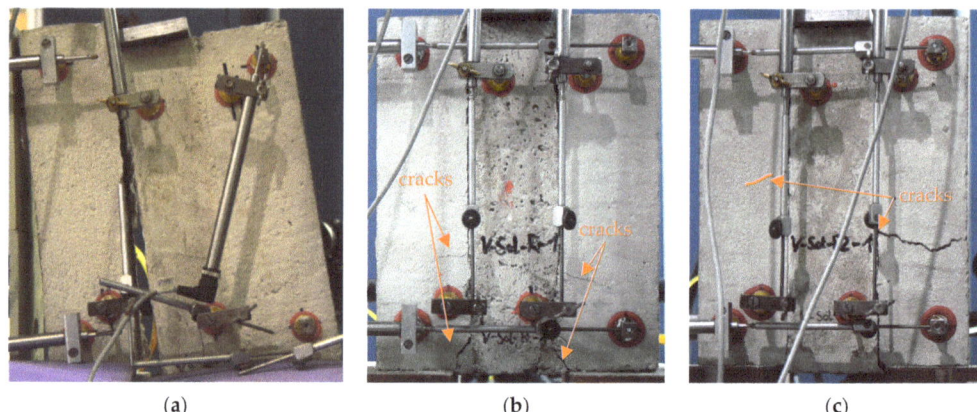

Figure 9. Push-off test: (**a**) unreinforced specimen with rough joint; (**b**) carbon-reinforced specimen (one reinforcement crossing the joint and rough joint); (**c**) carbon-reinforced specimen (two reinforcements crossing the joint and smooth joint).

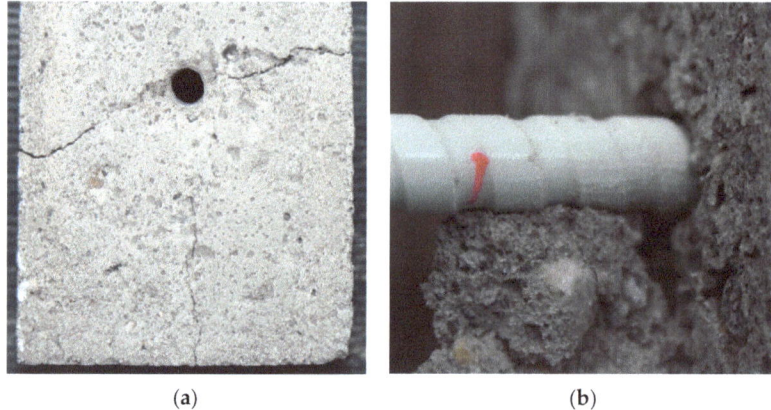

Figure 10. Details of failure for reinforcements embedded in concrete: (**a**) top view of a side concrete block with reinforcement (removed for photo); (**b**) damaged surface of glass reinforcement near the shear plane.

That the reinforcement has been exposed to shear is depicted in Figure 10b. The type of crack along the reinforcement near the shear plane resembles the crack obtained during the uniaxial shear test of the single bar (Figure 7b), but the cracks in Figure 10b are only in the surface of the reinforcement.

The brittle behavior of specimens without reinforcements is illustrated in a force-displacement diagram in Figure 11a,b. Roughening the surface does not impact the adhesion failure, as the curves have a significant drop as well. However, the ultimate force until failure of the joint is higher for the roughened surface. There is no drop near to 0 N for the reinforced specimens in Figure 11a, contrary to the unreinforced specimens. This cannot be shown for all unreinforced specimens because the machine stops the test when the force drops 95%. In Figure 11a, the ultimate force for the reinforced joint for two out of three specimens is slightly above that of the unreinforced joint. In Figure 11b, the opposite can be stated. The finding is unexpected, taking into account the fact that roughness typically results in a greater surface being available for adhesion forces, thus increasing the resistance to shear displacement. Comparing the calculation of resisting shear stress in joints according to [14], the rough surface has a higher resistance, since the design value of the adhesive bond (c_a in Equation (1)) is greater. One explanation for the test result could be that the reinforcement crossing the joint reduces the surface available for adhesion forces. A second explanation is offered by the erased material of the rough surface due to transport and reassembly, thus smoothening the joint before the second casting can take place. Another difference between the two diagrams is the behavior after the concrete joint fails. In Figure 11b, the rough surface with its hilly structure on a microscopic level becomes smoother with the further displacement of the joint. Hence, the force diminishes only slowly after the force drop. The contribution of reinforcement to the load-carrying capacity is evident in Figure 11a, which shows a new local maximum after an average machine path of 1.5 mm.

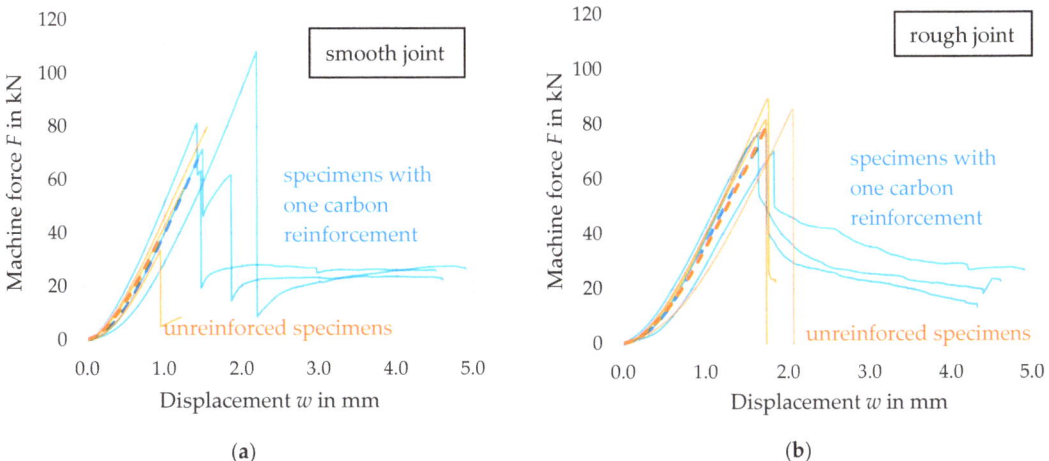

Figure 11. Force-displacement diagram comparing unreinforced and carbon-reinforced joints (dash line represents average value): (**a**) for smooth joints; (**b**) for rough joints.

The glass reinforced specimens achieved smaller ultimate loads compared to the carbon-reinforced specimens regarding the smooth joint for one and two reinforcements crossing the joint (see Table 5). This is not the case for the rough joint, where the glass reinforcement has a slightly higher ultimate load than the carbon fiber crossing the joint. The failure behavior is similar to that of the carbon-reinforced specimens, even though two significant drops in the curve take place in the case of the glass fiber reinforcement (Figure 12 light green line). This could be explained by an asymmetrical arrangement in the test machine, thus resulting in one joint being more loaded than the other, and as a result, one joint fails before to the other. A significant increase in the ultimate force is stated for two reinforcements crossing the joint (Figure 12, dark green line). This applies to both types of reinforcement. In addition, a new local maximum in the curve appears after the concrete failure of the joint. The local maximum of the highest curve represents an exception, as the machine recorded the real force of the reinforcement representing a resistance only after a short delay. The ultimate force for the specimens with two reinforcements crossing the joint could be even higher if the holes in the formwork are significantly greater and do not lead to the preliminary damage of the specimens, as described in Section 2.3.

Table 5. Push-off test results of reinforcements crossing a joint (ultimate force P_s in kN).

Joint Surface	Unreinforced		Carbon Reinforcement			Glass Reinforcement		
	Smooth	Rough	Smooth		Rough	Smooth		Rough
No. of reinforcements	0	0	1	2	1	1	2	1
Sample 1	57	81.7	108.1	92.6	70	36.8	74	76.2
Sample 2	79.7	85.5	81.2	109.5	76.9	60.1	92.2	68.5
Sample 3	33.7	89.2	71.4	62.3	75.9	64	56.2	82.9
Average value	56.8	85.5	86.9	88.1	74.3	53.6	74.1	75.9
Variation	529	14.1	361.1	571.9	13.9	216.3	324	51.9
Standard deviation	23	3.8	19	23.9	3.7	14.7	18	7.2

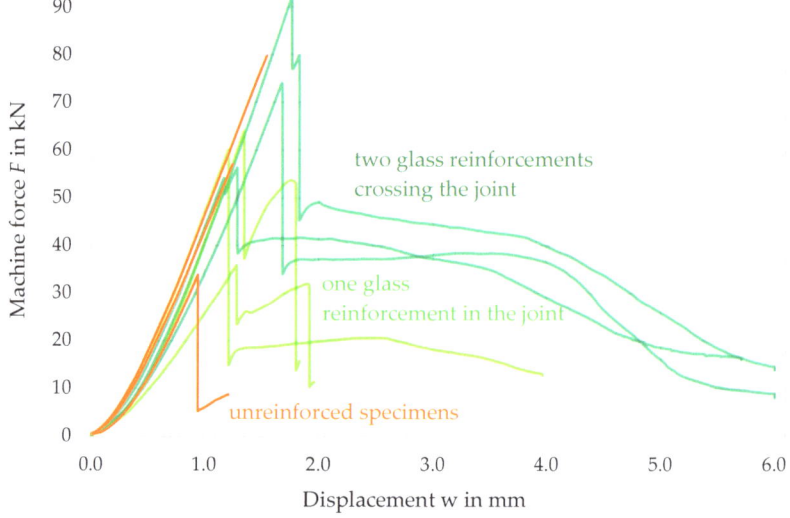

Figure 12. Force-displacement diagram comparing two glass reinforcements with one reinforcement crossing the joint and unreinforced specimens.

On average, the unreinforced samples with smooth joints achieved an average ultimate load of 56.8 kN, compared to 85.5 kN for the unreinforced samples with rough joints. A significant standard deviation was stated for the unreinforced specimens with smooth joints. Concerning the specimens with carbon reinforcement, the ultimate load was 86.9 kN for the smooth joint and 74.3 kN for the rough joint. Comparing these values with the glass fiber reinforcement, one can note that the ultimate load for the smooth joint with 53.6 kN was higher for the carbon-reinforced joints. Regarding the rough joint, the ultimate load for joints with glass rebars was slightly greater at 75.9 kN compared to carbon-reinforced joints. Throughout the smooth joints, a high standard deviation was noted. This can be explained in part by the inhomogeneity of the concrete joint. The details of the test results are listed in Table 5.

The findings correlate with the results of other authors. The ultimate shear force is a function of the reinforcement ratio, according to [30]. The shear force was determined by reinforcing the joint of two L-formed specimens with glass fiber-reinforced polymers (GFRPs). The test setup of two concrete blocks, casted one after the other and reinforcing the joint with GFRPs, showed that with a rising number of reinforcements crossing the joint, the bearing capacity increases.

A qualitative analysis of the bending behavior of steel, carbon and glass bars is shown in Figure 13. Steel reinforcement develops a plastic hinge at a distance of approximately one time the diameter from the joint (see Figure 13a) [7]. Comparing the tested non-metallic reinforcement as in Figure 13b, it can be observed that there is a slight deformation, for the carbon bar, and there are cracks in the profiling of the glass bar. In order to make the cracks visible, a red color spray was used. The photographs, as in Figure 13b, are the result of a different modulus of elasticity. The carbon fiber has a higher modulus of elasticity in the longitudinal direction, thus resulting in a greater resistance against cracks in the profiling. There were also some small fissures that could not be made visible using coloring spray.

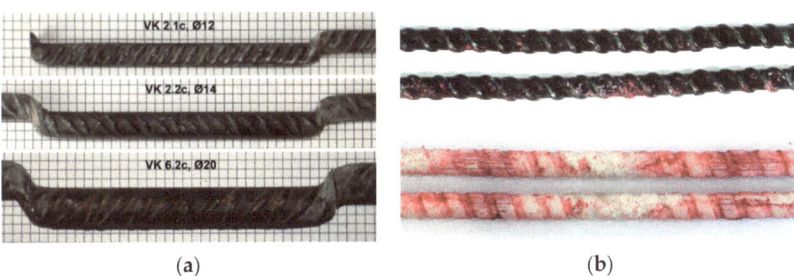

Figure 13. Details of failure for reinforcements crossing a joint: (**a**) steel reinforcement [7]; (**b**) carbon reinforcement (top) and glass reinforcement (bottom) for specimens with two reinforcements crossing the joint.

4. Conclusions

This paper investigated shear loads for single non-metallic bars and concrete specimens with non-metallic reinforcements crossing two joints. Reinforcement bars as carbon fiber-reinforced polymers (CFRPs) and glass fiber-reinforced polymers (GFRPs) with the same diameter have been used. The contribution of non-metallic reinforcement should not be negated, as in the German DAfStb-guideline for non-metallic reinforcement [6]. This paper gives a first insight into the shear capacity of non-metallic reinforcement.

The main conclusions are as follows:

1. Shear tests on single bars revealed shear strengths that exceeded manufacturer-provided values, attributed to a different test setup and the unmodified profile of the reinforcement bars. Initially, only force and machine displacement were recorded; future research will include strain measurements in the longitudinal direction, circumferential crack development and modifications to the test setup as per the ASTM D7617/D7617M standards [16].
2. Push-off tests of reinforcements crossing the joints in concrete specimens showed that the behavior after exceeding adhesion was ductile in comparison to joints without reinforcements, where the behavior was brittle.
    a. A drop from, on average, 87 kN to 16 kN was stated for the specimens with a smooth joint and one bar of CFRPs crossing the joint. A drop from 53 kN to 25 kN concerning one bar of GFRPs crossing the smooth joint existed compared to a drop from, on average 57 kN to 0 kN, for a specimen with a smooth joint and no reinforcement crossing the joint.
    b. Specimens with two reinforcements crossing the joint exhibited a higher ultimate shear force than those with a single reinforcement. The factor for the GFRP reinforcement crossing the joint was 1.38 on average. The CFRPs showed only a small factor of 1.01.
    c. CFRP-reinforced specimens demonstrated higher ultimate forces than GFRP-reinforced ones, as the ultimate tensile strength of the CFRP bars was 2.1 higher than those for the GFRP bars.
    d. Significant standard deviations highlighted the need for larger population size in future tests.

3. Further research should explore varying reinforcement diameters, the compressive strength of concrete, joint roughness and the reinforcement orientation (e.g., embedding at a 45° angle). A comprehensive examination of shear transfer mechanisms is essential for developing a robust assessment approach through additional modeling.

Author Contributions: Conceptualization, L.Z. and E.B.; methodology, L.Z. and E.B.; formal analysis, L.Z.; investigation, L.Z.; resources, S.M., D.S. and E.B.; writing—original draft preparation, L.Z.; writing—review and editing, L.Z., E.B. and D.S.; visualization, L.Z.; supervision, S.M. and E.B.; project administration, E.B. All authors have read and agreed to the published version of the manuscript.

Funding: This research was funded by the Federal Ministry of Education and Research of Germany via grant number 03LB3007C EDISON-rCF short fot he german project name "Verbundvorhaben: EDISON-rCF – Energieeffizientes werkstoffgerechtes Recycling von CFK durch einen innovativen Solvolyseprozess sowie die Entwicklung und Herstellung neuartiger quasiunidirektionaler Halbzeuge; Teilvorhaben: Bewehrungsentwicklung und Charakterisierung stabförmiger und flächiger Halbzeuge". Project period was from 2021 to 2023.

Data Availability Statement: The data presented in this study are available upon reasonable request from the corresponding author. The data are not publicly available due to this work being part of an ongoing study.

Acknowledgments: The authors would like to thank the staff of the OML for the support extended during the experimental works carried out at the laboratory.

Conflicts of Interest: The authors declare no conflict of interest.

References

1. Organisation Baunetzwissen Betonherstellung und Klimaschutz. Available online: https://www.baunetzwissen.de/beton/fachwissen/herstellung/betonherstellung-und-klimaschutz-7229519 (accessed on 21 February 2024).
2. Gouda, M.G.; Mohamed, H.M.; Manalo, A.C.; Benmokrane, B. Experimental investigation of concentrically and eccentrically loaded circular hollow concrete columns reinforced with GFRP bars and spirals. *Eng. Struct.* **2023**, *277*, 115442. [CrossRef]
3. Frenzel, M.; Baumgärtel, E.; Marx, S.; Curbach, M. The Cracking and Tensile-Load-Bearing Behaviour of Concrete Reinforced with Sanded Carbon Grids. *Buildings* **2023**, *13*, 2652. [CrossRef]
4. Frenzel, M.; Curbach, M. Shear strength of concrete interfaces with infra-lightweight and foam concrete. *Struct. Concr.* **2018**, *19*, 269–283. [CrossRef]
5. Wagner, J.; Curbach, M. Carbonstäbe im Bauwesen. *Beton-Und Stahlbetonbau* **2021**, *116*, 587–593. [CrossRef]
6. German Committee for Reinforced Concrete (DAfStb). *DAfStb-Richtlinie—Betonbauteile mit Nichtmetallischer Bewehrung: 2024-01*; Beuth Verlag GmbH: Berlin/Heidelberg, Germany, 2024.
7. Wingenfeld, D.R. Fügetechnische Konstruktionslösungen für Bauteile aus Ultrahochfestem Beton (UHPC). Ph.D. Dissertation, Technisches Universität München, Munich, Germany, 2013.
8. Lenz, P. Beton-Beton-Verbund. Ph.D. Dissertation, Technische Universität München, Munich, Germany, 2012. Available online: https://mediatum.ub.tum.de/604993?query=Potential+f%C3%BCr+Schubfugen+Lenz&show_id=1106588&srcnodeid=604993 (accessed on 7 March 2023).
9. Paulay, T.; Park, R.; Phillips, M.H. Horizontal Construction Joints in Cast-In-Place Reinforced Concrete. *Int. Concr. Abstr. Portal* **1974**, *42*, 599–616. Available online: https://www.concrete.org/publications/internationalconcreteabstractsportal/m/details/id/17303 (accessed on 7 March 2023).
10. Vintzēleou, E.N.; Tassios, T.P. Behavior of Dowels Under Cyclic Deformations. *ACI Struct. J.* **1987**, *84*, 18–30. [CrossRef]
11. Utescher, G.; Herrmann, H. *Versuche zur Ermittlung der Tragfähigkeit in Beton Eingespannter Rundstahldollen aus Nichtrostendem Austenitischem Stahl*; Deutscher Ausschuss für Stahlbeton: Berlin/Heidelberg, Germany, 1983; Volume 346, pp. 49–104.
12. Randl, N. Untersuchungen zur Kraftübertragung Zwischen Alt- und Neubeton bei Unterschiedlichen Fugenrauhigkeiten. Dissertation, Universität Innsbruck, Innsbruck, Austria, 2000. Available online: https://bibsearch.uibk.ac.at/AC02287162 (accessed on 27 June 2023).
13. Reinecke, R. Haftverbund und Rissverzahnung in Unbewehrten Betonschubfugen. Ph.D. Dissertation, Technische Universität München, Munich, Germany, 2002. Available online: https://mediatum.ub.tum.de/node?id=601067 (accessed on 23 June 2024).
14. Randl, N. Design recommendations for interface shear transfer in fib Model Code 2010. *Struct. Concr.* **2013**, *14*, 230–241. [CrossRef]
15. Gentry, R.; Bakis, C.; Harries, K.; Brown, J.; Prota, A.; Parretti, R. Development of ASTM test methods for FRP composite materials: Overview and transverse shear. In Proceedings of the 6th International Conference on FRP Composites in Civil Engineering, CICE 2012, Rome, Italy, 13–15 June 2012.
16. *ASTM D7617*; Test Method for Transverse Shear Strength of Fiber-reinforced Polymer Matrix Composite Bars. ASTM International: West Conshohocken, PA, USA, 2017.

17. Weber, A. Useful Shear Tests for FRP Dowels and Rebars. Available online: https://www.researchgate.net/publication/335136655_USEFUL_SHEAR_TESTS_FOR_FRP_DOWELS_AND_REBARS (accessed on 18 January 2024).
18. Bielak, J. On the role of dowel action in shear transfer of CFRP textile-reinforced concrete slabs. *Compos. Struct.* **2023**, *311*, 116812. [CrossRef]
19. Solidian. Technical Product Data Sheet—Solidian-REBAR-Dxx-CCE. 2023. Available online: https://solidian.com/wp-content/uploads/solidian-REBAR-Dxx-CCE-Technical-Product-Data-Sheets-v2303.pdf (accessed on 10 April 2024).
20. Schöck Bauteile. Technische Information Schöck Combar. 2019. Available online: https://www.schoeck.com/view/7725/Technische_Information_Schoeck_Combar___7725__.pdf/de (accessed on 21 May 2023).
21. *DIN EN 206*; Beton—Festlegung, Eigenschaften, Herstellung, Konformität. Beuth Verlag GmbH: Berlin/Heidelberg, Germany, 2021.
22. *DIN EN 1015-11:2020*; Prüfverfahren für Mörtel für Mauerwerk—Bestimmung der Biegezug- und Druckfestigkeit von Festmörtel. Beuth Verlag GmbH: Berlin/Heidelberg, Germany, 2020.
23. *DIN EN 206-1/DIN 1045-2*; Bestimmung von Korrelationen zwischen Würfel- und Prismendruckfestigkeit von Vergussmörteln für die Einordnung in Druckfestigkeitsklassen nach. Fraunhofer IRB Verlag: Stuttgart, Germany, 2009. Available online: https://www.irbnet.de/daten/rswb/08119015371.pdf (accessed on 23 April 2024).
24. *ISO 10406-1:2015*; Fibre-Reinforced Polymer (FRP) Reinforcement of Concrete—Test Methods—Part 1: FRP Bars and Grids. Beuth Verlag GmbH: Berlin/Heidelberg, Germany, 2015.
25. *DIN EN 12350-5:2019-09*; Prüfung von Frischbeton—Ausbreitmaß. Beuth Verlag GmbH: Berlin/Heidelberg, Germany, 2019.
26. *DIN EN 13036-1:2010-10*; Oberflächeneigenschaften von Straßen und Flugplätzen_-Prüfverfahren_-Teil_1: Messung der Makrotexturtiefe der Fahrbahnoberfläche mit Hilfe eines volumetrischen Verfahrens. Beuth Verlag GmbH: Berlin/Heidelberg, Germany, 2010.
27. *DIN EN 1766:2017-05*; Produkte und Systeme für den Schutz und die Instandsetzung von Betontragwerken_-Prüfverfahren_-Referenzbetone für Prüfungen. Beuth Verlag GmbH: Berlin/Heidelberg, Germany, 2017.
28. Bochmann, J. Carbonbeton unter Einaxialer Druckbeanspruchung. Ph.D. Dissertation, Technische Universität Dresden, Fakultät Bauingenieurwesen, Dresden, Germany, 2019.
29. Luo, Y.; Liao, P.; Pan, R.; Zou, J.; Zhou, X. Effect of bar diameter on bond performance of helically ribbed GFRP bar to UHPC. *J. Build. Eng.* **2024**, *91*, 109577. [CrossRef]
30. Alkatan, J. FRP Shear Transfer Reinforcement for Composite Concrete Construction. Master's Thesis, University of Windsor, Windsor, ON, Canada, 2016. Available online: https://scholar.uwindsor.ca/etd/5792 (accessed on 18 June 2023).

Disclaimer/Publisher's Note: The statements, opinions and data contained in all publications are solely those of the individual author(s) and contributor(s) and not of MDPI and/or the editor(s). MDPI and/or the editor(s) disclaim responsibility for any injury to people or property resulting from any ideas, methods, instructions or products referred to in the content.

Article

Effects of Different Fiber Dosages of PVA and Glass Fibers on the Interfacial Properties of Lightweight Concrete with Engineered Cementitious Composite

Haider M. Al-Baghdadi * and Mohammed M. Kadhum

Department of Civil Engineering, College of Engineering, University of Babylon, Babylon 51002, Iraq; eng.mohammed.mansour@uobabylon.edu.iq
* Correspondence: eng.haider.m@uobabylon.edu.iq

Abstract: The bond strength at the interface zone between two concrete sections plays a critical role in enhancing long-term durability, ensuring that both materials perform homogenously. Ensuring compatibility at the interfaces between repair and concrete materials is one of the most challenging aspects of constructing composite systems. Despite various studies, a comprehensive understanding of the engineered cementitious composite (ECC) bonding mechanism at the repair interface is still limited. The objective of this research is to identify the interfacial properties between lightweight concrete (LWC) and engineered cementitious composite (ECC) with varying fiber dosages of polyvinyl alcohol (PVA) and glass fibers under different surface roughness conditions. The study tested LWC-ECC specimens in direct shear using slant shear and bi-surface shear tests, recording the maximum shear stress at failure. Two grades of LWC—normal-strength lightweight concrete (NSLW) and high-strength lightweight concrete (HSLW)—were used as substrates, while the ECC overlays contained varying fiber dosages: 2% PVA, 1.5% PVA with 0.5% glass, 1.0% PVA with 1.0% glass, and 0.5% PVA with 1.5% glass. The surface conditions considered included grooved and as-cast substrates. The results indicated that the highest bond strength was achieved by specimens with 1.5% PVA and 0.5% glass fiber, with a maximum shear strength of 24.05 MPa for grooved HSLW substrates. Interface roughness had minimal impact on shear strength for NSLW substrates but significantly affected HSLW substrates, with bond strengths varying from 13.81 MPa to 24.05 MPa for grooved surfaces. This study demonstrates the critical role of fiber dosage and surface roughness in enhancing the bond performance of composite materials.

Keywords: engineered cementitious composite; polyvinyl-alcohol fiber; bi-surface shear test; roughness surfaces; slant shear test; substrate concrete

Citation: Al-Baghdadi, H.M.; Kadhum, M.M. Effects of Different Fiber Dosages of PVA and Glass Fibers on the Interfacial Properties of Lightweight Concrete with Engineered Cementitious Composite. *Buildings* **2024**, *14*, 2379. https://doi.org/10.3390/buildings14082379

Academic Editors: Bo Wang, Rui Guo, Muye Yang, Weidong He and Chuntao Zhang

Received: 24 May 2024
Revised: 22 July 2024
Accepted: 24 July 2024
Published: 1 August 2024

Copyright: © 2024 by the authors. Licensee MDPI, Basel, Switzerland. This article is an open access article distributed under the terms and conditions of the Creative Commons Attribution (CC BY) license (https://creativecommons.org/licenses/by/4.0/).

1. Introduction

The bond strength at the interface zone between two concrete sections plays a critical role in enhancing long-term durability and ensuring that both materials perform homogenously. This bond ensures the effective transfer of loads between the concrete layers in a composite section, enabling it to resist applied loads and stresses over an extended period [1]. The strength of the bonding area between two types of concrete is primarily influenced by four factors: the compressive strength of the weaker concrete, the stresses within the bonding area, the roughness of the bonding area, and the presence of shear reinforcement crossing the interface [2]. Additionally, bond strength is determined by cohesion and friction at the material boundary, which contributes to shear force and shear resistance [3].

Lightweight concrete (LWC) is known for its low density and high performance in terms of both strength and durability, particularly high-strength LWC. When dealing with composite structural members that combine lightweight and normal concrete, it is crucial to assess the material properties and behavior at the lightweight-to-normal concrete interface.

Differential shrinkage may not be a significant concern when LWC is used as a repair layer thanks to its self-curing properties, which reduce shrinkage [4].

Recent studies have highlighted the potential of using environmentally sustainable composites to enhance the properties of lightweight aggregate concrete (LAC). For instance, Suparp et al. [5] demonstrated that the axial load capacity of LAC can be significantly improved by using sustainable composites [5]. Similarly, Sirisonthi et al. [6] conducted a full-scale load test on precast post-tensioned continuous girders, providing insights into the load-deformation behavior of such systems under service and ultimate loading conditions [6]. The use of glass fiber reinforced polymer (GFRP) rods for the remediation of punching shear failure has also been investigated, showing promising results in enhancing the structural integrity of concrete slabs [7]. Moreover, Ullah et al. [8] explored the effect of partial replacement of e-waste as a fine aggregate on the compressive behavior of concrete specimens, highlighting the potential of alternative materials in improving concrete performance [8].

Ultra-high-performance fiber concrete is characterized by low porosity and excellent mechanical properties, including high tensile and compressive strength. Ultra-high-performance fiber concrete is ideal for rehabilitating and strengthening structures exposed to high mechanical and severe environmental loads, while normal structural concrete can be used for areas subjected to moderate exposure [9]. This combination enhances structural performance in terms of durability and life cycle costs [10]. Ensuring compatibility at the interfaces between repair and concrete materials is one of the most challenging aspects of constructing composite systems. This compatibility significantly impacts the overall structural behavior of the repaired structure [11]. Despite various studies, a comprehensive understanding of the bonding mechanism at the repair interface is still lacking [12]. For instance, studies on steel fiber-reinforced concrete have shown that steel fibers can improve interface bond strength and ductility in repaired structures, albeit slightly [13].

Engineered cementitious composite (ECC) are a new generation of high-performance fiber-reinforced cementitious materials with high flexibility and moderate fiber content. ECC materials using polyvinyl alcohol (PVA) fibers with a fiber volume fraction of no more than 2% have shown tensile strain capacities of between 3 and 5%. This high deformation capacity is due to the development of multiple microcracks rather than the continuous opening of a single crack [14]. The ECC's high fracture toughness and controlled crack width make it an ideal material for improving the serviceability and durability of infrastructure. ECC has been successfully applied in various structural elements, including embankments, bridge deck overlays, and connecting beams in high-rise buildings [15].

ECC's properties also make it suitable for repairing and modernizing concrete buildings. Potential applications include new structures requiring high energy absorption, impact resistance, and controlled crack width [16,17]. Studies have demonstrated that pseudo-ductile cement composite is an effective and durable material for concrete repair. The bond between pseudo-ductile cement composite and sufficiently cured concrete is strong, often leading to bond failure within the concrete substrate under shear or traction stresses [18].

Experimental studies on the bond strength between ECC layers and regular concrete substrates with varying surface textures have shown that ECC significantly enhances bond strength. This is particularly evident in compressive loading scenarios where ECC-coated concrete's bond strength is higher than that of substrate concrete [19]. ECC's ability to withstand tensile stresses makes it an excellent material for repairing structures prone to cracking due to low tensile strength [20].

Several researchers have explored the application of ECC in strengthening reinforced concrete structures, including beams, columns, and beam–column joints. The introduction of ECC into structural members results in the formation of multiple small cracks on tensile surfaces, leading to a ductile failure mode without reducing the load capacity of the member. These studies have shown that the bond strength between ECC and concrete, including interfacial tensile and shear strengths, is sufficient to transfer forces from the original reinforced concrete structure to the ECC layer [21]. The slant shear test (see Figure 1a), in particular, has proven to be an effective method for assessing bond strength, often yielding higher bond strength values compared to other tests [22]. To evaluate the bond performance between different concrete interfaces, various test methods exist, such as the pull-off test, splitting tensile test, direct shear test (see Figure 1b), and slant shear test [19,20].

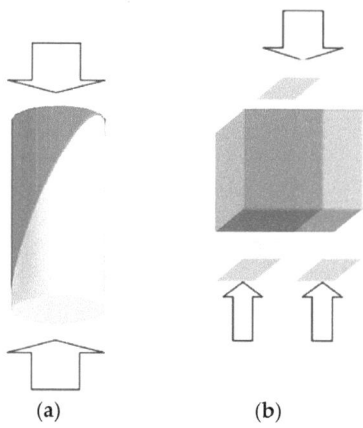

Figure 1. Specimen tests of bond strength: (**a**) slant shear and (**b**) direct shear (bi-surface).

2. Research Significance

This research addresses gaps in the use of polyvinyl alcohol (PVA) and glass fibers in engineered cementitious composite (ECC) for lightweight concrete (LWC) are addressed, offering significant contributions to both academic research and civil engineering practices. It shows how fiber dosages affect ECC's mechanical properties, provides new guidelines for optimizing composite interfaces, and introduces innovative, sustainable construction practices. The aim of this research is to identify the interfacial properties between LWC and ECC with different fiber dosages of PVA and glass fibers, along with varying degrees of roughness. In order to evaluate ECC effectiveness and suitability as repair materials for retrofitting reinforced concrete structures, LWC-ECC specimens were tested in direct shear using the slant shear test and bi-surface shear test, and the maximum shear stress at failure was recorded. Also, two grades of LWC (normal-strength lightweight concrete (NSLW) and high-strength lightweight concrete (HSLW)) were used as substrates, while for the overlay, ECC mixes contained various types and dosages of fibers (2% PVA, 1.5% PVA with 0.5% glass, 1.0% PVA with 1.0% glass, and 0.5% PVA with 1.5% glass fibers) were used. Grooved and as-cast substrate surface conditions were considered. Furthermore, this research investigates the relationship between surface roughness and bond strength for the LWC-ECC specimens with different types and dosages of fibers, along with two types of interface treatments—namely, as-cast and grooved surfaces.

3. Experimental Work

3.1. Apparatus and Measurement Techniques

The slant shear test was conducted using cylindrical molds with dimensions of 75 mm in diameter and 150 mm in height. The molds were cut at an inclination angle of 30 degrees to create the slant interface, as shown in Figure 2. This apparatus allowed compressive loads to be applied to the inclined surface, simulating real-life shear and compressive stresses at the bond interface. The elliptical area of the slant surface was calculated for stress determination. For the bi-surface shear test, cubic molds measuring 150 mm × 150 mm × 150 mm were used. The molds were designed to contain two-thirds of the LWC substrate and one-third of the ECC overlay material, as shown in Figure 2. Three thick steel plates with dimensions of 100 mm × 33 mm × 25 mm were used to facilitate direct shearing between the different concrete composites. The setup ensured symmetrical load application to simulate realistic shear stress conditions. Two surface preparation methods were employed to achieve the desired surface roughness: as-cast and grooved surfaces. Grooves inclined at approximately 45 degrees were created using special cutting discs designed for hard stone. This preparation ensured good interlock between the concrete surface and the ECC layer. The surfaces were cleaned with water and compressed air to remove any dust and debris and then left to dry before the ECC overlay application. A 2000 kN capacity digital compression testing machine was utilized to apply loads at a constant rate of 2 kN/s. This machine was employed for both the slant shear test and the bi-surface shear test to evaluate the bond strength between the LWC substrates and the ECC overlays. The machine's setup ensured precise load application and measurement throughout the testing process.

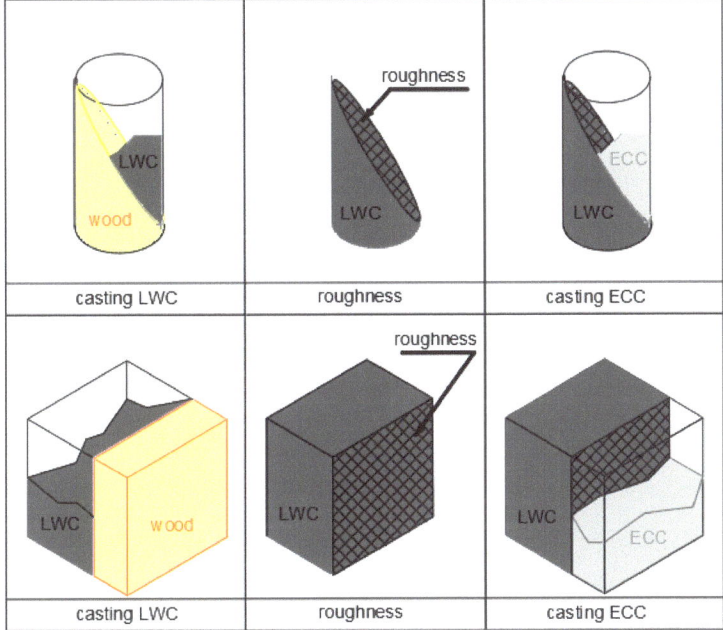

Figure 2. Preparation of composite specimens (slant shear and the bi-surface shear).

3.2. Materials

In all mixes (LWC and ECC), ordinary Portland cement was used in this study. The cement was tested to ensure conformity with No. 5/1984 standard [23]. The physical and chemical properties of the fine aggregate are listed in Table 1 in accordance with No. 45/1984 standard [24] and ASTM C128 standard [25]. Uniformly sized lightweight aggregates (0.475–1 cm) were used in this study. These aggregates are made from porous ceramic materials with uniformly small closed-cell pores and firmly sintered, durable exterior surfaces. The particle volume increases significantly during production due to swelling when the lightweight aggregates are produced from clay mineral raw materials, as shown in Figure 3. This process involves burning the materials in rotary kilns at temperatures between 1100 and 1200 °C. The sieve analysis of the lightweight aggregate was conducted to ensure it met the limits specified in ASTM C330-17a, 2017 [26], as shown in Table 2. Table 3 presents the physical and chemical properties of the lightweight aggregate tested by a manufacturer. Due to its high-water absorption capability, the lightweight aggregate was soaked in water for hours to prevent it from absorbing water during mixing. After soaking, the aggregate was spread out in laboratory air until the surface dried, resulting in a saturated surface dry state, as recommended by ACI 211.2-98 [27].

Table 1. Sand physical and chemical properties.

Size of Sieve (mm)	Cumulative Passing %	Limits of IQS No. 45/1984 as in (Zone 2)	Physical Properties	Test Results
10	100	100	Specific gravity	2.65
4.75	92	90–100	Absorption	0.92%
2.36	81	75–100	Fine material passing from the sieve (75 μm)	2.50%
1.18	73	55–90	Fineness modulus	2.68
0.6	55	35–59		
0.3	24	8–30	Chemical properties	
0.15	7	0–10	Sulfate content	0.35%

Figure 3. Lightweight aggregate.

Table 2. Grading of used lightweight aggregate.

Size of Sieve (mm)	Cumulative Passing (%)	Limits of ASTM C330-17a, 2017
12.5	100	100
10	99	80–100
8	66	-
6	44	-
4.75	7	5–40
2.36	2	0–20
1.18	0	0–10

Table 3. Physical and chemical properties of lightweight aggregate.

Chemical Properties	
Chemical Composition	Percentage by Weight%
CaO	3.78
SiO_2	61.58
Al_2O_3	16.99
Fe_2O_3	7.62
MgO	2.56
SO_3	0.19
Na_2O	1.03
Loss on Ignition	0.2
Physical Properties	
Properties	Test Results
Specific Gravity	1.2
Absorption	12%
Bulk density Kg/m^3	700

Silica fume consists of extremely tiny spherical particles with diameters ranging from 0.1 to 0.2 μm. In this research, silica fume was used as a partial replacement (15% by weight) for cement to enhance the microstructure of the cement paste, increasing its resistance to external influences. The technical data for the silica fume includes Form—Powder, Color—Grey, Density—0.55 to 0.7 kg/L, and Chloride content—<0.1% [28]. A high-performance concrete superplasticizer based on modified polycarboxylic ether, MasterGlenium 54, was also used. A proportion of 3.5% was employed to achieve acceptable flow properties for ECC. Table 4 provides the technical description of MasterGlenium 54 [29].

Table 4. Technical description of MasterGlenium 54 [29].

Chemical Basis	Aqueous Solution of Modified Polycarboxylic Ether
Color	Whitish to straw
Specific gravity	1.07
PH	5–7
Chloride content	None
Toxicity	Danger hazardous material.
Fire	Not fire-propagating

The PVA fibers used in this study are 12 mm in length and 39 μm in diameter. The fiber's nominal tensile strength, stiffness, and density are 1600 MPa, 40 GPa, and 1300 kg/m^3, respectively. Figure 4 displays the PVA fibers used in this research. Table 5 provides the technical properties of the PVA fibers according to a manufacturer. Another type of fiber used in this study is glass fiber (GF). The properties of the glass fibers are shown in Table 5. Figure 5 displays the glass fibers used in this research.

Table 5. Properties of PVA and glass fibers *.

Fiber	Fiber Length (mm)	Diameter μm	Tensile Strength (MPa)	Young's Modulus (GPa)	Fiber Elongation (%)	Density (kg/m^3)
PVA	12	39	1600	40	10-Apr	1.3
GF	12	-	2200	80	0–4	2.78

* Manufacturer Properties.

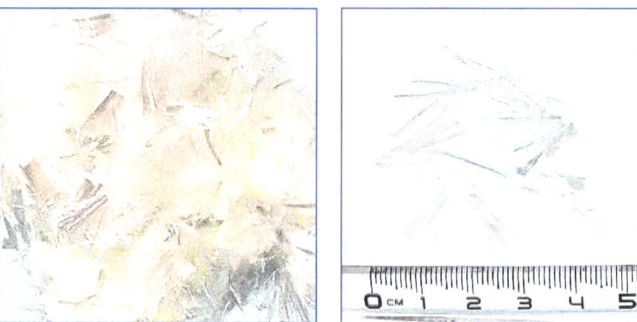

Figure 4. PVA fiber geometry.

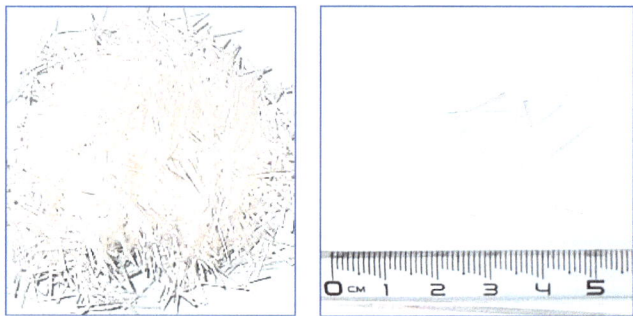

Figure 5. Glass fiber geometry.

Moreover, the fly ash used in this study complies with BS 3892-1 [30] and BS EN 450-1 [31]. The specific gravity of the fly ash is 2.70 g/cm^3, and its Blaine-specific surface area is 2970 cm^2/g. The chemical properties of the fly ash are presented in Table 6.

Table 6. Chemical analysis of fly ash (Type F).

Chemical Composition	% by Weight
Al_2O_3	24.62
SiO_2	48.53
CaO	9.49
Fe_2O_3	7.59
K2O	2.51
MgO	2.28
Na_2O	1.18
SO_3	2.48
Loss on ignition	1.69
28 d activity index	90%

3.3. Detailed Composition and Mixing Procedures

For substrate concrete mixes, the mix proportions for the NSLW and HSLW were designed to achieve a compressive strengths of 30 MPa and 50 MPa at 28 days, following ACI Committee 211.2-98 [32] and ACI 211.4R-08 [33] guidelines, respectively. To attain the necessary strength, a variety of trial mixes were tested; the specific mix ratios for NSLW and HSLW are listed in Table 7.

Table 7. LWC design mixes.

ID	Cement Kg/m^3	Sand Kg/m^3	Lightweight Aggregate Kg/m^3	Silica Fume Kg/m^3	Water Kg/m^3	SP (%) *	W/C
NSLW	478	667	440	0	153	1.1	0.32
HSLW	550	678	400	81	160	1.7	0.25

* SP by wt. of cm.

For overlay concrete mixes, ECC has been optimized using micromechanics to achieve high tensile ductility and tight micro-crack width while keeping the fiber content low (2% by volume) [34]. For ECC, the maximum grain size of 250 μm with an average size of 110 μm of fine aggregates was used to ensure a homogeneous mix. Many trial mixes were prepared to achieve the optimal properties for ECC mortars, with a mini-slump flow range of 240–260 mm, ensuring easy and complete penetration of the mortars through the concrete mold specified for repair.

Table 8 presents the typical mixture design of ECC adopted in this study, which exhibits self-consolidating casting properties. All proportions are given for materials in the dry state. Four different ECC mixes with various types and percentages of fibers (PVA and glass) were used as overlay materials. To account for material heterogeneity, a maximum fiber content of 2% by volume, which is greater than the calculated critical fiber content needed to achieve strain-hardening, is typically used in the mix design. Different percentages of glass fiber volume fractions were added to the ECC mix and PVA fibers, creating hybrid fiber composites. Therefore, the ECC-P2.0-G0.0, ECC-P1.5-G0.5, ECC-P1.0-G1.0, and ECC-P0.5-G1.5 mixes included 2% PVA fiber, 1.5% PVA + 0.5% glass fiber, 1.0% PVA + 1.0% glass fiber, and 0.5% PVA + 1.5% glass fiber, respectively.

Table 8. ECC Mix ID.

Mix ID	C/C	F/C	S/C	W/C	HRWR/C	PVA Fiber (%)	Glass Fiber (%)	Total (%)
ECC-P2.0-G0.0	1	1.2	0.8	0.56	0.012	2.0	0.0	2
ECC-P1.5-G0.5	1	1.2	0.8	0.56	0.012	1.5	0.5	2
ECC-P1.0-G1.0	1	1.2	0.8	0.56	0.012	1.0	10.	2
ECC-P0.5-G1.5	1	1.2	0.8	0.56	0.012	0.5	1.5	2

C: cement, F: fly ash, S: sand, W: water.

3.4. Specimen Preparation and Curing Regimes

Composite specimens consisting of NSLW and HSLW substrates with various ECC overlay materials were prepared and tested. Slant shear and bi-surface shear tests were used to evaluate the interface bond strength at 28 days. For the slant shear test, the specimens were 75 × 150 mm cylinders with an inclination angle (α) of 30°, measured relative to the vertical axis [22]. For the bi-surface shear test, the specimens were 150 mm cubes, with two-thirds of the cube consisting of the NSLW and HSLW substrates and one-third consisting of the ECC overlay material [35].

To prepare the LWC substrate concrete specimens, wooden cylinders were cut at the specified slanted dimension, and wooden cubes were cut to a height of 150 mm with a base size of 50 × 150 mm, as shown in Figure 6. First, the cylindrical molds were filled halfway, and the cubic molds were filled two-thirds of the way with the cast LWC substrate. The fresh LWC mixtures were left in the molds for 24 h. Afterward, the samples were de-molded and cured in water for 28 days.

(a)   (b)

Figure 6. Wood slices for (**a**) slant shear and (**b**) bi-surface shear test.

To obtain quantitative surface roughness parameters for the LWC surfaces of the specimens, both grooved and as-cast surface methods were used to create a rough surface that ensures a strong bond between the concrete substrate and the repair material. This is crucial to ensure that the ECC repair material contributes to bearing part of the stresses. Special discs for cutting hard stones were used to make grooves inclined in opposite directions at an angle of approximately 45 degrees, ensuring good interlock between the concrete surface and the ECC layer. After the roughening process was completed, a final cleaning process was performed using water extrusion followed by compressed air to ensure no dust remained. The specimens were then left to dry before the ECC repair material was applied, as shown in Figure 7. NSLW and HSLW were used as reference interfaces, as listed in Table 9.

(a)   (b)

Figure 7. Interface roughness (**a**) half cylinder grooved and as cast surface and (**b**) two-thirds of the cube grooved and as cast surface.

Beushausen et al. [36] emphasized the necessity of reaching saturated surface dry conditions at the LWC-substrate interface before overlay application by soaking the interface. The purpose of this is to allow the mixed water from the fresh layer to pass into the dry LWC substrate when a new layer is applied. The LWC substrate overlay surfaces were moistened and subsequently dried using a damp towel. The saturated surface dry slanted cutting specimens were positioned in cylinder molds with the bevel side facing upward, and the overlay materials were poured on top of the LWC substrate concrete to complete these cylinders. Similarly, the saturated surface dry bi-surface cutting specimens were placed in cube molds and covered with overlay material on one-third of the cubes. A total of 16 groups of specimens are listed in Table 9, and in each group three specimens were prepared.

Table 9. Specimen identification.

Specimen ID	Substrate Concrete Mix	Overlay Concrete Mix	Interface Roughness
N-E-P2.0-G0.0-G	NSLW	ECC-P2.0-G0.0	Grooved surface
N-E-P1.5-G0.5-G	NSLW	ECC-P1.5-G0.5	Grooved surface
N-E-P1.0-G1.0-G	NSLW	ECC-P1.0-G1.0	Grooved surface
N-E-P0.5-G1.5-G	NSLW	ECC-P0.5-G1.5	Grooved surface
H-E-P2.0-G0.0-G	HSLW	ECC-P2.0-G0.0	Grooved surface
H-E-P1.5-G0.5-G	HSLW	ECC-P1.5-G0.5	Grooved surface
H-E-P1.0-G1.0-G	HSLW	ECC-P1.0-G1.0	Grooved surface
H-E-P0.5-G1.5-G	HSLW	ECC-P0.5-G1.5	Grooved surface
N-E-P2.0-G0.0-A	NSLW	ECC-P2.0-G0.0	As cast surface
N-E-P1.5-G0.5-A	NSLW	ECC-P1.5-G0.5	As cast surface
N-E-P1.0-G1.0-A	NSLW	ECC-P1.0-G1.0	As cast surface
N-E-P0.5-G1.5-A	NSLW	ECC-P0.5-G1.5	As cast surface
H-E-P2.0-G0.0-A	HSLW	ECC-P2.0-G0.0	As cast surface
H-E-P1.5-G0.5-A	HSLW	ECC-P1.5-G0.5	As cast surface
H-E-P1.0-G1.0-A	HSLW	ECC-P1.0-G1.0	As cast surface
H-E-P0.5-G1.5-A	HSLW	ECC-P0.5-G1.5	As cast surface

3.5. Load Application and Monitoring

Slant shear tests were performed according to ASTM C882 [22]. Specimens were tested using a standard compression apparatus, as shown in Figure 8a. The loading was applied as recommended in ASTM C39 [37]. The maximum load values were recorded, and the applied stress (σo) required to produce bond rupture was determined using Equation (1):

$$\sigma o = P/Ae. \qquad (1)$$

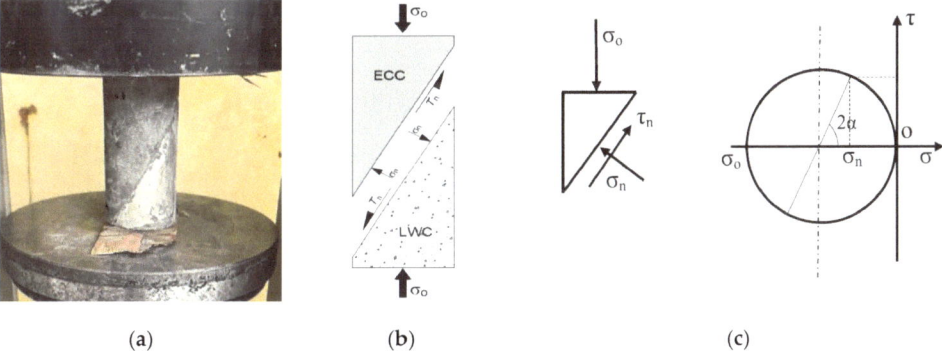

Figure 8. Slant shear test (a) and compression test (b) stresses developed at the interface, and (c) Mohr–Coulomb circle.

In Equation (1), P is the maximum applied load, and Ae is the elliptical area of the slant surface [38]. The maximum applied stress is a combined measure of the shear and compressive strength of the bond (Figure 8b), which are represented by Equation (2) and Equation (3), respectively.

$$\tau n = \sigma o \cos\alpha \qquad (2)$$

Equation (2) calculates the normal shear stress acting on the interface. This shear stress component is derived from the overall stress and the angle of inclination, factoring in the geometric orientation of the bond interface.

$$\sigma n = \sigma o \sin 2\alpha \qquad (3)$$

Equation (3) determines the normal compressive stress at the interface. This equation emphasizes the influence of the inclination angle on the compressive stress experienced by the interface, which is critical for evaluating the capacity of the interface to resist compression. In Equations (2) and (3), $\alpha = 30°$, while τn and σn are the shear and compressive stresses acting on the bond plane, respectively. τn and σn are related in form, as in Equation (4) [39].

$$\tau n = c + \mu\, \sigma n = c + \tan\phi \cdot \sigma n, \tag{4}$$

where c, μ, and ϕ are cohesion, coefficient of friction, and internal friction angle of the bond, respectively. Equation (4), known as the Mohr–Coulomb failure criterion, links the shear stress with the compressive stress. The parameters c (cohesion) and μ (coefficient of friction) are material properties that describe the adhesive and frictional contributions to the interface strength. This equation integrates the effects of both shear and compressive stresses to provide a comprehensive view of the failure mechanics at the interface.

To determine the average shear bond strength between two types of concrete, Momayeza et al. [21] proposed the bi-surface shear test method used in this research. This test is straightforward to perform and provides consistent results. The samples comprised two-thirds LWC and one-third repair materials (ECC), as shown in Figure 9. Three thick steel plates with dimensions of $100 \times 33 \times 25$ mm were used to facilitate direct shearing between the interfaces of the different concrete composites. A 2000 kN digital testing machine was used to apply a constant loading rate of 2 kN/s to all bi-surface specimens. The bi-surface shear strength values were determined using Equation (5).

$$\tau v = Pv/(2 \times Av), \tag{5}$$

where τv = bi-surface shear bond strength (MPa), Pv = ultimate load indicated by the testing machine (N), Av = area of interface in shear (mm^2). The average of two shear force values was used to evaluate the direct shear strength. Figure 9 shows the testing machine and test setup.

Figure 9. Bi-surface shear test and test setup.

4. Results and Discussion

4.1. Concrete Machinal Properties of LWC and ECC

For evaluating the material properties of NSLW and HSLW at 28 days, nine samples were prepared for each type of concrete. Three samples were used to determine the compressive strength, three for flexural strength, and three for splitting tensile strength. The mechanical properties of NSLW and HSLW are summarized in Table 10. For ECC materials with different types and dosages of fibers, the average compressive strength at 28 days was obtained from the average results of three cube samples (70 mm $\times$ 70 mm $\times$ 70 mm) tested for compression. Additionally, four prismatic samples of each type of ECC were tested to determine the flexural strength at 28 days. Micromechanics is a branch of mechanics applied at the material constituent level, capturing the mechanical interactions between the fiber, mortar matrix, and fiber/matrix interface. The matrix heterogeneities in ECC, including defects, sand particles, cement grains, and mineral admixture particles, range in size from the nanoscale to the millimeter scale. The ingredients and mix proportions have been optimized to satisfy multiple cracking criteria. Specifically, the type, size, amount of fiber, matrix ingredients, and interface characteristics are tailored for multiple cracking and controlled crack width. ECC incorporates fine silica sand with a sand-to-binder ratio (S/B) of 0.36 to maintain adequate stiffness and volume stability. The ECC mix has a water-to-binder (W/B) ratio of 0.26 to balance fresh and hardened properties. The binder system is defined as the total amount of cementitious material (i.e., cement and Type F fly ash) in ECC. The particle size of all matrix components is properly graded to achieve self-consolidating fresh properties [40]. The mechanical properties of the ECC materials with various types and dosages of fibers are summarized in Table 10.

Table 10. Mechanical properties of substrate and overlay concretes.

Mix ID	Compressive Strength (MPa)	Flexural Strength (MPa)	Splitting Tensile Strength (MPa)
NSLW	29.3	3.8	3.6
HSLW	56.2	6.8	4.5
ECC-P2.0-G0.0	78.7	21.5	14.1
ECC-P1.5-G0.5	81.3	23.3	15.7
ECC-P1.0-G1.0	76.0	20.7	15.0
ECC-P0.5-G1.5	59.5	19.8	13.6

4.2. Bond Strength—Slant Shear Test and Bi-Surface Test

Previously, various test methods have been reported to estimate bond strength [41–44]. These tests can be classified into the following categories: (1) direct tension tests, (2) direct shear tests (bi-surface tests), (3) indirect tensile (split tensile tests), (4) shear and compression tests (inclined shear tests), (5) withdrawal tests, and (6) three-point bending tests. Among these, the direct shear and oblique cut tests are widely used due to their ease of preparation, cost-effectiveness, and consistent results [42,43]. Many standards and codes also approve slant shear and Bi-surface tests to define bond strength [38,45]. Therefore, in this study, these methods were employed to determine the bond strength between different concrete substrates and overlays. The bond strength values of the composite LWC substrate-ECC overlay specimens were determined using slant shear and bi-surface shear tests. These values represent the average bond strengths, as listed in Tables 11–14.

Table 11. Slant shear bond strength and failure modes for NSLW concrete substrate.

Specimen ID	σ_o (MPa)	Ave. σ_o (MPa)	COV (%)	τ_n (MPa)	σ_n (MPa)	Failure Mode
N-E-P2.0-G0.0-G	10.8					C
N-E-P2.0-G0.0-G	9.5	10.58	7.6	9.16	1.7	C
N-E-P2.0-G0.0-G	11.44					C
N-E-P1.5-G0.5-G	11.95					B
N-E-P1.5-G0.5-G	12.83	12.75	4.9	11.04	1.96	B
N-E-P1.5-G0.5-G	13.47					D
N-E-P1.0-G1.0-G	10.12					C
N-E-P1.0-G1.0-G	8.89	9.5	5.3	8.22	1.48	C
N-E-P1.0-G1.0-G	9.49					C
N-E-P0.5-G1.5-G	5.81					C
N-E-P0.5-G1.5-G	6.45	6.25	5.0	5.41	1.06	C
N-E-P0.5-G1.5-G	6.49					C
N-E-P2.0-G0.0-A	9.12					C
N-E-P2.0-G0.0-A	9.45	9.05	4.0	7.83	1.48	C
N-E-P2.0-G0.0-A	8.58					C

Table 11. *Cont.*

Specimen ID	σo (MPa)	Ave. σo(MPa)	COV (%)	τn (MPa)	σn (MPa)	Failure Mode
N-E-P1.5-G0.5-A	11.02					B
N-E-P1.5-G0.5-A	10.25	10.81	3.7	9.36	1.72	C
N-E-P1.5-G0.5-A	11.16					B
N-E-P1.0-G1.0-A	6.85					C
N-E-P1.0-G1.0-A	7	7.23	6.0	6.26	1.21	C
N-E-P1.0-G1.0-A	7.84					C
N-E-P0.5-G1.5-A	4.89					C
N-E-P0.5-G1.5-A	5.35	5.4	8.1	4.67	0.93	C
N-E-P0.5-G1.5-A	5.96					C

Table 12. Slant shear bond strength and failure modes for HSLW concrete substrate.

Specimen ID	σo (MPa)	Ave. σo (MPa)	COV (%)	τn (MPa)	σn (MPa)	Failure Mode
H-E-P2.0-G0.0-G	18.12					B
H-E-P2.0-G0.0-G	17.5	17.58	2.3	15.22	2.49	C
H-E-P2.0-G0.0-G	17.12					C
H-E-P1.5-G0.5-G	23.75					B
H-E-P1.5-G0.5-G	24.15	24.05	0.9	20.82	3.26	B
H-E-P1.5-G0.5-G	24.25					B
H-E-P1.0-G1.0-G	14.81					B
H-E-P1.0-G1.0-G	15.2	15.3	2.9	13.24	2.57	B
H-E-P1.0-G1.0-G	15.89					C
H-E-P0.5-G1.5-G	11					B
H-E-P0.5-G1.5-G	10.85	11.25	4.1	9.74	1.81	B
H-E-P0.5-G1.5-G	11.9					B
H-E-P2.0-G0.0-A	9.63					B
H-E-P2.0-G0.0-A	8.58	9.39	6.3	8.13	1.55	B
H-E-P2.0-G0.0-A	9.96					B
H-E-P1.5-G0.5-A	12.95					B
H-E-P1.5-G0.5-A	14.12	13.81	4.5	11.95	2.27	C
H-E-P1.5-G0.5-A	14.36					C
H-E-P1.0-G1.0-A	7.85					B
H-E-P1.0-G1.0-A	8.55	8.23	3.5	7.13	1.56	B
H-E-P1.0-G1.0-A	8.29					B
H-E-P0.5-G1.5-A	4.75					B
H-E-P0.5-G1.5-A	5	4.86	2.1	4.21	0.96	B
H-E-P0.5-G1.5-A	4.83					B

Table 13. Bi-surface bond strength and failure modes for NSLW concrete substrate.

Specimen ID	σo (MPa)	Ave, σo	COV (%)	Failure Mode
N-E-P2.0-G0.0-G	7.1			B
N-E-P2.0-G0.0-G	5.8	6.8	10.4	C
N-E-P2.0-G0.0-G	7.5			B
N-E-P1.5-G0.5-G	7.6			B
N-E-P1.5-G0.5-G	8.1	7.9	2.4	D
N-E-P1.5-G0.5-G	7.9			D
N-E-P1.0-G1.0-G	6.3			C
N-E-P1.0-G1.0-G	5.8	5.9	4.4	B
N-E-P1.0-G1.0-G	5.7			B

Table 13. Cont.

Specimen ID	σo (MPa)	Ave, σo	COV (%)	Failure Mode
N-E-P0.5-G1.5-G	4.0			C
N-E-P0.5-G1.5-G	5.6	4.3	23.9	C
N-E-P0.5-G1.5-G	3.2			C
N-E-P2.0-G0.0-A	5.5			C
N-E-P2.0-G0.0-A	5.6	6.0	9.5	C
N-E-P2.0-G0.0-A	6.8			B
N-E-P1.5-G0.5-A	7.0			B
N-E-P1.5-G0.5-A	6.8	6.9	1.7	C
N-E-P1.5-G0.5-A	6.9			B
N-E-P1.0-G1.0-A	5.1			C
N-E-P1.0-G1.0-A	4.5	4.9	5.9	C
N-E-P1.0-G1.0-A	5.0			C
N-E-P0.5-G1.5-A	3.7			C
N-E-P0.5-G1.5-A	4.5	3.7	15.8	C
N-E-P0.5-G1.5-A	3.0			C

Table 14. Bi-surface bond strength and failure modes for HSLW concrete substrate.

Specimen ID	σo (MPa)	Ave σo	COV (%)	Failure Mode
H-E-P2.0-G0.0-G	9.8			B
H-E-P2.0-G0.0-G	9.1	10.0	8.0	B
H-E-P2.0-G0.0-G	11.0			B
H-E-P1.5-G0.5-G	14.0			B
H-E-P1.5-G0.5-G	13.2	13.1	6.5	B
H-E-P1.5-G0.5-G	12.0			B
H-E-P1.0-G1.0-G	9.8			B
H-E-P1.0-G1.0-G	10.5	10.3	3.5	B
H-E-P1.0-G1.0-G	10.6			B
H-E-P0.5-G1.5-G	7.4			C
H-E-P0.5-G1.5-G	8.1	7.3	10.2	B
H-E-P0.5-G1.5-G	6.3			C
H-E-P2.0-G0.0-A	5.8			C
H-E-P2.0-G0.0-A	6.7	6.2	6.0	B
H-E-P2.0-G0.0-A	6.1			C
H-E-P1.5-G0.5-A	9.4			B
H-E-P1.5-G0.5-A	9.0	9.1	2.4	B
H-E-P1.5-G0.5-A	8.9			B
H-E-P1.0-G1.0-A	6.1			B
H-E-P1.0-G1.0-A	5.4	6.2	11.8	B
H-E-P1.0-G1.0-A	7.2			B
H-E-P0.5-G1.5-A	3.4			C
H-E-P0.5-G1.5-A	4.7	3.9	15.8	B
H-E-P0.5-G1.5-A	3.5			C

From the slant shear test results, the applied stress (σo) on the interfacial bond was calculated by dividing the maximum force at bond failure obtained from compression loading by the elliptical area (Ae) using Equation (1). For the bi-surface shear test, the bonding shear strength was calculated by dividing the maximum applied force by the bonded surface area using Equation (5). Figure 10 presents the average slant shear strengths, and Figure 11 presents bi-surface shear strengths for grooved and as-cast surfaces for NSLW and HSLW at 28 days of age.

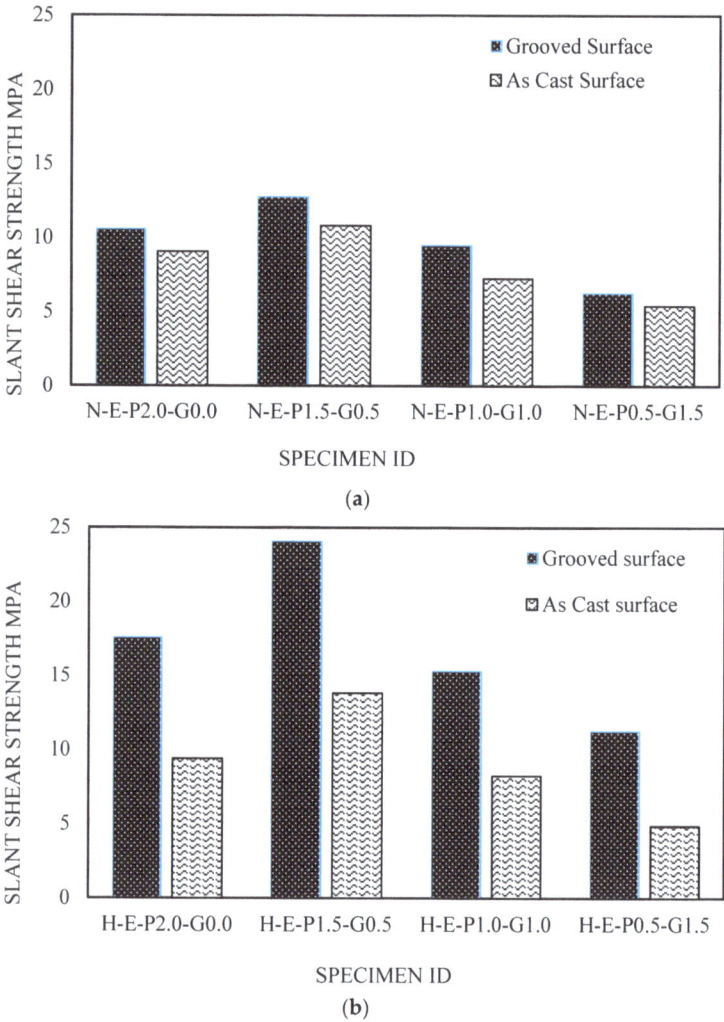

Figure 10. Slant shear bond strengths with grooved and as-cast surfaces for (**a**) NSLW substrate and (**b**) HSLW substrate.

The figures indicate that the bond strengths measured by the slant shear method were higher than those measured by the bi-surface shear method for all ECC repair materials and surface preparations for both NSLW and HSLW substrates. This is attributed to the higher friction forces and interlock resulting from the high compressive stresses in the slant shear test, which increases the shear failure load [21].

Among all the NSLW and HSLW substrate-ECC specimens, the N-E-P1.5-G0.5 and H- H-E-P1.5-G0.5 specimens exhibited the highest bond strengths at 28 days. The bond quality decreased in the following order: N-E-P2.0-G0.0, N-E-P1.0-G1.0, N-E-P0.5-G1.5, H-E-P2.0-G0.0, H-E-P1.0-G1.0, and H-E-P0.5-G1.5. The decrease in bond strength with higher glass fiber content can be primarily attributed to the formation of voids within the ECC matrix. Increasing the volume of glass fibers tends to disrupt the uniformity of the cementitious matrix due to their rigid and non-pliable nature compared to more flexible fibers like polyvinyl alcohol (PVA). Glass fibers, when added in larger quantities, can create localized areas of weak matrix density due to poor particle packing and increased air entrapment during mixing. The formation of these voids is critical as they act as stress concentrators within the composite material, significantly weakening the interfacial transition zone between the ECC and the LWC substrate. Voids reduce the effective contact area through which load and stresses can be transferred between the ECC overlay and the concrete substrate. As a result, the mechanical interlocking necessary for optimal bond strength is compromised, leading to reduced overall adhesion.

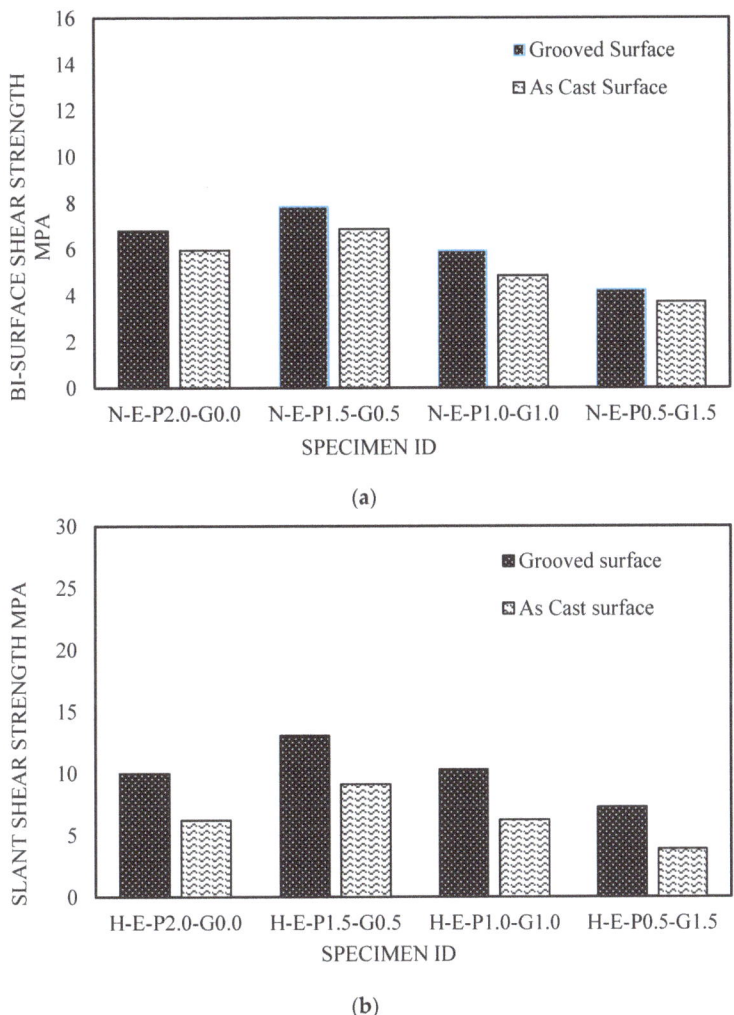

Figure 11. Bi-surface shear bond strengths with grooved and as-cast surfaces for (a) NSLW substrate (b) HSLW substrate.

The increase in PVA fibers leads to a stronger bond with the cement matrix due to the fibers' interaction with the $Ca(OH)_2$ compound. This interaction forms complex clusters with metal hydroxide, which can create a consistency that reduces the ability to fill the pores on the substrate surface [46]. Conversely, an increase in glass fibers results in the formation of voids in the cement matrix, reducing adhesion and bond strength between the overlay and the original concrete.

Figures 10a and 11a show that the quality and roughness of the adhesion surface of the ECC repair material minimally affect NSLW. The results were similar for grooved and as-cast surfaces in both slant shear and bi-surface shear tests, respectively. The adhesion strength of ECC was such that failure occurred in the NSLW (substrate layer). Several studies have concluded that the bond strength between two materials is highly influenced by the degree of roughness of the substrate surfaces [47–49]. In this study, to obtain quantitative surface roughness parameters for the LWC surfaces of the specimens, both grooved and as-cast surface methods were used to create a rough surface that ensures a strong bond between the concrete substrate and the repair material. This is crucial to ensure that the ECC repair material contributes to bearing part of the stresses. The effect of surface roughness on bond strength is significant for NSLW in both tests. Figures 10 and 11 show that grooved specimens have a markedly higher bond strength than as-cast surface specimens. This is due to the better interlock provided by the grooved surfaces between the ECC and the LWC substrate, especially in HSLW.

Based on the investigation of PVA and glass fibers on the interfacial properties of lightweight concrete with engineered cementitious composite, a comparison of the test results from the present study with past research is warranted. Reference [50] by Chuanqing Fu et al. focused on corrosion's impact on bond strength and cracking in cement mortar and PVA-ECC, highlighting the complex interactions between material degradation and mechanical properties under corrosive conditions. This complements this study on the mechanical impact of fiber content in non-corrosive environments. Hui et al. [51] examined the dynamic compressive strength of steel fiber-reinforced concrete under cyclic conditions, underscoring the importance of fiber content, similar to these findings on how fiber variations influence ECC's bond strength and integrity. Hui et al. [52] explored the mechanical properties of polypropylene fiber cement mortar under different loading speeds, relevant to our interest in fiber dosage effects under static conditions, suggesting a further investigation into dynamic behaviors. These studies collectively enrich the understanding of fiber-reinforced composites' performance across various conditions, aligning with and expanding upon our research on fiber dosage's structural impacts.

4.3. Failure Mode—Slant Shear Test and Bi-Surface Test

The bond quality between the lightweight concrete (LWC) substrate and the engineered cementitious composite (ECC) overlay can be effectively evaluated by analyzing the locations and types of failure. These failures, documented through visual inspections, are categorized into four distinct modes. Type A: Interfacial Bond Failure—This type of failure, where the bond interface itself fails without material adherence from either the substrate or overlay, was not observed in our experiments. Type B: Interface Failure with Thin Layer Detachment—This mode is characterized by the detachment of thin layers of the LWC substrate that remain adhered to the ECC, indicative of moderate bond strength. Representative images are displayed in Figures 12 and 13. Type C: Interfacial Failure with Thick Layer Detachment—Here, a substantial layer of the LWC substrate remains attached to the ECC, suggesting a robust bond. This type is illustrated in Figures 14 and 15. Type D: Complete Substrate Failure—This most substantial bond type results in the complete structural failure of the substrate, while the overlay remains intact. This scenario is depicted in Figures 16 and 17. Most failures, particularly in the normal-strength lightweight concrete (NSLW) during both slant shear and bi-surface tests, occurred within the substrate itself, showing no cracks in the ECC overlay across both as-cast and grooved surfaces. However, the failures in high-strength lightweight concrete (HSLW) with as-cast surfaces generally stemmed from inadequate friction at the interface between the substrate and the ECC.

In specific cases, such as with the N-E-P1.5-G0.5 and H-E-P1.5-G0.5 samples, interface failures followed internal substrate fractures or occurred wholly within the substrate, demonstrating an exceptionally strong bond strengths of 12.75 MPa and 24.05 MPa, respectively. These observations highlight the critical role of mechanical interlocking and substrate conditions in influencing bond efficacy.

Figure 12. Slant shear typical failure mode B—Interface failure with a thin layer on the substrate.

Figure 13. Bi-surface typical failure mode B—Interface failure with a thin layer on the substrate.

Figure 14. Slant shear typical failure mode C—Interfacial failure with a thick layer of the substrate.

Figure 15. Bi-surface typical failure mode C—Interfacial failure with a thick layer of substrate.

Figure 16. Slant shear typical failure mode D—Complete substrate failure.

Figure 17. Bi-surface typical failure mode D—Complete substrate failure.

In this study, it was observed that higher concentrations of fibers, particularly glass fibers, are associated with increased porosity within the interfacial transition zone. This increase in porosity is attributed to the disruption caused by the fibers to the packing density of the cement matrix, resulting in the formation of voids and gaps at the interface. Conversely, it was found that optimal dosages of polyvinyl alcohol (PVA) fibers enhance the density of the interfacial transition zone by promoting better interactions between the fibers and the matrix, thereby reducing porosity and enhancing bond strength.

Regarding surface roughness, it was found that higher roughness levels lead to a more mechanically interlocked interfacial transition zone. Rougher substrate surfaces provide greater mechanical anchorage points for the ECC, which decreases the porosity of the interfacial transition zone through improved contact and interlocking at the micro-level. This enhancement in physical bonding contributes to reduced porosity at the interface, which is correlated with increased shear transfer capacity and higher overall bond strength.

4.4. Friction Coefficient

The calculated values of the coefficient of friction (μ) for the studied specimens are summarized in Table 15. The results indicate an increase in the μ values for the grooved surface specimens compared to the as-cast specimens for both NSLW and HSLW substrates. For HSLW, the μ values ranged from 2.101 to 1.375 for grooved specimens and 1.246 to 0.361 for as-cast specimens. These results suggest that the increase in μ is due to the type of substrate surface preparation and the interlock effects between the concrete substrate and the overlay. The highest μ values were observed in specimens H-E-P1.5-G0.5 (2.383) and N-E-P1.5-G0.5 (1.626) with grooved surfaces, which is consistent with the previously mentioned adhesion strength results.

Table 15. Coefficient of friction (μ).

Specimen ID	μ	Specimen ID	μ
N-E-P2.0-G0.0-G	1.4	H-E-P2.0-G0.0-G	2.1
N-E-P1.5-G0.5-G	1.6	H-E-P1.5-G0.5-G	2.4
N-E-P1.0-G1.0-G	1.5	H-E-P1.0-G1.0-G	1.1
N-E-P0.5-G1.5-G	1.1	H-E-P0.5-G1.5-G	1.4
N-E-P2.0-G0.0-A	1.3	H-E-P2.0-G0.0-A	1.2
N-E-P1.5-G0.5-A	1.4	H-E-P1.5-G0.5-A	1.3
N-E-P1.0-G1.0-A	1.2	H-E-P1.0-G1.0-A	0.5
N-E-P0.5-G1.5-A	1.0	H-E-P0.5-G1.5-A	0.4

The fiber dosage within the ECC mix plays a pivotal role in modifying the friction coefficient by directly affecting the surface texture and the microstructure of the ECC when it interfaces with the LWC substrate. Increasing the dosage of fibers, particularly polyvinyl alcohol (PVA) fibers, contributes to an involved and more irregular surface topology at the microscopic level. These fibers, due to their high tensile strength and stiffness, create micro-anchoring points within the ECC, which protrudes slightly at the interface, enhancing mechanical interlocking with the LWC substrate. As fiber content increases, these micro-anchoring points become more prevalent, effectively increasing the roughness at the microscale, even if the macroscopic surface appears smooth. This increase in microscopic roughness enhances the friction coefficient by providing more points of resistance against sliding, thereby improving the shear transfer capacity across the interface.

The differences in the μ values are significant and can be attributed to the effect of surface roughness on the adhesion surface of HSLW compared to NSLW models. For HSLW, the impact of surface roughness is considerable, as shown in Figures 18 and 19, whereas, for NSLW, the effect of surface roughness is slight.

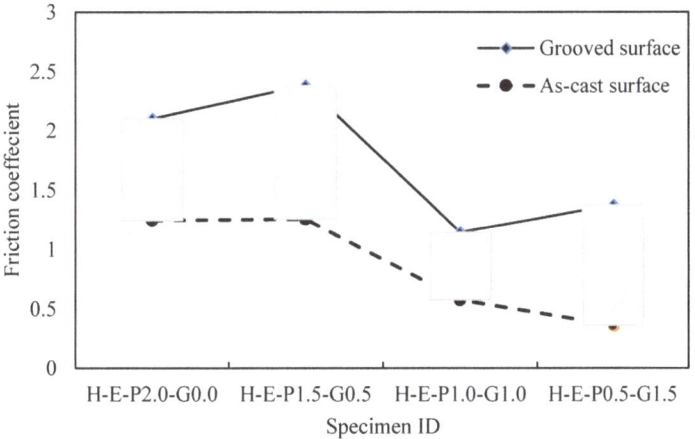

Figure 18. Friction coefficient values between grooved and as-cast surface for HSLW substrate.

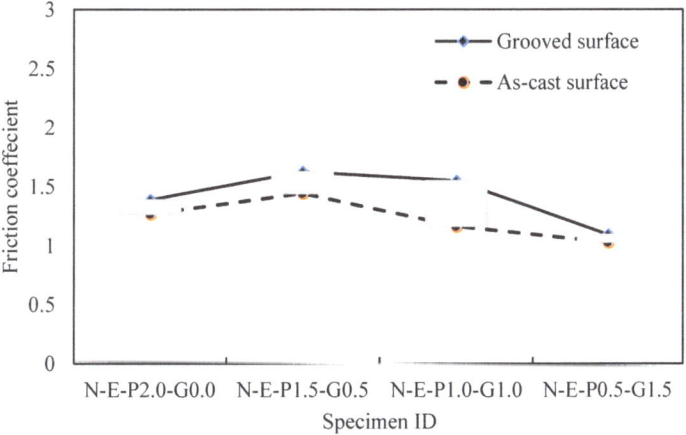

Figure 19. Friction coefficient values between grooved and as-cast surface for NSLW substrate.

This study observed a clear correlation between increased surface roughness and enhanced bond strength. Specifically, as surface roughness parameters increased, bond strength was proportionally enhanced. This relationship is attributed to the improved mechanical interlocking facilitated by more pronounced surface textures, which increases the effective contact area and frictional resistance between the Engineered Cementitious Composite (ECC) overlay and the LWC substrate. Statistical analyses, including regression models, indicated a positive correlation, showing that an increase in surface roughness parameters significantly enhanced bond strength across various samples and test conditions. Additional data and graphical representations illustrating this relationship will be detailed in the revised manuscript, providing comprehensive insight into how surface roughness directly impacts the bond performance in fiber-reinforced composites.

5. Conclusions

In this research, each analyzed sample consisted of two types of concrete: the substrate LWC and the overlay ECC material. The ECC contained various types and percentages of fibers, specifically 2% PVA fiber, 1.5% PVA + 0.5% glass fiber, 1.0% PVA + 1.0% glass fiber, and 0.5% PVA + 1.5% glass fiber. Multiple factors were considered, including the testing methods and the roughness of the substrate surfaces at 28 days of age. The investigation aims to comprehensively understand how these variables influence the bond performance between the repair materials and LWC substrates. Based on the experimental investigation, the following conclusions were reached:

1. The absence of a high rough surface (as-cast surface) did not significantly affect the adhesion strength between the ECC matrix and the LWC, preventing interface failure. This strength can be attributed to the chemical interaction between active silicon dioxide from supplementary cementitious materials in ECC and $Ca(OH)_2$ in mature concrete, forming secondary C–S–H. This phenomenon has been demonstrated in several prior studies, indicating that the material has a strong adhesion capacity, which varies depending on surface roughness.
2. Increasing the proportion of glass fibers beyond 0.5% while reducing PVA fibers in the ECC matrix reduces adhesion strength for both slant shear and bi-surface shear tests across different surface roughness levels (as-cast and grooved surfaces).
3. The bond strength is greatly affected by the test method employed. The bond strengths obtained from the slant shear test were significantly greater in both NSLW and HSLW types than those obtained from the bi-surface shear test. It is crucial to select bond tests that can accurately represent the shear stress experienced by structures in real conditions.
4. For the HSLW substrate, surface roughness significantly impacts the bond strength with the repair material (ECC). This effect is observed in both the slant shear and bi-surface shear tests.
5. The coefficient of friction is influenced by the properties and texture of the adhesive surface. When the surface was grooved, there was an increase in the coefficient of friction between both NSLW and HSLW types and the repair material ECC.

It is recommended that the initial findings related to curing age and fiber orientation in composite concrete systems be further explored. The effects of extended curing periods on the bond strength between normal-strength and high-strength lightweight concrete, when combined with engineered cementitious composite (ECC), are to be further investigated. Additionally, the influence of the alignment of polyvinyl alcohol (PVA) and glass fibers within the ECC matrix on mechanical properties and stress transfer should be explored. Further experimental and computational studies are also recommended to quantify the influence of porosity and microstructural characteristics of NSLW on the bonding behavior with ECC. Advanced imaging techniques like scanning electron microscopy (SEM) and micro-computed tomography (μCT) could provide deeper insights into the interfacial properties and clarify the role of surface roughness in different substrate types.

Author Contributions: Data curation, H.M.A.-B.; formal analysis, H.M.A.-B. and M.M.K.; investigation, H.M.A.-B.; methodology, H.M.A.-B. and M.M.K.; project administration, M.M.K.; resources, M.M.K.; supervision, M.M.K.; writing—original draft, H.M.A.-B.; writing—review and editing, M.M.K. All authors have read and agreed to the published version of the manuscript.

Funding: This research received no external funding.

Data Availability Statement: The data presented in this study are available on request from the corresponding author.

Conflicts of Interest: The authors declare no conflicts of interest.

Abbreviations

LWC	Lightweight Concrete
ECC	Engineered Cementitious Composite
PVA	Polyvinyl Alcohol
NSLW	Normal-Strength Lightweight Concrete
HSLW	High-Strength Lightweight Concrete
GFRP	Glass Fiber Reinforced Polymer
CFRP	Carbon Fiber Reinforced Polymer
ASTM	American Society for Testing and Materials
MPa	Megapascal
mm	Millimeter
COV	Coefficient of Variation
E-waste	Electronic Waste

References

1. Carbonell Muñoz, M.A.; Harris, D.K.; Ahlborn, T.M.; Froster, D.C. Bond performance between ultrahigh-performance concrete and normal-strength concrete. *J. Mater. Civ. Eng.* **2014**, *26*, 401–4031. [CrossRef]
2. Santos, P. Assessment of the Shear Strength between Concrete Layers. Ph.D. Thesis, University of Coimbra, Coimbra, Portugal, 2009.
3. Farzad, M.; Shafieifar, M.; Azizinamini, A. Experimental and numerical study on bond strength between conventional concrete and Ultra High-Performance Concrete (UHPC). *Eng. Struct.* **2019**, *186*, 297–305. [CrossRef]
4. Costa, H.; Júlio, E.; Lourenço, J. Lightweight Aggregate Concrete–Codes Review and Needed Corrections. Codes in Structural Engineering—Developments and Needs for International Practice, IABSE-Fib, Dubrovnik. 2010. Available online: https://www.researchgate.net/publication/268817379_Lightweight_Aggregate_Concrete_-_Codes_Review_and_Needed_Corrections (accessed on 6 May 2023).
5. Suparp, S.; Ali, N.; Al Zand, A.W.; Chaiyasarn, K.; Rashid, M.U.; Yooprasertchai, E.; Hussain, Q.; Joyklad, P. Axial Load Enhancement of Lightweight Aggregate Concrete (LAC) Using Environmentally Sustainable Composites. *Buildings* **2022**, *12*, 851. [CrossRef]
6. Sirisonthi, A.; Suparp, S.; Joyklad, P.; Hussain, Q.; Julphunthong, P. Experimental study of the load-deformation behaviour of the precast post-tensioned continuous girder for straddle monorail: Full-scale load test under service and ultimate loading conditions. *Case Stud. Constr. Mater.* **2021**, *15*, e00666. [CrossRef]
7. Yooprasertchai, E.; Dithaem, R.; Arnamwong, T.; Sahamitmongkol, R.; Jadekittichoke, J.; Joyklad, P.; Hussain, Q. Remediation of Punching Shear Failure Using Glass Fiber Reinforced Polymer (GFRP) Rods. *Polymers* **2021**, *13*, 2369. [CrossRef] [PubMed]
8. Ullah, S.; Qureshi, M.I.; Joyklad, P.; Suparp, S.; Hussain, Q.; Chaiyasarn, K.; Yooprasertchai, E. Effect of partial replacement of E-waste as a fine aggregate on compressive behavior of concrete specimens having different geometry with and without CFRP confinement. *J. Build. Eng.* **2022**, *50*, 104151. [CrossRef]
9. Harris, D.K.; Sarkar, J.; Ahlborn, T.M. Interface Bond Characterization of Ultra-High Performance Concrete Overlays. In Proceedings of the Transportation Research Board 90th Annual Meeting, Washington, DC, USA, 23–27 January 2011.
10. Al-Osta, M.A.; Ahmad, S.; Al-Madani, M.K.; Khalid, H.R.; Al-Fakih, A. Performance of bond strength between ultra-high-performance concrete and concrete substrates (concrete screed and self-compacted concrete): An experimental study. *J. Build. Eng.* **2022**, *51*, 104291. [CrossRef]
11. Li, V.C.; Wang, S.; Wu, C. Tensile strain-hardening behavior of polyvinyl alcohol engineered cementitious composite (PVA-ECC). *ACI Mater. J.* **2001**, *98*, 483–492.
12. Li, M.; Li, V.C. Influence of material ductility on performance of concrete repair. *ACI Mater. J.* **2009**, *106*, 419–428.
13. Li, V.C.; Wu, C.; Wang, S.; Ogawa, A.; Saito, T. Interface Tailoring for Strain-Hardening Polyvinyl Alcohol-Engineered Cementitious Composite (PVA-ECC). *ACI Mater. J.* **2002**, *99*, 463–472.
14. Lim, Y.M.; Li, V.C. Durable repair of aged infrastructures using trapping mechanism of engineered cementitious composites. *J. Cem. Concr. Compos.* **1997**, *19*, 373–385. [CrossRef]
15. Zhu, H.; Leung, C.K.Y.; Cao, Q. Preliminary Study on the Bond Properties of the PDCC/Concrete Repair System. *J. Mater. Civ. Eng.* **2011**, *23*, 1360–1364. [CrossRef]
16. Sahmaran, M.; Yücel, H.E.; Yildirim, G.; Al-Emam, M.; Lachemi, M. Investigation of the bond between concrete substrate and ECC overlays. *J. Mater. Civ. Eng.* **2014**, *26*, 167–174. [CrossRef]
17. Shang, X.Y.; Yu, J.T.; Li, L.Z.; Lu, Z.D. Strengthening of RC Structures by Using Engineered Cementitious Composites. *Sustainability* **2019**, *11*, 3384. [CrossRef]

18. Eyre, J.; Campos, E.S. Upper bounds in the slant shear testing of perfectly plastic joints in concrete. *Mag. Concr. Res.* **1996**, *48*, 181–188. [CrossRef]
19. Omer, G.; Kot, P.; Atherton, W.; Muradov, M.; Gkantou, M.; Shaw, A.; Riley, M.L.; Hashim, K.S.; Al-Shamma'a, A. A nondestructive electromagnetic sensing technique to determine chloride level in maritime concrete. *Karbala Int. J. Mod. Sci.* **2021**, *7*, 61–71. [CrossRef]
20. Mirmoghtadaei, R.; Mohammadi, M.; Samani, N.A.; Mousavi, S. The impact of surface preparation on the bond strength of repaired concrete by metakaolin containing concrete. *Constr. Build. Mater.* **2015**, *80*, 76–83. [CrossRef]
21. Momayeza, A.; Ehsani, M.R.; Ramezanianpour, A.A.; Rajaie, H. Comparison of methods for evaluating bond strength between concrete substrate and repair materials. *Cem. Concr. Res.* **2005**, *35*, 748–757. [CrossRef]
22. *ASTM C882/C882M*; Standard Test Method for Bond Strength of Epoxy-Resin Systems Used with Concrete by Slant Shear. National Laboratories: Pittsburgh, PA, USA, 2005.
23. IQS. *Iraqi Specifications No. (5), 1984 for Portland Cement*; IQS: Barcelona, Spain, 1984.
24. IQS. *Iraqi Specifications No. (45), 1984 for Aggregates of Natural Resources Used for Concrete and Construction*; IQS: Barcelona, Spain, 1984.
25. *ASTM C128*; Standard Test Method for Density, Relative Density (Specific Gravity), and Absorption of Coarse Aggregate. ASTM: West Conshohocken, PA, USA, 2004; pp. 1–5.
26. *ASTM C330*; Standard Specification for Lightweight Aggregates for Structural Concrete. ASTM: West Conshohocken, PA, USA, 2017; Volume 552. [CrossRef]
27. *ACI 211.1-98, A*; Standard Practice for Selecting Proportions for Normal, Heavyweight, and Mass Concrete. ACI: Tokyo, Japan, 1998.
28. Available online: https://assets.construction-chemicals.mbcc-group.com/en-mne/masterroc-ms-610-tds.pdf (accessed on 13 July 2022).
29. Available online: https://assets.construction-chemicals.mbcc-group.com/en-mne/masterglenium-54-tds.pdf (accessed on 2 July 2022).
30. *BS 3892-1*; Pulverized-Fuel Ash. Specification for Pulverized-Fuel Ash for Use with Portland Cement. British Standards Institution: London, UK, 1997.
31. *BS EN 450-1*; Fly Ash for Concrete—Definition, Specifications and Conformity Criteria. British Standards Institution: London, UK, 2012.
32. *ACI. 211*; ACI Standard: Standard Practice for Selecting Proportions for Structural Lightweight Concrete. ACI: Tokyo, Japan, 1982; Volume 4, p. 20. [CrossRef]
33. *ACI. 211.4R-08*; Guide for Selecting Proportions for High-Strength Concrete Using Portland Cement and Other Cementitious Materials. ACI Committee 211; ACI: Tokyo, Japan, 2008.
34. Li, V.C. From Micromechanics to Structural Engineering the Design of Cementitious Composites for Civil Engineering Applications. *Doboku Gakkai Ronbunshu* **1993**, *1993*, 1–12. [CrossRef] [PubMed]
35. Momayez, A.; Ramezanianpour, A.A.; Rajaie, H.; Ehsani, M.R. Bi-surface shear test for evaluating bond between existing and new concrete. *Mater. J.* **2004**, *101*, 99–106.
36. Beushausen, H.; Höhlig, B.; Talotti, M. The influence of substrate moisture preparation on bond strength of concrete overlays and the microstructure of the OTZ. *Cem. Concr. Res.* **2017**, *92*, 84–91. [CrossRef]
37. *ASTM C39/C39M-20*; Standard Test Method for Compressive Strength of Cylindrical Concrete Specimens. ASTM: West Conshohocken, PA, USA, 2020.
38. *BS 6319*; Part 4 Slant Shear Test Method for Evaluating Bond Strength of, Epoxy Systems. BSI: Yokohama, Japan, 1984.
39. Zanotti, C.; Rostagno, G.; Tingley, B. Further evidence of interfacial adhesive bond strength enhancement through fiber reinforcement in repairs. *Constr. Build. Mater.* **2018**, *160*, 775–785. [CrossRef]
40. Fisher, G.; Li, V.C. Proc. Int. RILEM Workshop on High-Performance Fiber-Reinforced Cementitious Composites (HPFRCC) in Structural Applications. RILEM Publications S.A.R.L., Bagneux (France), PRO 49. 2006. Available online: https://orbit.dtu.dk/en/publications/international-rilem-workshop-on-high-performance-fiber-reinforced (accessed on 28 May 2023).
41. Yıldırım, G.; Sahmaran, M.; Al-Emam, M.K.M.; Hameed, R.K.H.; Al-Najjar, Y.; Lachemi, M. Effects of compressive strength, autogenous shrinkage, and testing methods on bond behavior of high-earlystrength engineered cementitious composites. *ACI Mater. J.* **2015**, *112*, 409–418.
42. Austin, S.; Robins, P.; Pan, Y. Shear bond testing of concrete repairs. *Cem. Concr. Res.* **1999**, *29*, 1067–1076. [CrossRef]
43. Geissert, D.G.; Li, S.E.; Frantz, G.C.; Stephens, J.E. Splitting prism test method to evaluate concrete-to-concrete bond strength. *ACI Mater. J.* **1999**, *96*, 359–366.
44. Abu-Tair, A.I.; Rigden, S.R.; Burley, E. Testing the bond between repair materials and concrete substrate. *ACI Mater. J.* **1996**, *9*, 553–558.
45. *546.3R-14*; Guide to Materials Selection for Concrete Repair. ACI: Tokyo, Japan, 2014.
46. Deshpande, U.L.; Murnal, P.B. Ductile Concrete Using Engineered Cementitious Composites. *Int. J. Eng. Res.* **2016**, *5*, 756–776.
47. Omar, B.; Fattoum, K.; Maissen, B. Influence of the roughness and moisture of the substrate surface on the bond between old and new concrete. *Contemp. Eng. Sci.* **2010**, *3*, 139–147.
48. Duan, K.; Hu, X.; Mai, Y.W. Substrate constraint and adhesive thickness effects on fracture toughness of adhesive joints. *J. Adhes. Sci. Technol.* **2004**, *18*, 39–53. [CrossRef]
49. Espeche, A.D.; Le, J. Estimation of bond strength envelopes for old-to-new concrete interfaces based on a cylinder splitting test. *Constr. Build. Mater.* **2011**, *25*, 1222–1235. [CrossRef]

50. Fu, C.; He, R.; Wang, K. Influences of corrosion degree and uniformity on bond strength and cracking pattern of cement mortar and PVA-ECC. *J. Mater. Civ. Eng.* **2023**, *35*, 04023134. [CrossRef]
51. Chen, H.; Zhou, X.; Li, Q.; He, R.; Huang, X. Dynamic Compressive Strength Tests of Corroded SFRC Exposed to Drying–Wetting Cycles with a 37 mm Diameter SHPB. *Materials* **2021**, *14*, 2267. [CrossRef] [PubMed]
52. Chen, H.; Huang, X.; He, R.; Zhou, Z.; Fu, C.; Wang, J. Mechanical properties of polypropylene fiber cement mortar under different loading speeds. *Sustainability* **2021**, *13*, 3697. [CrossRef]

Disclaimer/Publisher's Note: The statements, opinions and data contained in all publications are solely those of the individual author(s) and contributor(s) and not of MDPI and/or the editor(s). MDPI and/or the editor(s) disclaim responsibility for any injury to people or property resulting from any ideas, methods, instructions or products referred to in the content.

MDPI AG
Grosspeteranlage 5
4052 Basel
Switzerland
Tel.: +41 61 683 77 34
www.mdpi.com

Buildings Editorial Office
E-mail: buildings@mdpi.com
www.mdpi.com/journal/buildings

Disclaimer/Publisher's Note: The statements, opinions and data contained in all publications are solely those of the individual author(s) and contributor(s) and not of MDPI and/or the editor(s). MDPI and/or the editor(s) disclaim responsibility for any injury to people or property resulting from any ideas, methods, instructions or products referred to in the content.